"This may be one of the most important books
poisoned, and this book is sounding a well-infor
and then join the thousands rising up against th̶
the health of our bodies and our earth."
—**Eve Ensler**, *New York Times* bestselling author of *I Am an Emotional Creature:
The Secret Life of Girls Around the World*, *The Vagina Monologues*,
and *In the Body of the World*

"Activism and science need one another to stay grounded in reality. Few environmental activists have done more than the poet and science writer Mitchel Cohen to
connect with scholars across a multitude of disciplines in his tireless campaign to
keep the natural world from turning into a toxic hell. This book, with its remarkably varied group of expert contributors, is both a monument to Cohen's ongoing
efforts and a resource for those who will be inspired by it to join forces with him."
—**Stuart Newman, Ph.D.**, Professor of Cell Biology and Anatomy,
New York Medical College

"This book delivers the goods. Mitchel Cohen and his coauthors have thoroughly
and effectively indicted Monsanto, Syngenta, and other Big Food corporations for
poisoning our soil, our water, and our genomes."
—**Clifford D. Conner**, author of *A People's History of Science*, and the
forthcoming *Tragedy of American Science: From Truman to Trump*

"*The Global Fight Against Monsanto's Roundup* is an absolutely vital book for so
many who have been diagnosed with diseases their doctors say were caused by
vague 'environmental factors.' Mitchel Cohen explains precisely what are these
'environmental factors.' Read it as if your life depended on it, because it does."
—**George Caffentzis**, author of *No Blood for Oil: Essays on Energy,
Class Struggle and War 1998-2016*

"Few battles are as important today across the planet as that to free the earth, the
seas, and woods from the poisons that companies like Monsanto are pouring in
them. *The Fight Against Monsanto's Round up: The Politics of Pesticides* is a great
resource in that struggle. Read it and give it to all your friends and comrades."
—**Silvia Federici**, author of *Caliban and the Witch*

"Evidence of cataclysmic climate change and environmental collapse are all around
us, wherever you look: the raging fires and recurring storms, poisoned water and
filthy air, droughts and floods of Biblical proportion, climbing rates of extinction,
stressed-out birds and fading bees, frazzled fish and misshapen frogs. It takes

some bizarre combination of self-interest, privilege, cynicism, ideology, corruption, dogma, or chutzpah for supine politicians to lie passively on their backs denying the facts in favor of cloud-cuckoo-land fantasies and the interests of the filthy extractors and the billionaire barbarians from ExxonMobil, Monsanto, and Chevron. Seriously engaging the environmental catastrophe, which this important collection edited by Mitchel Cohen does brilliantly, and taking the necessary steps to solve it, will mean—I'll spit it out—overthrowing capitalism. This is the real choice in front of us: the end of capitalism or the end of the habitable earth, saving the system of corporate finance capital or saving the system that gives us life. Which will it be?"

—**Bill Ayers**, author of *Demand the Impossible!*

"Pesticides not only wreak havoc with our health, they accelerate global warming by destroying carbon-fixing soil ecosystems. They are the linchpin of the chemicalized capitalist agriculture that must be pulled out if we are to make the transition to organic agroecologies that draw carbon out of the atmosphere and into the biosphere in restored soil ecosystems. Read this collection to arm yourself with knowledge for the fight to ban pesticides and build a sustainable and climate-friendly agriculture."

—**Howie Hawkins**, three-time Green Party candidate for Governor of New York

"In this age of corporate greed and its effects on the environment, this book provides a valuable critique of corporate environmental destruction as well as suggestions for a way out of this global destruction."

—**Irene Javors**, author of *Culture Notes: Essays on Sane Living* and advocate for optimal living www.ijavorsoptimalliving.com

"This book reveals the triumph of the collective scholarship, humanity, and humor over the hubris of the mammoth pesticide-chemical industry aligned with insensitive bureaucrats."

—**Joel R Kupferman, Esq.**, New York Environmental Law & Justice Project, co-Chair of National Lawyers Guild - Environmental Justice Committee, and co-counsel to the No Spray Coalition in its victorious lawsuit against New York City's spraying of toxic pesticides.

"Our poor planet is being attacked on so many sides by so many issues that it has become increasingly difficult to give any one of them the attention they deserve. This is especially true of the spread of pesticides, to which most of its victims are blind . . . until it's too late. Mitchel Cohen has brought together a collection of powerful essays on this subject—including several excellent ones by him—that has just the right combination of the most important facts, scholarly analysis, outrage (not to be neglected), and other solutions to the problems that pesticides are supposed to address . . . *The Fight Against Monsanto's Round up: The Politics of Pesticides* makes a major breakthrough in this life and death discussion. Highly recommended!"

—**Bertell Ollman**, Professor of Politics, New York University, author of *Alienation: Marx's Concept of Man in Capitalist Society, Dialectical Investigations,* and *Dance of the Dialectic*

"If you care about the planet, what you eat, your health and our whole, fragile ecosystem, then this book is for you. It's jam-packed with up-to-date information about pesticides, politics, protest. and the monster known as Monsanto. All of the contributors write with style, savvy, and with a sense of humor, too. Each author carves out his or her own part of the big picture. Together they've created a book that's greater than its individual parts, and that will help organizers and activists who want a planet that's free of poisons and lies."

—**Jonah Raskin** is the author of 16 books, including *The Mythology of Imperialism*, plus biographies of Jack London, Allen Ginsberg and Abbie Hoffman.

"This is a superb collection of essays on pesticides by a superb collection of independent experts. Importantly, they're not only highly knowledgeable about how pesticides kill but fully understand the politics and economics of the pesticide push over the many years. They are fully aware, too, of the safe, green alternatives—in harmony with nature—to these deadly poisons. Moreover, the work is edited and many of the essays written by a brilliant Green and social activist, Mitchel Cohen. I have known Mitch for more than 50 years—starting with my covering him as a reporter when at Stony Brook University he led student protests against the Vietnam War and other outrages. Carefully done and comprehensive, this book is in the great tradition of Rachel Carson's *Silent Spring*. It is an essential learning and teaching tool."

—**Karl Grossman**, professor of journalism at the State University of New York/ College at Old Westbury, author of *The Poison Conspiracy*, and host of "Enviro Close-Up with Karl Grossman."

"I've known Mitchel Cohen, the editor of this remarkable volume, for over 30 years. Mitchel is brilliant, a dedicated activist, and deeply involved in opposing the mass use of pesticides since 1999 and the politics that promote it at the expense of human health and the environment. In *The Fight Against Monsanto's Roundup: The Politics of Pesticides*, Mitchel informs and guides us."

—**Elizabeth Liberty** is a poet, feminist writer, and former Executive Director of the European branch of International P.E.N.

"In an era when Monsanto seeks to poison the earth with its pesticide Roundup, Mitchel Cohen has done his own round up, assembling an amazing array of scientists, journalists, and activists to strategically fight back with the truth. There's no better place to learn about and join that fight than with *The Fight Against Monsanto's Roundup: The Politics of Pesticides*.

The movement against the routinized use of poisons in agriculture has taken on global proportions. This book examines how activists around the world are now fighting back against Monsanto, the manufacturer of the best-selling, cancer-causing pesticide glyphosate, better known as 'Roundup.' The new book, edited by longtime Green activist Mitchel Cohen, explores not only the scientific dangers of glyphosate, but the nitty-gritty of the grassroots movements organizing to ban it.

The book's contributing scientists and activists detail how corporations such as Monsanto, Bayer, Dow and DuPont scuttle attempts to regulate the pesticides they manufacture. Moreover, in an age where banned pesticides are simply replaced with newer and more deadly ones, the book also explores the best strategy to win the struggle for healthy foods and a clean environment."

—**Jack Shalom**, an educator in the New York City public school system for over 25 years, has worked with Brooklynites Against Apartheid, Brooklyn For Peace, and the New Sanctuary Coalition, and produces segments for WBAI radio in New York City.

"Mitchel Cohen, author, poet and activist is a force of nature. For decades he has tilted against the purveyors of poison who profit from the conversion of our complex, diverse, and interdependent biosphere into sterile, unsustainable monoculture. In his new book, Cohen, and the dozen or so other authors who round out this anthology, share insights as to how we got into the mess we're in, and wise guidance as to how we might challenge the poison-laden paradigm Monsanto and company are foisting upon us."

—**Mark Haim** is Director of Mid-Missouri Peaceworks and has been active around peace, justice, sustainability and climate concerns for many decades.

"The catchphrase is disaster capitalism, but we all know capitalism IS the disaster! No one makes this point more eloquently than Mitchel Cohen. No topic exposes this truth more clearly than the destruction of the environment. 99% of human history has been spent without this toxic system....we can't allow the 1% to control a minute more of our future."

—**Margaret Stevens**, Antiwar veteran of US Army Nat'l Guard, teacher in Newark, NJ, and author of *Red International, Black Caribbean: Communist Organizations in New York, Mexico and the West Indies, 1919 to 1939.*

"This volume is a 'must read' for anyone who cares about life on this planet. It is a remarkable collection, providing both breadth and depth of understanding of not only the current state of the environment and protecting it but also the cast of characters who seem bent on destroying it to serve their drive for greater profits. You won't be able to put it down."

—**Marilyn Vogt-Downey** is a journalist and Russian translator, who translated the *Writings of Leon Trotsky* series and *Notebooks for the Grandchildren.*

"*The Fight Against Monsanto's Roundup: The Politics of Pesticides* examines Monsanto's history and misuse of science, as well as our relationship to nature. It picks up from where Rachel Carson left off in her deep concern about the devastating effects of chemical pollution for all of us."

—**Marilyn Berkon** is a retired NYC high school English teacher, whose major interest is in Greek classical literature.

"Mitchel Cohen is one of the most dedicated, brilliant activists that I have ever had the pleasure and honor to work with over the years, and *The Fight Against Monsanto's Roundup: The Politics of Pesticides* is a passionate and highly informative treatise for protecting public health and the environment."

—**Aton Edwards**, Executive Director of the International Preparedness Network, winner of the Grio 2012 100 history maker award, Today Show, MSNBC contributor, author, environmental & social activist.

"With levity, poetry, art, and a solid lineup of informed and dedicated contributors, Mitchel Cohen brings to our attention the uncomfortable reality of today's silent spring: the invisible persistent killer pesticide, Glyphosate. Read, learn, and enjoy Cohen's creative efforts to tell us what we need to know. An informed citizen is our best defense against the chemical vultures of our time."

—**Linda Zises**, Co-founder and president of Women in City Government United

THE

FIGHT AGAINST

MONSANTO'S ROUNDUP

THE POLITICS OF PESTICIDES

WRITTEN AND EDITED BY MITCHEL COHEN

Foreword by VANDANA SHIVA

Skyhorse Publishing

Skyhorse Publishing books may be purchased in bulk at special discounts for sales promotion, corporate gifts, fund-raising, or educational purposes. Special editions can also be created to specifications. For details, contact the Special Sales Department, Skyhorse Publishing, 307 West 36th Street, 11th Floor, New York, NY 10018 or info@ skyhorsepublishing.com.

Skyhorse Publishing® is an imprint of Skyhorse Publishing, Inc.®, a Delaware corporation.

Visit our website at www.skyhorsepublishing.com.

10 9 8 7 6 5 4 3 2 1

Library of Congress Cataloging-in-Publication Data is available on file.

Cover design by Brian Peterson

Hardcover ISBN: 978-1-5107-3513-2
Paperback ISBN: 978-1-5107-6829-1
Ebook ISBN: 978-1-5107-3514-9

Printed in the United States of America

This book is dedicated to
Allan Sicignano, chiropractor extraordinaire
and
Lawrence Gamble, acupuncturist
and
Cathryn Swan, aromatherapy
and
Dr. Khabir Bhasin, Electro-Cardiologist
and
Hannah Reimann, Musician and songwriter

who each generously gave of their time and expertise to help my brain and
body heal from serious health issues while I was writing this book (and
who don't realize how much their efforts meant to me and helped me to
heal)
and
Malika Moro-Cohen
whose very existence is something of an ongoing miracle

With much love and thanks to the efforts, insights and encouragement of
Robin Esser

In loving memory
Valerie Sheppard (1953–2004), founding board member, No Spray
Coalition against pesticides
Annette Averette (1945–2020), founding board member, No Spray
Coalition against pesticides
DayStar Chou (1961–2020), representative of the Queens Greens to the
board, No Spray Coalition against pesticides
Maria Kuriloff (1949–2009), representative of the Central Nassau Greens
to the board, No Spray Coalition against pesticides
Steve Tvedten (1942–2018), former pesticides applicator who became a
heroic anti-pesticides activist
Joel Kovel (1936–2018), former psychiatrist who turned against the abuses
of psychiatry, became a founder of Eco-socialism, and supported the No
Spray Coalition and anti-pesticides efforts across the country

And

To those friends, teachers, and activists who participated with, created, or supported the work of the No Spray Coalition against pesticides, and who have now passed. In loving and powerful memory:

Barri Boone, Frank Carr, Therese Chorun, Binny Ipcar Correll, David Crowe, Enid Dame, Peggy Dye, Bryna Eil, Dr. Samuel Epstein, Seth Farber, Liz Fink, Glen Ford, Arly Fox-Daly, Peter Freund, John "Tito" Gerassi, Barbara Goldberg, Ibrahim Gonzalez, David Graeber, Corey Gregory, Dick Gregory, Saralee Hamilton, Fred Ho, Mae-Wan Ho, Connie Hogarth, Robert Knight, Bill Koehnlein, Ray Korona, Bud Korotzer, Jim Krivo, Frank Lefever, Faith Legier, Carl Lesnor, Virginia Lerner, Donald Lev, Richard Levins, Grandpa Al Lewis, Bernice Linton, Bill Livant, Carl Makower, Manning Marable, Julius Margolin, John McCarthy, Bernie McFall, Bob McGlynn, Will Miller, Dr. Luc Montagnier, John Moran, Mike Pahios, Gloria Pasin, Ed Pearl, Dorothy Williams Pereira, Lillian Pollak, Carol Randeros, Michael Ratner, Mark Rausher, Pete Seeger, Michael Shenker, Daniel Simidor, Kent Smith, Ann Snitow, Albert Solomon, Ernest J. Sternglass, Lynne Stewart, Gene Vanderport, Len Weinglass, Brad Will, Freda Zames, Kevin Zeese, Bob Zink

CONTENTS

Note: Chapters and Poems in this book not attributed to an author are contributed by Mitchel Cohen.

Rachel Carson and Moppet, photographed in 1962 by Alfred Eisenstaedt

Remember the lessons of Rachel Carson and Silent Spring *as evidence mounts that we are in a very precarious point, as the push for pesticide dependence and the drive for corporate profits take precedence over people's lives and our environment.*
—Carey Gillam, author of *Whitewash: The Story of a Weed Killer, Cancer and the Corruption of Science*

We can now prove that all Monsanto's claims about glyphosate's safety were myths concocted by amoral propaganda and lobbying teams. Monsanto has been spinning its lethal yarn to everybody for years and suborning various perjuries from regulators and scientists who have all been lying in concert to American farmers, landscapers and consumers.

—Robert F. Kennedy Jr.

Foreword

by Vandana Shiva

The story of Roundup told by activists, scientists, and academics in *The Fight Against Monsanto's Roundup: The Politics of Pesticides* is an important contribution to the history of how a toxic war against nature and people has been unleashed over the last fifty years through industrial agriculture by the Poison Cartel.

The Poison Cartel began as IG Farben in Hitler's Germany and has shaped a century of ecocide and genocide. Bayer was an important part of IG Farben. Bayer and Monsanto worked as MOBAY earlier, and have merged again. Monsanto's Roundup is now Bayer's Roundup. The multiple cases in U.S. courts against Monsanto's cancer-causing Roundup have already led to a 30 percent fall in Bayer's share values. Today the Poison Cartel has been reduced to a cartel of three, with common ownership and cross-licensing arrangements on patents and technologies. Besides Bayer merging with Monsanto, Dow has merged with Dupont, and Syngenta has merged with Chem China.

The World Health Organization (WHO) classified Roundup (glyphosate) as *probably carcinogenic to humans* (Group 2A).

This book discusses the historic case of Dewayne Johnson, a forty-six-year-old former groundskeeper who used to spray Roundup to maintain the grounds of a school. On Aug. 10, a jury in California ruled that Monsanto's Roundup weed killer had caused Johnson's cancer and ordered Monsanto pay $289 million to the victim. Eight thousand more cases have been filed in the U.S.A by cancer victims.

Monsanto sells Roundup, which causes cancer. Bayer, now merged with Monsanto, sells patented cancer drugs. And both Monsanto and Bayer are still trying to undermine patent laws of countries like India, which prohibit patents on seeds and on medicines that are already being produced as generics.[1]

This book is not just about the hazards of Roundup to our health and the environment, but also about the political hazard to democracy and freedom when corporations like Monsanto control our agriculture, knowledge, science, and the media, and try to control and manipulate our scientific and regulatory institutions.

As Mitchel Cohen, the editor of the book, writes:

"This book focuses not so much on examining the dangers of each and every pesticide *du jour*, but on the processes by which corporations such as Monsanto, Bayer, Dow, DuPont, Syngenta, Novartis, BASF Corporation and the other pesticide and pharmaceutical manufacturers are allowed to mask the truth about their products. They are facilitated by the complicity of federal (and global) regulatory agencies, allowing them to intentionally thwart the development and congealing of educated and effective opposition."

The very existence of chemicals like Roundup is based on thinking of agriculture not as care for the land and co-creation with biodiversity, but as a war against the earth and the diversity of her species, including insects and plants. The monarch butterfly has declined by 90 percent because the spread of Roundup and Roundup Ready crops has destroyed the milkweed.

Roundup is not just a toxic chemical. It embodies the violent and distorted worldview of Monsanto and the Poison cartel that developed poisons to kill humans and later introduced their toxic chemicals into agriculture.

As a chemical which kills everything green, Roundup is key to the destruction of the biodiversity that allows small farms practising chemical-free farming to regulate pests and weeds. Pests and weeds are symptoms of non-sustainable farming that works against nature and the ecological functions that biodiversity provides.

In the 1990s, when the United Nations Convention on Biodiversity was being negotiated, a Monsanto representative described Round Up Ready GMO crops as a "technology" that "prevents the weeds from stealing the sunshine."

This is war.

The sun shines on all in abundance and generosity. In India we greet the sun daily as the 'Dispeller of darkness' ('*Om Suryaya Namaha*'), including the darkness of untruth in the mind.

We call on the sun (*Aditya*) to illuminate our intellect. "*Om Bhaskaraya Vidmahe Mahadyudikaraaya Dhimahi Tanno Aditya Prachodayaat*" (I learn about the one who is the source of light and meditate on the one who is so effulgent. Let Lord *Aditya* illuminate my intellect).

The perception of scarcity of sunshine is a darkness of the mind that cannot see light. It is this unscientific and distorted and dark world view rooted in the militaristic, mechanistic, militarised mind that encourages use of violent tools like Roundup.

The darkness creates a distorted worldview that sees us at war with biodiversity, and diverse species at war with each other, instead of seeing the earth as a common home for diverse life forms co-evolving in harmony with each other, sharing the sunshine and transforming it into life on earth through photosynthesis.

The sun blesses all beings and allows biodiversity to flourish. Biodiversity is not "weeds" to be exterminated by Roundup.

I write this foreword at the Navdanya Biodiversity Conservation farm where we grow more than 2,000 varieties of crops, including 750 rice varieties.

There is, in addition, abundant biodiversity that is wild, biodiversity we did not plant. We just counted more than 56 medicinal plants and uncultivated edibles that grow on their own, that we do not sow. This biodiversity is nature's pharmacy and pantry, healing us and feeding us. Biodiversity of insects and plants is a sign of healthy ecosystems. All insects are not pests. Biodiversity is not weeds. Some insects and some plants become pests and weeds as a result of the toxic war against biodiversity and destruction of the ecological balance that is the result.

Toxics like Roundup are at the root of the pest and weed problem in agriculture, not a solution. Toxics create pests and weeds. Corporations like Monsanto first create a problem, then use the problem to create a new market.

First industrial agriculture based on chemical monocultures created the problem of weeds dominating agriculture. Then they introduced Roundup. To expand their market and maintain their monopoly, they introduced Roundup Ready GMO crops.

"Since 1974 in the U.S., over 1.6 billion kilograms of glyphosate active ingredient have been applied, or 19% of estimated global use of glyphosate (8.6 billion kilograms). Globally, glyphosate use has risen almost 15-fold since so-called "Roundup Ready," genetically engineered glyphosate-tolerant crops were introduced in 1996. Two-thirds of the total volume of glyphosate applied in the U.S. from 1974 to 2014 has been sprayed in just the last 10 years. The corresponding share globally is 72%. In 2014, farmers sprayed enough glyphosate to apply ~1.0 kg/ha (0.8 pound/acre) on every hectare of U.S.-cultivated cropland and nearly 0.53 kg/ha (0.47 pounds/acre) on all cropland worldwide."[2]

Now they are creating superpests and superweeds with the spread of GMO Bt crops and Roundup Ready crops.

Amaranth is a sacred plant in India. Its grains are used in fasting. Its leaves are rich in Vitamin A and iron and are a major source of nutrition. Diverse species of amaranth grow on the Navdanya farm and are among our favourite vegetables.

With the intensive use of Roundup Ready crops in the U.S., palmer amaranth has become a superweed that cannot be destroyed by Roundup. The military mind of DARPA and Bill Gates now wants to use gene drives to exterminate amaranth. The extermination logic that began in concentration camps and gas chambers has now spread to our farms and forests.[3]

Roundup has spread across the world, but not because it was needed or people asked for it. In industrialised countries it has spread by destroying small family farms and making "farms" so large that they cannot but be managed through instruments of war. In countries like mine, where small, biodiverse farms dominate, Roundup has spread illegally, by deceiving farmers, corrupting regulatory institutions, and attacking science and scientists.

I recently wrote to the Prime Minister of my country drawing attention to the illegal spread of Roundup and Roundup Ready crops in total violation of our biosafety regulations. These issues are in Indian courts.

Over the last few years, with the forcing by globalization and so-called "Free Trade," while our farmers have not been able to sell the diverse pulses (the edible seeds of plants in the legume family) they grow, India has increasingly become dependent on imports of Roundup-sprayed pulses, just like we were made dependent on imports of edible oils after 1998, the forced shutting down of our "ghanis" and indigenous cold pressed healthy edible oils.

Imported pulses also have Roundup residues. The spread of Roundup through imports as well as through illegal spread of Roundup Ready Bt Cotton in India is a major threat to the health of the Indian people.

Roundup symbolizes not just a war against the planet and its biodiversity. It is a war against our bodies, and the gut microbiome which is key to our health. When Monsanto declared that Roundup is safe for humans, it did not just distort the science that showed that it was a carcinogenic, it also distorted the science that recognizes that the microbes in our gut have the shikimate pathway which Roundup disrupts, thus triggering the diverse chronic diseases that are emerging in epidemic form. Health is a continuum from the biodiversity in our farms to the biodiversity in our gut. And Roundup is an instrument of war against the biodiversity outside us and within us.

It is simultaneously a war against democracy, knowledge, and a science that shines the light of truth. The fight against Monsanto's Roundup is a non-violent response to this multidimensional war. This is why we organized the Monsanto tribunal and People's Assembly in the Hague in October 2016.[4]

The Fight Against Monsanto's Roundup: The Politics of Pesticides gives details of how this process evolved in the U.S. The book gives details of how Monsanto manipulated and falsified science, attacked scientific institutions like the World Health Organization's International Agency for Cancer Research, attacked scientists whose scientific research provided evidence of harm, did ghostwriting for journalists and journals, corrupted regulatory agencies and undermined laws.

In spite of the manipulation of scientific evidence and institutions, the evidence is finally reaching the courts. In spite of the desperate attempt to dismantle laws and regulations that protect the public, Roundup is on trial everywhere, in courts, in institutions, and in communities. The trial is between people's rights and corporate greed and irresponsibility, between life and death, between truth and falsehood.

The Fight Against Monsanto's Roundup: The Politics of Pesticides provides the detailed story of the ongoing trial and the politics behind Roundup, Monsanto (now Bayer), and the mass use of pesticides.

The struggles described in this book are the contemporary *Satyagraha* —the word Gandhi used for the fight for truth and justice. It is a *Satyagraha* to "Roundup Roundup" and sows the seeds for poison-free food and farming everywhere.

Preface

The Monsanto Corporation—now owned by Bayer—manufactures the most widely used herbicide in the world. Roundup causes cancer. It causes neurological diseases. It destroys all plant life it touches. How could this product ever have been authorized by U.S. government regulatory agencies?

Lawsuits filed by some of those sickened by Roundup—deadly cancers, neurological damage, even arsenic poisoning—reveal scandalously unethical and corrupt corporate and governmental behavior. Monsanto paid off politicians and pressed government and world health regulatory agencies to bypass critical safety protocols. The Food and Drug Administration, charged with overseeing and reviewing pharmaceutical companies' submissions, stacks its board of directors with pesticide industry hacks.[1] This is what has become known as the "revolving door" between industry and government regulatory bodies. In this regard, it doesn't matter which political party controls the executive branch of government[2]—former Monsanto executives write the rules for the FDA on what products the company is allowed to bring to market and how to "regulate" them. The agencies not only "looked the other way" in approving the manufacture and use of Roundup, but actually ghostwrote Monsanto's applications for the herbicide, which the company then submitted for approval to those same agencies that illegally helped draft them.[3] The practices have been so deceptive, and the corporation so *rewarded* for a process of development and marketing devoid of all precaution, that one would be justified in first repairing the failed oversight before approval of any new product is even considered.

A year after the publication of *The Fight Against Monsanto's Roundup: The Politics of Pesticides* in 2019, the world became engulfed in the Covid-19 pandemic, and the pharmaceutical corporations rushed to develop new "vaccines," promoted initially as "magic bullets" that would save us all. The head of Bayer/Monsanto's Pharmaceutical division (hereafter, "Bayer"), Stefan Oelrich, enthusiastically lauded his company's plans to develop what he calls "solid gene therapy," which attempts to create treatments for autoimmune and other diseases by, controversially, altering genetic sequences. "Ultimately, the mRNA vaccines," Oelrich proclaims without the slightest trepidation, "are an example for that solid gene therapy."[4] Many scientists, though, reject the classification of mRNA vaccines as "gene therapy" because, they say, Pfizer's and Moderna's vaccines are not designed to change an individual's DNA. So, says Oelrich, Bayer plans to gain approval for therapies that *do* change DNA by tying them illegitimately to the public's wary acceptance of the Moderna and Pfizer mRNA technology.

"If we had surveyed two years ago the public: 'Would you be willing to take gene or cell therapy and inject it into your body?' we would have probably had a 95 percent refusal rate," Oelrich noted in a speech in October 2021 at the World Health Summit in Berlin. "I think this pandemic has also opened many people's eyes to innovation in the way that was maybe not possible before."[5] Bayer has apparently decided to piggyback its new and highly controversial gene technology on the Moderna and Pfizer precedents, where the FDA allowed an emergency bypass of the required long-term testing of the mRNA technology to enable the pharmaceutical corporations to receive official authorization to produce what Oelrich is calling "solid gene therapies." Bayer is losing no time pairing with Mammoth Biosciences and hopes to "harness . . . the diversity of life to power the next generation of CRISPR [gene editing] products."[6] Given the FDA's dreadful record with regard to its approval of Monsanto's product Roundup, and its acceptance of Monsanto's submission of falsified or intentionally inadequate data, concern over Bayer's *new* gene-altering products would be more than warranted.

With the pandemic deep into its third year, an important and unexpected link is becoming evident between the widespread use of Monsanto's Roundup and an individual's susceptibility to the SARS-CoV-2 virus. Thanks to researchers who have courageously stood up against the pharmaceutical corporations' powerful influence over our health care system, we are now learning that the risk of contracting Covid-19 and the severity of the illness are linked to low levels of Vitamin D in the body.[7] The incidence of Covid has been observed to be greater among people who are deficient in Vitamin D.

Beneficial bacteria (E. coli) in the large intestine help produce Vitamin D. Roundup was designed to kill plant cells that had not been genetically engineered to withstand the herbicide by disrupting the production of specific amino acids—phenylalanine, tyrosine, and tryptophan—and targeting the Shikimate biological pathway, which is found only in plants and some microorganisms, but not in animals. Thus, Monsanto presented glyphosate as being perfectly safe for humans (not even considering other ways in which a particular chemical or genetic alteration might have unanticipated consequences apart from the ones expected). Roundup has now been shown experimentally to suppress cytochrome P450 (CYP) enzymes in the liver, which without the presence of glyphosate would otherwise activate Vitamin D.[8] (CYP enzymes in the liver convert vitamin D to 25-hydroxy-vitamin D, which is the form that is usually measured in a blood test.) It kills gut bacteria, disrupting the microflora composition in the GI tract. MIT researcher Stephanie Seneff has examined research showing that this may be an important factor in the widespread epidemic we face today of vitamin D deficiency, an outcome that should have been obvious (and it probably *was* obvious to Monsanto's scientists) but one that Monsanto intentionally obscured. The widespread use of Roundup may be killing off the bacteria in the gut needed to raise vitamin D levels, increasing the risk of Covid (and other) infections.[9]

Vitamin D discrepancies would also explain, in part, the greater threat of Covid to those lacking exposure to sunlight, where industrial pollution, frequent overcast skies, and sometimes skin pigmentation, reduces vitamin D production by blocking ultraviolet light.[10] Other factors—diet, obesity, diabetes, poor access to nutrition and health care, chronic exposure to toxins, stress, and pathogens, and being frontline health workers and caregivers in hospitals—drive the death rate from Covid for Black people to more than twice the national average than for non-Blacks, with the hospitalization rate for Blacks almost five times higher than for others.[11]

One study, published in the *Journal of the American Medical Association* on March 19, 2021, found that levels of 30 ng/ml of Vitamin D, which are usually considered sufficient in everyday life, are too low for protecting against Covid. Most African Americans do not test higher than 25 ng/ml at any time of the year,[12] and 40 ng/ml of Vitamin D in the blood or greater would cut the Covid risk by half,[13] and improve overall health.[14] Monsanto's

Roundup, and its concentration in many prepared foods—even in healthy ones like Quaker oats, where Roundup is used as a drying agent (desiccant)—may be responsible for the destruction of the gut biome and hindering of the body's ability to produce Vitamin D, thus increasing susceptibility to and severity of the SARS-CoV-2 virus.

Rather than welcoming the growing research of Roundup's effects on "friendly" bacteria in the human gut needed for the body to make Vitamin D, Bayer/Monsanto obscured the data as it pressured governments to approve its product. How has Monsanto been able to get away with it? Why have the U.S. regulatory agencies continued to downplay and even lie about Roundup's effects on human health and the environment? How has a corrupt governmental approval process downplayed and even falsified the "science" and collaborated with Monsanto to further the interests of U.S. foreign and domestic policies? Why haven't officials ordered the mass distribution of Vitamin D pills and zinc, especially in Black communities?

Monsanto's Roundup, and its genetically engineered "Roundup Ready" crops, are designed for saturation spraying, preserving the genetically modified plant while obliterating weeds and unengineered plants in the soil. Monsanto benefits mightily in terms of corporate profits from such an approach, while sacrificing the soil and nutritional quality of the foods made from them. Even more important than immediate profits to those in power is the longer-term control of the world's food supply, which is as much a part of political and imperialist strategies as it is a mechanism for achieving immediate economic benefit. Billionaires Bill and Melinda Gates, having purchased $27 million of Monsanto's stock in 2010, announced their hearty support for genetically engineered agriculture. They have become the largest individual owners of farmland in the U.S., currently owning 269,000 acres of farmland in eighteen states.[15]

Abroad, with the help of the biotechnological production of the world's agriculture and the massive spraying of Bayer/Monsanto's Roundup, corporate behemoths are significantly removing small family farmers from their lands and food supply, enabling the corporate takeover of those lands for the planting of GMO export crops. This new colonization of the small farmer and the world's agriculture is a hallmark of neoliberalism—itself a poor description for what should be more accurately categorized as global "neo-feudalism." The World Economic Forum's "great reset"[16] calls for uprooting workers who once farmed those lands to feed themselves and their communities, and removing them to so-called "export zones." As WEF Chair Klaus Schwab notes: "The pandemic represents a rare but narrow window of opportunity to reflect, reimagine, and reset our world to create a healthier, more equitable, and more prosperous future."[17] Welcome to the New World Order, where mass-produced GMO crops are heavily sprayed and assembled into material for clothing and fuel in addition to food, and sold abroad to consumers in the U.S. and wealthier countries.

Even prior to the Covid pandemic, a European Parliament investigation reported in January 2019 that EU regulators had based their decision to re-license Roundup on a supposedly independent assessment that was shadow-written by Monsanto. The scandal that ensued over Monsanto and its product Roundup in Europe brought to light so much corruption that a number of governments at last introduced legislation restricting the use of Roundup, and in some cases banning it altogether. But the regulatory bodies in the U.S. continue to deflect responsibility for the falsified studies leading to the government's administrative approval of Roundup.

In Chapter 17 of this book, "When Rights Collide: Genetic Engineering & Preserving Biocultural Integrity," Dr. Martha Herbert explores how developing a systemic awareness of the collusion between government and corporations leads to new ways of seeing the ramifications of such collusion. Dr. Herbert examines the role that GMOs play in the larger geopolitical context. The genetic engineering of agriculture is key to the consolidation of corporate control of the world's farmlands and seeds. With thousands now recognizing the larger geopolitical purposes, that awareness could and should change the nature and demands of our *environmental* movements, and their philosophies, programs, and strategies.

In 2019, New York City, as a result of a decades-long battle by environmental justice activists, enacted a new anti-pesticides law. (It went into effect in 2021). The new law is well-intended, but—as is customarily the case when movements rely on legislation they've achieved but don't accelerate and expand the fight—it contains certain "flaws of omission". Here, those include the exclusion of golf courses from the environmental requirement to reduce pesticide applications—Golf courses like the one owned by Donald Trump at the foot of the Whitestone Bridge in the Bronx. This is a very big deal since golf courses are major users of herbicides that then leach into surrounding neighborhoods and the public water supply, and increase the possibilities of cancer for golfers on those courses as well as in the surrounding communities. The new law does, however, look to greatly reduce the amounts of pesticides used in the city overall, particularly in and around schools where children congregate and who are especially vulnerable to the ravages of environmental toxins. Even small quantities of pesticides can have major long-term effects. The government has also released tens of thousands of genetically engineered mosquitoes[18] for the first time into the East Coast environment despite the new law, and continues to spray to kill "nuisance" mosquitoes as well as those thought to be transmitting diseases such as West Nile Virus.

Nevertheless, many activists in New York City remain hopeful that the worst excesses of the applications of pesticides, including Roundup, will be ameliorated by the new law. However, given ecological activists' experience with the NYC government over its ongoing pesticide spraying despite the No Spray Coalition's court victory against the City on this issue fifteen years ago (see, herein, Mitchel Cohen's "Poisoning the Big Apple—Forgotten History in the Lead-Up to 9/11"), environmental activists are not exactly holding our breath waiting for a flawed law to bring to an end the mass use of pesticides. On the other hand, given the ongoing spraying, holding our breath would indeed be more than appropriate.

Mitchel Cohen, Brooklyn
April 2022

Introduction

O, pardon me, thou bleeding piece of earth,
That I am meek and gentle with these butchers![1]
—William Shakespeare, Julius Caesar, Act 3, Scene 1

A battle royale is ripping through every country against the Monsanto Company's carcinogenic chemical glyphosate, the primary active ingredient in its most profitable pesticide, "Roundup." (Monsanto is now owned by the drug and agrochemical multinational corporation Bayer.)

Advocates of pesticides claim that they increase crop yields, protect the public from insect-borne diseases, and save labor costs by chemically killing "weeds." Opponents counter that synthetic pesticides harm human beings, animals, beneficial insects, wildlife, and plants;[2] they pollute drinking water and food chains, increase health-related costs, and represent a contemptuous and colonizer's approach to life and nature. The production and application of pesticides for corporate profit ignores and, in fact, assaults genuine health needs and the ecological balance of the natural environment.

For every environmental movement success in pressuring governments to ban an egregious pesticide, the industry spits out a new one and the cycle begins again. Victories on individual pesticides are undermined by a methodology that examines each chemical in isolation from the others; each corporate polluter is seen as an exception, a "bad apple" in an otherwise benevolent system. Thus, arsenic begat DDT, DDT begat organophosphates, the first wave of organophosphates begat pyrethroids and glyphosate, and now glyphosate begets dicamba.

Jonathan Latham, co-founder and Executive Director of the Bioscience Resource Project and the Editor of *Independent Science News*, points out in his chapter "Unsafe at any Dose? Glyphosate in the Context of Multiple Chemical Safety Failures," that although stopping the applications of glyphosate will be a significant victory, it will not be enough; the chemical corporations' policies will remain unchanged. They will simply substitute another poison.

Movements concerned with stopping the mass applications of pesticides need to go deeper, beyond the usual concerns about a particular chemical or corporation. If each pesticide, banned after years of struggle, is thought of as the exception to the rule, then the system itself is assumed to be fundamentally stable and beneficial, save for those few rotten apples.

The system, though, is fundamentally unstable, unsustainable, and harmful. It reflects—and regurges—an approach to nature and to human life in which life is denigrated, and maximization of corporate profits is par for the course. (Golf courses are one of the prime abusers of pesticides in urban areas.)

The continued use of chemical fertilizers and pesticides degrades the soil, killing off the biodiversity needed to rebuild and enrich it. It's a destructive and unstable system perpetuated by the ravages of corporate capitalism as it engages in its relentless drive to expand, consolidate smaller companies, centralize production, and exert monopolistic control. We will never succeed in saving ourselves, our children, and the environment by opposing one pesticide (or pipeline, or corporation) at a time. Consideration of more radical frameworks and actions is therefore essential if ecological activists are to build upon our limited victories and save complex life on this planet.

In 2015, rocker Neil Young wrote and belted out the lead song for an album titled, *The Monsanto Years*:

You never know what the future holds in the shallow soil of Monsanto,
    Monsanto
The moon is full and the seeds are sown while the farmer toils for Monsanto,
    Monsanto
When these seeds rise they're ready for the pesticide
And Roundup comes and brings the poison tide of Monsanto, Monsanto
The farmer knows he's got to grow what he can sell, Monsanto, Monsanto
So he signs a deal for GMOs* that makes life hell with Monsanto,
    Monsanto
Every year he buys the patented seeds
Poison-ready they're what the corporation needs, Monsanto

—Lyrics and song by Neil Young,[3] "The Monsanto Years," (2015)
*GMOs = genetically modified organisms

Monsanto officials attempted to discredit Neil Young. They investigated him, monitored his communications, and posted internal memos about his social media activity and music. The company did the same with Reuters' senior correspondent Carey Gillam, who published devastating investigations of the company's weedkiller and its links to cancer.[4]

And then Dewayne "Lee" Johnson, a forty-six-year-old groundskeeper and pest-control manager at Benicia Public School District in Solano County, California, sued Monsanto. Johnson was diagnosed with non-Hodgkin's lymphoma after he had repeatedly sprayed Roundup, as directed, on school grounds (and thereby unwittingly jeopardized the lives of 5,000 young students in that district as well as his own). After a four-week trial ending on August 10, 2018, a unanimous jury awarded Johnson $289 million.

While San Francisco Superior Court judge Suzanne Bolanos upheld the jury's verdict finding that Roundup indeed caused Johnson's cancer, she cut its unprecedented punitive damage award by seventy-five percent.[5]

Some jurors were so upset by the prospect of having their verdict thrown out that they wrote to Bolanos: "I urge you to respect and honor our verdict and the six weeks of our lives that we dedicated to this trial," wrote juror Gary Kitahata. Another, Robert Howard, said that the jury had paid "studious attention" to the evidence and that any decision to overturn its verdict would shake his confidence in the judicial system.[6]

"The cause is way bigger than me," Johnson said. "Hopefully this thing will start to get the attention that it needs to get right."[7] Johnson's was the first of what has mushroomed into 125,000 lawsuits against Monsanto over cancers, injuries and deaths caused by Roundup.[8] In another legal victory against Monsanto, the Court of Appeal for the State of California affirmed the lower court's decision, ruling that "Monsanto acted with willful disregard for the safety of others ... Monsanto's conduct evidenced reckless disregard of the health and safety of the multitude of unsuspecting consumers it kept in the dark. This was not an isolated incident; Monsanto's conduct involved repeated actions over a period of many years motivated by the desire for sales and profit."[9]

Dewayne Johnson's victory against Monsanto—a David vs. Goliath battle—snapped attention to the mass use of toxic pesticides and the horrors that thousands of workers are subjected to on a daily basis. It also shined a spotlight onto the corruption of regulatory agencies charged with protecting the health of those workers, and on the Bayer corporation's sordid history of medical "experiments" done first on workers at the Mexico-US border and then on the transmission of that "research" over the Atlantic, to use on the Jewish minority population in Nazi Germany's concentration camps and gas chambers leading into World War II.[10]

"We didn't cross the border, the border crossed us."

Along the U.S. southern border, agribusiness conglomerates recruit farmworkers from Mexico and Central America to pick crops for low pay, under wretched conditions. There are two million workers on U.S. farms who grow and harvest the fruits, vegetables, grains, dairy, and meat that the rest of the country depends on for its daily food—almost all of it, like the workers, soaked with pesticides. American folksinger Woody Guthrie's song, "Pastures of Plenty," depicts those conditions:

> It's a mighty hard row that my poor hands have hoed
> My poor feet have traveled a hot dusty road
> Out of your Dust Bowl and westward we rolled
> And your deserts were hot and your mountain was cold
>
> I worked in your orchards of peaches and prunes
> Slept on the ground in the light of your moon
> On the edge of the city you'll see us and then
> We come with the dust and we go with the wind[11]

Every year, there are ten to twenty thousand cases of farmworker poisonings reported in the US, with different chemicals impacting different areas of the body. For example, citrus industry workers have higher incidence of gastric cancer, which is caused by phenoxyacetic acid herbicide 2,4-D; the organochlorine insecticide chlordane; and the herbicide triflurin.[12]

That spraying is meant to kill insects and rodents; all too often the health of the workers is a secondary concern, when it is considered at all. We can trace the U.S. government's mass spraying of migrant workers at least as far back as 1917, when, under the guise of protecting the country from the threat of typhus, U.S. Customs agents began delousing Mexicans who were legally crossing the border at the El Paso-Juarez international bridge and into areas of the U.S. that were formerly *part of* Mexico.

In 1917 alone, 127,000 workers crossing the border were forced to strip naked and given pesticide showers. By the 1920s, the chemical used at the border switched to the notorious cyanide-based Zyklon B, manufactured by the German chemical conglomerate IG Farben. Border agents tested the gas on Mexican workers, with the results sent to the German affiliate where it was used by the Nazis to exterminate Jews in the gas chambers. Like the Mexican workers, the Jews in Germany were similarly considered "vermin."

Historian/musician David Dorado Romo remembers how his great Aunt, Adela Dorado, would tell his family about the humiliation of having to go through the delousing every eight days "just to clean American homes in El Paso. All immigrants from the interior of Mexico, and those whom U.S. Customs officials deemed 'second-class' residents of Juarez, were required to strip completely, turn in their clothes to be sterilized in a steam dryer and fumigated with hydrocyanic acid, and stand naked before a Customs inspector who would check his or her 'hairy parts'—scalp, armpits, chest, genital area—for lice. Those found to have lice would be required to shave their heads and body hair with clippers and bathe with kerosene and vinegar."[13] She recalled how on one occasion "the U.S. Customs officials put her clothes and shoes through the steam dryer and her shoes melted."[14]

In a moment now forgotten from history, one day in 1917 a 17-year-old female migrant Carmelita Tores, working as a maid in Juarez, crossed the border as she did every day to clean houses and refused to undress and be showered in pesticides. By noon, she was joined by several thousand "refusers" at the border bridge. Carmelita Tores became the Rosa Parks of what would be dubbed "the Bath Riots."

The Orchestration of Disease. Manufacturing Fear of Immigrants.

At the time of the Bath Riots, the mainstream (corporate) press did everything it could to sensationalize the typhus "threat" from Mexican migrants. That disease devastated the Russian working class at the time of the 1917 revolution, along with tens of thousands of Austrian prisoners from World War I in Serbia.[15]

But in the U.S. that year, there was the small total of only thirty-one typhus cases overall, and only three typhus-related fatalities in El Paso. While public health was certainly a concern, officials used fear of Typhus as a vehicle for fomenting repressive migrant and anti-working-class policies. Today, Typhus—caused by a bacterium transmitted by infected body lice—is readily treated with oral Ivermectin and clean clothing.

As one of the founding corporations in the IG Farben consortium, Bayer had no compunction about testing drugs on unwilling human subjects—prisoners, soldiers, and migrants. In El Paso the company was presented with a golden opportunity and, for the next two decades, it found other opportunities to continue its human "experimenting", particularly on Jews and others imprisoned in Nazi concentration camps. The U.S. Holocaust Museum describes Bayer's involvement in Nazi "medical experiments" on Jews and other prisoners, who were deliberately infected against their will with tuberculosis, diphtheria, and other diseases at the Dachau, Auschwitz, and Gusen concentration camps. Nazi physician Helmuth Vetter, appointed as the German Reich's chief doctor by Heinrich Himmler, coordinated the experiments.

Bayer paid Vetter to test Rutenol and other sulfonamide drugs on deliberately infected prisoners. The Holocaust Museum notes that in Buchenwald, "physicians infected prisoners with typhus in order to test the efficacy of anti-typhus drugs, resulting in high mortality among test prisoners."[16] Bayer was particularly active in Auschwitz. "A senior Bayer official oversaw the chemical factory in Auschwitz III (Monowitz). Most of the experiments were conducted in Birkenau in Block 20, the women's camp hospital. There, Vetter and Auschwitz physicians Eduard Wirths and Friedrich Entress tested Bayer pharmaceuticals on prisoners who suffered from and often had been deliberately infected with tuberculosis, diphtheria, and other diseases."[17] Following World War II, Vetter was convicted by an American military tribunal at the Mauthausen Trial, and was executed at Landsberg Prison in February 1949.

Some employees of Bayer, writes the Holocaust Museum, "appeared in the IG Farben Trial, one of the Nuremberg Subsequent Tribunals under U.S. jurisdiction. Among them was Fritz ter Meer, who helped to plan the Monowitz camp (Auschwitz III) and IG Farben's Buna Werke factory at Auschwitz, where medical experimentation had been conducted and where 25,000 forced laborers were deployed. Ter Meer was sentenced to seven years, but was released in 1950 for good behavior. One positive outcome of these subsequent Nuremberg Trials was the establishment of the Nuremberg Code, a product of the Nuremberg Doctors' Trial which codified prohibitions against the kinds of involuntary experimentation conducted by Bayer in the concentration camp system.[18]

In the immediate postwar period, the victorious allies divided the IG Farben conglomerate into individual companies, but still allowed them to function. Bayer, along with BASF and Hoechst—all part of the IG Farben conglomerate and supporters of the Nazis in World War II—re-emerged as one of the world's largest pharmaceutical companies.[19]

Bayer, however, "did little to come to terms with its Nazi past," the Holocaust Museum notes, and adds this tidbit: "Fritz ter Meer, convicted of war crimes for his actions at Auschwitz, was elected to Bayer AG's supervisory board in 1956, a position he retained until 1964."[20]

IG Farben's Max Faust, center, accompanies Heinrich Himmler, second from left, leader of the SS, during an inspection of the plant at the Monowitz concentration camp in Poland in July 1942. Faust worked as an engineer for BASF prior to playing a lead role in building the plant. Credit: Galerie Bilderwelt/Getty Images

At the U.S. border with Mexico, the U.S. government policy of spraying Mexican workers with toxic pesticides continued apace thru the late 1950s and into the 1960s. The organochloride insecticide DDT became the pesticide of choice, until the movement inspired by Rachel Carson's book *Silent Spring*, forced out DDT after a decade of crescendoing protests. (More on Rachel Carson and the movement she inspired, later.)

Acute symptoms of exposures to pesticides include dizziness, nausea, vomiting, headaches, shortness of breath, and inflammation of eyes, nose, throat, and skin. Such exposure could lead to convulsions and death among those workers, their families and children, who suffered immediate and long-term health damage from pesticide-drift exposures at their nearby homes, playgrounds, and schools. As Patricia Wood details in Chapter 6 ("Children and Pesticides"), children have experienced not only the acute symptoms listed, but also learning disabilities, autism, and chronic respiratory diseases, among other negative long-term outcomes.

A key culprit in these poisonings is chlorpyrifos, a strong neurotoxin widely used to kill insects in orange, almond, grape, broccoli, and many other crop fields, where they mix with Roundup in a toxic soup. Even tiny levels of exposure to chlorpyrifos can damage the development of children's brains, a fact that resulted in its ban in 2000 for household uses. Yet the Environmental Protection Agency allowed it to be used in agriculture. In California, Latinx children are 91 percent more likely than others to attend schools near heavily sprayed areas. In one year alone, more than 750 pounds of chlorpyrifos pesticides were sprayed within a quarter mile of four different public schools in Tulare County, in the heart of California's Central Valley, where much of America's fruits and nuts are grown.[21]

Decades of pressure from environmental groups and the growing body of research documenting chlorpyrifos, Roundup, and other pesticides' harmful impacts on agricultural workers and surrounding communities, their destruction of soil microbes, and hazards to aquatic organisms came to a head in 2015, and resulted in former President Obama's Environmental Protection Agency proposal for a total ban on agricultural use of chlorpyrifos. In an article published in *The Ecologist* in September 2015, Richard Gale and Gary Null publicized documents that showed that Monsanto had, for decades, carried out detailed studies of glyphosate and Roundup toxicity.

Monsanto knew just how toxic its products were all along, while claiming they were 'safe as lemonade'.[22]

In March 2017, former President Trump's Environmental Protection Agency announced a reversal of the proposed ban on chlorpyrifos. It allowed its use in agricultural settings, after Dow Agrosciences, manufacturer since the 1960s of the pesticide under the brand name Lorsban, and two other chemical corporations, sent letters to the White House declaring the proposed ban to be ill-advised and harmful to the economic interest of growers. (They said nothing, of course, about the health effects on workers, nor on consumers abroad as well as here at home.) The corporations claimed that the science used to justify the ban was "flawed." Dow had also notoriously written a $1 million check to support Trump's presidential inauguration. Scott Pruitt, the first head of the EPA under Trump, said he rejected the ban, in order to provide "regulatory certainty to the thousands of U.S. farms that rely on chlorpyrifos."[23]

But a U.S. federal Appeals Court ruled in favor of a lawsuit brought against the EPA by the grassroots enviro group EarthJustice, and ordered the agency to ban Chlorpyrifos.[24] EarthJustice celebrated the ruling as a huge victory for children and agricultural communities across the country, who will finally be spared the needless poisonings and lifelong learning disabilities. Taking heart from such court decisions, a coalition of agricultural workers and environmentalists have filed a lawsuit that seeks to reverse the Environmental Protection Agency's approval of glyphosate.[25]

An Aspirin the Size of the Sun

In June 2018, the U.S. Justice Department approved the $66 billion purchase of Monsanto by the German pharmaceutical corporation Bayer, making Bayer-Monsanto the most powerful agribusiness entity on the planet. Bayer—the leading manufacturer of neonicotinoid pesticides responsible for the mass slaughter of the world's bees—now owns and controls more than 25 percent of the world's seeds, and more than a third of global herbicide sales.

The Bayer-Monsanto consolidation—one of the rotting legs of what physicist and world ecology advocate Vandana Shiva calls "The Poison Cartel"—came on the heels of the merger of the agricultural divisions of Dow and Dupont (now called Corteva Agriscience), and Syngenta's merger with ChemChina. (Syngenta itself was the outcome of the consolidation of part of Novartis with AstraZeneca.)

As a result of its acquisition of Monsanto, Bayer—no stranger to protecting itself from condemnations of its dreadful record when it comes to human rights and environmental justice—now has to decide how to proceed with the torrent of lawsuits, which have resulted in penalties and fines soaring into many billions of dollars.[26] Bayer could not alleviate this headache even by gulping down the adult-size dose of its other famous drug, packaged in its familiar yellow and brown box. To do so would take, in poet Roque Dalton's verse, "an aspirin the size of the sun".

Bayer gambles that its vast control of seeds will eventually offset the costs of the corporation's acquisition of Monsanto/Roundup's liabilities. (Monsanto has also reportedly been looking to develop a genetically engineered marijuana seed, a technology that may be part of Bayer's calculations, especially as marijuana reform movements succeed in legalizing marijuana in many states in the U.S.[27])

Bayer—on the hook for an estimated $1 trillion—is attempting to settle existing lawsuits and pay for future ones for a total bargain price of $14.5 billion while making no admission of liability or wrongdoing. A lesser amount was nixed by a U.S. federal judge. Bayer's CEO Werner Baumann apparently expected that ruling: "We have an alternative course of action: we are in charge and in control now. We continue to pursue a comparable solution. There are different ways to skin a cat,"[28] Baumann delicately put it.

Following the jury's verdict in the Dewayne Johnson case, the company's $63 billion stock value plummeted by almost 40 percent. But the decline in value of Bayer/Monsanto's stocks has since steadied and bounced half-way back, making the injuries, mass poisonings and deaths—skinning cats, Bayer put it—"acceptable risks," the costs and benefits of doing business under capitalism. Bayer maintains it will continue selling its weed killer, and stands by the company's assertions that it's safe when used as directed, despite findings to the contrary by international scientific bodies and U.S. courts.

But by July 2021, the financial pressures on Bayer forced the company to begin pulling Roundup from the shelves of such companies as The Home Depot and Lowe's; as of 2023, Roundup will no longer be sold to individual gardeners in the U.S.,[29] a historic victory for enviro-activists and the environment.[30] (It is, unfortunately, still being allowed on vast agriculture plantations and fields, industrial and municipal applications, and in parks and forests.) Bayer will replace its glyphosate-based products with "new formulations that rely on alternative active ingredients beginning in 2023, subject to a timely review by the U.S. Environmental Protection Agency (EPA) and state counterparts."[31]

Andrew Kimbrell, executive director of the Center for Food Safety (CFS), described Bayer's decision to end U.S. residential sales of Roundup, as "a historic victory for public health and the environment," But "as agricultural, large-scale use of this toxic pesticide continues," he added, "our farmworkers remain at risk. It's time for EPA to act and ban glyphosate for all uses."[32]

As is typical under American jurisprudence, Bayer denied admitting to any of the health and safety claims made in tens of thousands of outstanding lawsuits against the company with regard to Roundup and chose only to emphasize its "bottom line." With its transparently false claims regarding Roundup's safety, Bayer's lack of morality hastens workers' mortality, an immorality that even Monsanto's losing the flood of lawsuits cannot wash clean.

Bayer is also facing a new wave of lawsuits, over Monsanto's follow-up pesticide called dicamba (also known as XtendiMax). In Feb. 2020, Jurors awarded $265 million to a Missouri farmer who blamed the herbicide for destroying his peach farm.[33]

Lawsuits have become an important tool enabling the modern ecology movement to gain victories against Monsanto's grave miscarriages of justice and ecological destruction. In one lawsuit filed by the Center for Food Safety, the U.S. Environmental Protection Agency was forced to admit it had failed to account for glyphosate's risk to Monarch butterflies and other endangered species.[34] But rather than reversing its approval of Roundup, the EPA told the court it wanted even more time to reassess. In the meantime, it would leave this pesticide on the market, despite EPA's admission that it didn't look carefully enough at Roundup's danger to butterflies and other species, and failed to consider Roundup's financial and social costs to farmers and farmworkers. In fact, the U.S. government regulators accepted bribes and falsified studies submitted by Monsanto.[35]

U.S. President Joe Biden continues the awful legacy of his predecessor. In June 2019, Donald Trump issued an Executive Order directing federal agencies to exempt many GMOs from regulation. Trump publicly aligned himself with pesticide-seed companies to promote the cultivation of GMO crops.[36] The Biden administration follows suit in defense of Monsanto and its attempt to control agriculture throughout the globe via genetically engineering the world's food supply, for which its herbicide, Roundup, has been so destructively designed.

Biden has gone so far as to appoint "Mr. Monsanto"—Tom Vilsack—*again*, as Agriculture Secretary. Vilsack served in that same capacity for eight years in the Obama administration, despite much protest from ecology activists.[37] This same Poison Cartel, under Republican as well as Democratic presidents, has dictated U.S. government food policy while accumulating trillions of dollars from its manufacturing of pharmaceutical drugs to "save us" from the cancers and neurological diseases their pesticides are causing. The "revolving door" spins freely between corporate interests and regulatory agency apparatchiks no matter which party is in power.[38] The agencies that are supposed to be looking out for the public's health not only "looked the other way" but have now been exposed for having ghost-written Monsanto's applications for Roundup and other chemicals—submitted for approval to those very same agencies that illegally helped draft them.[39]

In December 2017, the National Marine Fisheries Service issued a 3,700-page report warning that three widely-used pesticides pose a threat, through run-off into rivers and oceans, to salmon, sturgeon, orca and dozens of other endangered and vulnerable aquatic species. The report's release on the three organophosphate pesticides—chlorpyrifos, diazinon and malathion—was the result of a long legal battle by environmental groups to extract the report from history's dust bin and win its publication. The Trump administration, following heavy lobbying by the chemical manufacturers who claimed the report was flawed, then pushed for a two-year delay.

Glen Spain, northwest regional director of the Pacific Coast Federation of Fishermen's Associations, told the *Guardian* that the $1 billion industry has lost thousands of jobs as salmon numbers have plummeted. "Pacific salmon is the life-blood of our industry . . . [and] the Trump administration is supposed to be about jobs, isn't it?," Glen Spain noted.[40] The growth, reproduction, and swimming ability of salmon can be impaired by even low levels of pesticide run-off and, according to Joseph Bogard, executive director of the Save Our Wild Salmon Coalition, the Pacific Northwest's wild salmon population has now shrunk to only five to six percent of historic levels. And of the resident orca population in Puget Sound, Bogard warns, "There are only 76 individuals in this orca community and they lost eight in the last 18 months, which is a devastating rate. The population is starving and is being poisoned by these toxins."[41]

The entire planet is awash in chemical pollutants, pesticides and other effluents from multiple industrial sources that poison our drinking water, food and soil, human breast milk,[42] animals, and ecosystems. The U.S. government has a long and pathetic history of groveling before and collaborating with the titans of industry, whose propaganda machines promote their self-interested assurances that their products are "safe" and environmentally friendly. They know that truthful information, under the right circumstances, can move people to rebel. So they contaminate the truth, just as they pollute the natural environment, to foster public acceptance of pesticides and genetic engineering of the world's food supply.

Will the growing awareness of pesticides and their effects on our lives broaden into critical challenges on related issues such as habitat fragmentation and destruction, climate chaos, agribusiness based on monocropping, genetic engineering, patenting of seeds, and the loss of biodiversity? Will activists in the United States and other industrial countries be able to force their governments to reverse course? Will they succeed in challenging the corporate quest for ever-increasing profits and control? To do so requires those reading this book to be reborn as ecology activists who strive to win society to a different way of looking at human interactions with nature—no easy task, in current circumstances—and to take action based on that transformed consciousness. The consumer preference demonstrated by the growth of the organic and non-GMO food movement shows that many people are paying attention to the degradation of the food supply and the failure of regulators to act in the public interest.

Julian Assange, Wikileaks, and U.S. Food Policy

Key pieces of information regarding the U.S. government's worldwide advocacy (including the threatened use of its military) on behalf of Monsanto's patented seeds exploded onto the internet via thousands of cables published by current political prisoner Julian Assange. Those cables revealed massive U.S. government attempts on behalf of Monsanto, and its patents, to arm-twist countries throughout the world, along with its attempts to squelch opposition to GMOs. The cables showed U.S. diplomats applying financial, diplomatic, and even military pressure on behalf of Monsanto and other biotech corporations.

In a 2007 cable marked "confidential," Craig Stapleton, then U.S. Ambassador to France, advised the U.S. to prepare for economic war with countries unwilling to introduce Monsanto's GM corn seeds. He called for retaliation, to "make clear that the current path has real costs to EU interests and could help strengthen European pro-biotech voices. In fact, the pro-biotech side in France . . . [has] told U.S. retaliation is the only way to begin to turn this issue in France."[43] The U.S. diplomatic team recommended "that we calibrate a target retaliation list that causes some pain across the EU since this is a collective responsibility, but that also focuses in part on the worst culprits."[44]

The idea of U.S. government officials wanting to "cause some pain" to other countries is hardly a revelation. Ambassador Stapleton and his team's remark echoed a similar directive by Richard Nixon and Henry Kissinger vis à vis Chile in 1970-73. Nixon's deputy CIA director, Thomas Karamessines, wrote in a secret memo: "It is firm and continuing policy that [the elected President of Chile, Salvador] Allende be overthrown by a coup. . . . It is imperative that these actions be implemented clandestinely and securely so that the USG [the U.S. government] and American hand be well hidden." Under Henry Kissinger's managerial influence, President Nixon ordered the CIA to "make the economy scream" in Chile to "prevent Allende from coming to power or to unseat him."[45]

In another cable, this one from Macau and Hong Kong, a U.S. Department of Agriculture director requested $92,000 in U.S. public funds for "media education kits" to combat growing public resistance to GMO foods. It portrays attempts to mandate the labeling of GMOs as a "threat" to U.S. interests, and seeks to "make it much more difficult for mandatory labeling advocates to prevail."

The cables released by Wikileaks revealed that officials in the Obama administration, particularly in Hillary Clinton's State Department, intervened at Monsanto's request "to undermine legislation that might restrict sales of genetically engineered seeds."[46] Under Hillary Clinton, the U.S. State Department was so gung-ho to promote GMOs that *Mother Jones* writer Tom Philpott called it "the de facto global-marketing arm of the ag-biotech industry, complete with figures as high-ranking as former Secretary of State Hillary Clinton mouthing industry talking points as if they were gospel."[47] The *New York Daily News* reported that State Department officials under Hillary Clinton were actively using taxpayer money to promote Monsanto's controversial GMO seeds around the world.[48]

The fight against GMOs and Roundup is partly a propaganda war; U.S. officials recommended pro-biotech and bio-agriculture DVDs be sent to every high school in Hong Kong.[49] The cables reveal the joint strategic planning of Monsanto and the U.S. government. In one series, Monsanto and the U.S. government concluded that northern Thailand would be an ideal location to cultivate genetically engineered corn for export to other countries, due to the area's very low labor and infrastructure costs.

In this cable, one country, Peru, is designated as recipient, and the U.S. suggests that even with transportation expenses across two oceans included it would nevertheless be more profitable to grow and ship GMO corn from northern Thailand to Peru than from neighboring Argentina or Brazil. U.S. "diplomatic efforts" would be used to drive down the cost of production in northern Thailand and would press Thailand to drop its opposition to GM cultivation. In that scenario, Vietnam, the Philippines, and Indonesia lose out in the competition for those investments (though they gain in terms of human and ecological health).[50] The cables provide a fascinating (and terrifying) glimpse into the mechanisms of global imperialism on a very localized level, with Monsanto invoking the U.S. government's formidable economic and military power to impose its strategies. (See "Genetic Engineering, Pesticides, and Resistance to the New Colonialism" (Chapter 15), which examines the revolving door of U.S. government corporate, regulatory, and foreign policy officials, eager to serve Monsanto's interests and mold U.S. foreign policy around them.[51])

Among the most revelatory documents Wikileaks "acquired" and published were the searchable and unabridged texts and database of the secret 2015 TransPacific Partnership, Transatlantic Trade and Investment Partnership, and Trade in Services Agreement.[52] As the cables' publisher, Julian

Assange exposed the U.S. government's pressure on other countries to purchase and plant Monsanto's patented genetically engineered seeds, which required the concomitant purchase of Monanto's patented pesticides in order to grow.

The documents reveal in excruciating detail the tight interplay between government and corporate strategies on the most expansive but also on the most minuscule of levels. The treaties limited the ability of one country to legally challenge environmental depredation in trade with another, making it abundantly clear that environmental issues could not be successfully addressed in piecemeal fashion, but must be seen as integrated political, technological, economic, and scientifically packaged warfare. To succeed, movements would be compelled to not only examine the dangers of each pesticide *du jour* but, just as important, the underlying mechanisms by which corporations such as Monsanto, Bayer, Dow, DuPont, Syngenta, Novartis, BASF and other pesticide and pharmaceutical manufacturers come to determine government policies, while masking the truth about their products.

Environmental activists have always exposed the collaboration between government and corporate expansion, but the details exposed by Wikileaks' documents are nothing short of astounding. They reveal the need for ecological movements to develop far more radical strategies for dealing with the immense destruction by capitalism in practice, and not just in theory. For this largely unknown contribution by Julian Assange, ecological activists, along with antiwar radicals motivated by Wikileaks' "collateral damage" video (obtained from Chelsea Manning), owe him a debt of gratitude that can never be fully repaid.

Today, Julian Assange is locked away in a British prison, despite judicial findings in his favor, and is fighting for his life. The U.S. government seeks to bring this Australian citizen back to the United States for a show trial and then lock him up forever, if they don't assassinate him en route.[53] The sacrifices Julian Assange has made are profound, and his contribution to ecological as well as antiwar movements are enormous. It is incumbent on all to demand an end to his incarceration and torment by the U.S. and British governments.[54]

* * *

Throughout *The Fight Against Monsanto's Roundup: The Politics of Pesticides*, herbicides (for weeds) and insecticides (for insects) are seen as subsets of the overarching category, pesticides. The Federal Insecticide, Fungicide, and Rodenticide Act (FIFRA) is the statute that governs the registration, distribution, sale, and use of pesticides in the United States. With certain exceptions, a pesticide is any substance or mixture of substances intended for preventing, destroying, repelling, or mitigating any "pest," or for use as a plant regulator, defoliant, desiccant, or any nitrogen stabilizer.

While the chemical glyphosate and Monsanto/Bayer's Roundup are treated interchangeably, they are not the same. Roundup's formula of ingredients contains more than glyphosate. Roundup was never tested prior to being granted government approval as a composite formula; only glyphosate was examined as a weed-killer. A proper submission should include data on the entire formula, including so-called "inert" ingredients like arsenic, surfactants, and Perfluoroalkyl and Polyfluoroalkyl substances (PFAS).

Despite worldwide exposures of glyphosate's dangers and its designation as a "probable carcinogen," only a handful of governments throughout the world joined with environmental activists and health professionals in banning it. We—people who want to breathe clean air, drink pure water, preserve what's left of the old-growth forests, protect the many species that share this earth with us, and escape from the epidemics of cancer and neurological disorders—need to grasp why government officials ever approved Roundup, and ask profoundly radical questions about the misuses of science and our relationship to nature. For example:

- What is stopping officials from effectively opposing Roundup and other pesticides?
- Since it has never been determined and convincingly reported that Roundup is safe, why was Roundup allowed to be applied for so many decades, contaminating so much of the earth's soil?
- In fact, why did government officials, and international as well as domestic agencies, ever allow it at all?
- How did Monsanto and other companies thwart attempts by various government agencies to regulate Roundup and glyphosate?

Next comes a question about strategy for anti-pesticide activists:

- How significant is the fight to ban individual pesticides, since the industry releases new and equally dangerous ones into the environment, to replace the ones being banned or withdrawn?[55]

Finally,

- What lessons can ecology and social justice movements draw from victories and losses of the global struggle to ban Monsanto's Roundup and the fight to abolish the manufacture and use of chemical pesticides?

* * *

The Fight Against Monsanto's Roundup: The Politics of Pesticides encourages readers to think about human beings' relationship to nature, and the weaknesses and contradictions of the process used to approve pesticides. The purported experts have been proven wrong on so many occasions that we'd be fools to take their acceptance of Roundup at face value, especially since many researchers conceal their financial arrangements with corporate funders, thereby biasing the outcome and reporting of their research.[56]

The 2016 occupation and blockade of an underground oil pipeline under construction at the Standing Rock Indian Reservation in North Dakota offered a wider vision for how to construct effective social movements. The Dakota Access pipeline was to carry oil 1,172 miles from the Bakken oil fields for distribution in the U.S. Midwest. More than 10,000 participants took part in an occupation lasting months, and, unlike the politicians in D.C., heroically refused to be divided by the false assurances of those in power.

But even in the face of unprecedented, united opposition to the pipeline among Native American tribes and environmental activists, corporate power would not yield; it bulls forward, striving to add ever-newer and equally destructive pipelines, pesticides, and genetically engineered plants.[57] And budget-conscious officials—many of them in thrall to the pesticide industry—have decided that it is more cost-effective to lay off workers and replace their much safer weeding-by-hand with chemical herbicides like Roundup. In the short-run, chemical pesticides appear to reduce public expenditure . . .

until the outsized health and environmental costs are factored in, and the bill comes due.

Clean water, soil, and air remain as necessary today as they were in 1962, when Rachel Carson issued her call-to-arms against chemical pesticides such as DDT and, at the same time, against radioactive isotopes from atomic bomb tests. Monsanto's Roundup and the genetic engineering of agriculture have now significantly poisoned the planet's soil; the battles against those inter-connected technologies are indeed taking place throughout the world—as are the related environmental battles against monocropping, hydrofracking, climate chaos, huge dams, mountaintop removal, nuclear power plants and weapons, oil and gas pipelines, pollution, factory farming, EMF pollution, and the destruction of wetlands and the resulting floods. Sixty years ago, *Silent Spring* understood that industrial capitalism was (and remains) anti-ecological at its core. It bequeathed today's movements a legacy of insights, efforts, victories, and sacrifices upon which to draw in combining the fight against individual pesticides with opposition to the systemic wars promul-gated by the same corporations and governments for labor, land, resources, geopolitical control, and private corporate profits . . . to save the world, and to change it.

Genetic Enginearrings

with apologies to Joyce Kilmer

I think that I have never seen
A tree as lovely as the Neem
Whose twigs when chewed prevent disease
Keep bugs away by grinding seeds
A tree that may in summer wear
A nest of vipers in her hair
Who folds her leafy arms to pray
That no one steal her DNA
Patent her soul. What sick disgrace
Would profit W. R. Grace
And Company? Forgive my rant, O
But Grace is owned now by Monsanto.

1

Roundup the Usual Suspects

MONSANTO STANDS EXPOSED! ... ROUNDUP THE USUAL SUSPECTS!

Illustration by Haideen Anderson

As imperceptibly as grief
the Summer lapsed away
—Emily Dickinson

It was not a silent spring. For decades, ecology activists have been targeting the Monsanto Company, its herbicides, and genetically engineered crops as the embodiment of an industry that will do anything it can get away with for profit, regardless of how much environmental damage it causes and havoc it wreaks on human health.

In March 2017, the state of California, following in the footsteps of the United Nation's World Health Organization, declared its intent to list the widely used chemical glyphosate—the main active ingredient in the Monsanto Company's herbicide Roundup—as a cancer-causing agent. Monsanto challenged the state's designation in California State Supreme Court. And, in a substantial defeat for the corporation in July 2017, Monsanto lost.[1]

And then on April 19, 2018, a California appellate court delivered a crushing blow to Monsanto, rejecting the company's appeal of the earlier decision and reaffirming that Monsanto's glyphosate pesticide can be listed as a known carcinogen under California's Proposition 65, which "requires notification and labeling of all chemicals known to cause cancer, birth defects or other reproductive harm, and prohibits their discharge into drinking waters."[2]

Grassroots activists had been advocating that action since the early 1990s. *What took California so long?* Small farmers as well as gigantic industrial agricultural companies had for decades applied greater and greater amounts of Roundup, especially once Monsanto's genetically modified Roundup Ready corn and soy crops had been engineered to withstand the herbicide. It's a lucrative technological fix, where an herbicide is patented and plants are engineered to "tolerate" that chemical.

When the court released heretofore secret Monsanto documents to public scrutiny, ecology activists' decades-old suspicions were confirmed. The secret Monsanto documents revealed that those active in anti-pesticide protests, along with those opposing the World Trade Organization in Seattle in 1999 and hundreds of lesser-known actions, indeed had the good sense to give voice to their outrage. According to the *New York Times*, "The records suggested that Monsanto had ghostwritten research that was later attributed to academics." Moreover, they "indicated that a senior official at the Environmental Protection Agency had worked to quash a review of Roundup's main ingredient, glyphosate, that was to have been conducted by the United States Department of Health and Human Services."[3]

The documents show that both industry and regulators "understood the extraordinary toxicity of many chemical products and worked together to conceal this information from the public and the press," according to the Bioscience Resource Project, which publishes groundbreaking and topical analyses of the science underlying food and agriculture systems in the peer-reviewed academic literature and elsewhere.[4] "These papers will transform our understanding of the hazards posed by certain chemicals on the market and the fraudulence of some of the regulatory processes relied upon to protect human health and the environment."[5]

Dr. Jonathan Latham, executive director of the project, explains:

> These documents represent a tremendous trove of previously hidden or lost evidence on chemical regulatory activity and chemical safety. What is most striking about them is their heavy focus on the activities of regulators. Time and time again regulators went to the extreme lengths of setting up secret committees, deceiving the media and the public, and covering up evidence of human exposure and human harm. These secret activities extended and increased human exposure to chemicals they knew to be toxic.[6]

Revelations in 2016 about the spraying of glyphosate by the Quaker Oats Company on its pre-rolled oats as a drying agent,[7] the discovery that 90 percent of the samples tested of "socially responsible" Ben & Jerry's ice cream contained glyphosate (traced to the candy added, not to the milk),[8] and, shortly after, findings of glyphosate in leading orange juice brands[9] provided impetus for heated global debate. And now new findings reveal that children's vaccines[10] as well as popular brands of wine and beer are contaminated with the cancer-causing pesticide.[11]

A report by Carey Gillam, widely disseminated by EcoWatch, claimed that Monsanto intentionally suppressed information about the potential dangers of its Roundup herbicide and relied on corrupt U.S. regulators to cover it up.[12] The evidence of government's collusion with Monsanto infuriated the jury in the case of Dewayne Johnson v. Monsanto. In the summer of 2018, it voted unanimously to award Johnson, the school groundskeeper and plaintiff, a quarter-of-a-billion dollars, much of it as part of punitive damages against the company. Currently, thousands of people are suing Monsanto, alleging that Roundup "caused them or their family members to become ill with non-Hodgkin['s] lymphoma, a type of blood cancer. . . . The documents, which were obtained through court-ordered discovery in the litigation, are also available as part of a long list of Roundup court-case documents compiled by the consumer group U.S. Right to Know."[13]

Attorney Brent Wisner was stunned at the revelations found in the documents. "This is a look behind the curtain," he said. "These [documents] show that Monsanto has deliberately been stopping studies that look bad for them, ghostwriting literature and engaging in a whole host of corporate malfeasance. They [Monsanto] have been telling everybody that these products are safe because regulators have said they are safe, but it turns out that Monsanto has been in bed with U.S. regulators while misleading European regulators."[14]

Robert F. Kennedy Jr., one of the plaintiff's lawyers, explains that this trove of released documents shows Monsanto executives:

> colluding with corrupted EPA officials to manipulate and bury scientific data to kill studies when preliminary data threatened Monsanto's commercial ambitions, bribing scientists and ghostwriting their publications, and purchasing peer review to conceal information about Roundup's carcinogenicity, its toxicity, its rapid absorption by the human body, and its horrendous risks to public health and the environment.
>
> We can now prove that all Monsanto's claims about glyphosate's safety were myths concocted by amoral propaganda and lobbying teams. Monsanto has been spinning its lethal yarn to everybody for years and suborning various perjuries from regulators and scientists who have all been lying in concert to American farmers, landscapers and consumers. It's shocking no matter how jaded you are! These new revelations are commiserate with the documents that brought down big tobacco.[15]

Especially indicting in the Monsanto documents are messages from some of the corporation's own scientists who questioned the safety claims Monsanto was making, but who were either obstructed or ignored. "In one email," reports EcoWatch, "Monsanto scientist Donna Farmer writes, 'you cannot say that Roundup is not a carcinogen . . . we have not done the necessary testing on the formulation to make that statement. The testing on the formulations are not anywhere near the level of the active ingredient.'"[16] Monsanto nevertheless disregarded its own scientist's warning; it went ahead and claimed that Roundup is not a carcinogen and is perfectly safe.

Nor is cancer the only concern. Some researchers have also linked glyphosate and other pesticides to what appears to be a major upsurge in autism in children,[17] and to Monsanto's marketing of its pesticide to farmers planting genetically engineered seeds, which are frequently sprayed with much greater doses of pesticides than non-GMO crops. The company aggressively promotes its genetically engineered Roundup Ready corn and soy and markets its Roundup herbicide as designed specifically to unlock the secrets engineered into its crops. As a consequence, Roundup saw sales of more than $4.7 billion in 2016, triple that of a few years earlier. In that same year, Monsanto raked in more than $12 billion in sales of its genetically engineered products.

Monsanto, in collusion with lawmakers throughout the country, pressured growers, landscapers, and municipalities to deny consumers the right to even know whether their food is genetically engineered. It has fought tooth and nail against attempts to require labeling of products containing genetically engineered ingredients. "The present-day chemical companies have manufactured a nano planet, a second unseen world. We can't see it. We don't notice it. But we notice the epidemic of cancer deaths. We notice our dying loved ones. We vaguely blame modern life. We continue to hesitate to defend ourselves against Monsanto and Bayer, Syngenta, Dow and DuPont, BASF, and the big Chinese outfits," intones Rev. Billy Talen, exorcising the "Monsanto devil" at a performance with the Church of Stop Shopping outside Monsanto's headquarters in St. Louis, Missouri.

"They are a vanishing act. Their chemicals enter the world as dark magic. The sprays and vapors and seed coatings vanish into the eco-systems, and into our bodies. Our regulators are corrupted, and we are left with the surgical strikes from Big Chem's marketing departments, fake scientists, and bribed politicians."[18] Roundup, and another pesticide, 2,4-D,[19] have been deployed by the United States and other governments since the mid-1970s to defoliate entire forests, eradicate unwanted plants, facilitate extraction of minerals, and clear land for monocropped and pesticide-saturated export crops. In the country of Colombia, where people indigenous to the coca-growing areas have for millennia harvested coca plants as part of their culture and local economy, the U.S. government's "war on drugs" has funded—and U.S. citizens have piloted—airplanes spraying massive amounts of glyphosate over the tens of thousands of acres of coca fields, part of the U.S. government's "Plan Colombia."[20]

In Argentina, the same toxic brew is sprayed over miles of monocropped genetically modified soy.[21]

The Parks Department in New York City, as well as hundreds of agencies in towns and villages throughout the United States, applies Roundup and other herbicides in public parks and sidewalks for "cosmetic reasons." Under the spell of TV images of grassy suburban homes and Monsanto's propaganda depicting what a happy lawn looks like, homeowners spray Roundup and 2,4-D on their lawns and gardens to kill what they deem unsightly weeds, like dandelions and crabgrass.[22] Neil Gentzlinger, writing in the *New York Times Book Review,* points to the power of the industry's advertising even among intelligent people who care about their children's health and the environment: "Plenty of lawn-obsessed people read the paper, have college degrees, support the Nature Conservancy; they cannot possibly think the chemicals they dump on their grass are good for their children or wildlife or groundwater, yet they dump them anyway."[23]

A tiny dose of just ten *micrograms* of glyphosate is all that it takes to shrivel and kill a plant that has not been engineered to withstand it or has not yet grown resistant to it.[24] The proliferation of weeds *could be* controlled through a variety of methods, including permaculture, crop rotation, horticulture, interspersed rows of complementary plants, friendly insects, and organic (though labor-intensive) means.[25] But with monocropping, companion plants that repel weeds and insect "pests" are eliminated, prompting resistance to pesticides among such "pests." Thus, heavier and more frequent doses of pesticides are required, and we become trapped on the chemical treadmill. As a result, the aggregate dosages applied to farmland in the United States are enormous, with profound and destructive effects on our environment, and, consequently, on our health.[26] By 1999, agribusiness corporations (85 percent) and homeowners (15 percent) were together dumping more than 556 million pounds of toxic herbicides on U.S. farmlands each year,[27] which often ended up in drinking water, where they exceeded federal safety levels. By 2011, total pesticide volume applied in the United States rose to 1.1 billion pounds annually.[28]

According to the direct action group Greenpeace, which in 1997 published an early assessment of Roundup, glyphosate is one of the world's most toxic herbicides, belonging to a family of chemicals known as organophosphates.[29] Glyphosate can be more damaging to wild flora than many other herbicides. Aerial spraying with glyphosate, writes Greenpeace, has been detected drifting for two thousand five hundred feet beyond its intended target area, and ground spraying has been shown to damage sensitive plants up to three hundred feet away from the field sprayed.[30] Roundup is also extremely toxic to fish, earthworms, insects (including those beneficial to protecting crops), birds and some mammals, and tadpoles and frogs.[31]

Roundup does not attack a plant just through glyphosate. Additional extremely toxic compounds are built into Roundup and other herbicides. "In particular most contain surfactants known as polyoxyethyleneamines (POEA). Some of these are much more toxic than glyphosate,"[32] and can account for many of the worker-reported health effects. They are serious irritants of the respiratory tract, eyes, and skin. In addition, POEAs "are contaminated with dioxane (not dioxin), which is a suspected carcinogen."[33]

Greenpeace's 1997 *Glyphosate Fact Sheet* was followed by several additional critical reports. In July 2000, *Organic Gardening* reported that glyphosate "made bean plants more susceptible to disease, and reduce[d] the growth of beneficial soil-dwelling mycorrhizal fungi. In rabbits exposed to glyphosate, sperm production was diminished by 50%,"[34] and the herbicide has been shown to cause genetic damage to the livers and kidneys of mice.[35]

In fact, a U.S. Geological Survey concluded that a large percentage of waterways and streams throughout the United States, including those in urban and non-farming areas, are flooded with environmentally destructive chemicals[36] that have severe impacts on animal and aquatic life. And a 2005 study by the Centers for Disease Control found that even those living in urban and non-farming areas in the United States now carry in their bodies dangerously high levels of pesticides and pesticide residues.[37]

Neither the corporations nor the agencies assigned to regulate the toxic onslaught have been required to demonstrate glyphosate's safety. In his chapter in this book, "Glyphosate on Trial: The Search for Toxicological Truth," Sheldon Krimsky examines the procedures used by U.N. agencies to analyze differing interpretations of the existing research on glyphosate. The U.S. government failed to require the pesticide industry to use the "precautionary principle" throughout Roundup's approval process, which would require corporations to submit proof that the chemicals to be released into the environment are safe for humans and animal life—individually and in conjunction with other chemicals—before allowing the pesticides to be approved. In this case, as in others, government agencies actually promoted the herbicide throughout the approval process. As a result, the companies are manufacturing new and dangerous synthetic chemicals to kill "weeds" and "pests," substituting them for the existing pesticides as they are banned. Given this ongoing cycle, grassroots activists need to rethink strategies for ending the widespread use of toxic pesticides altogether.

In 2007, the No Spray Coalition—the grassroots group I co-founded in 1999 to oppose New York City's spraying of malathion (another organophospate) and pyrethroids to kill mosquitoes said to be carrying West Nile virus—met with Julius Spiegel, then commissioner of New York City parks for Brooklyn. Also present was Jean Halloran of Consumers Union, and a biologist working for the New York City Parks Department. The No Spray Coalition and Halloran urged the city officials to stop spraying glyphosate not only in the parks but also on New York City's sidewalks around schoolyards and playgrounds. The coalition's representatives handed Spiegel several studies on the disastrous effects pesticides have, especially on children, the elderly, and people with compromised immune systems—a solid majority of the people who live in Brooklyn. Spiegel's parents were Holocaust survivors, and of all the government officials I'd met over the years, Spiegel seemed truly concerned with the spraying of the public parks he presided over and of the surrounding sidewalks, which often abutted children's playgrounds. It was Spiegel who first explained to me that more than 70 percent of the Parks Department workforce had been laid off over the preceding decade as a cost-cutting measure, which meant that chemical herbicides were now replacing city workers who had previously weeded public parks and sidewalks by hand.

Of course, spraying Monsanto's Roundup and other toxins for cosmetic reasons is unnecessary and should simply be stopped. Even small amounts of pesticides are poisonous to children.[38] Further, what some call "weeds" in today's urban environment were once upon a time medicinal or nutritious plants used by American Indians and healers. Connie Lesold, a community activist, and I both noted in our conversation with Spiegel that the dyes the city had added to the herbicides expressly to warn the public to "stay away" were actually having an unintended opposite effect. Lesold lived along Eastern Parkway, a main traffic artery cutting across the heart of Brooklyn. We had observed children there attracted to the bright yellow and green areas, which they saw as play zones. I also witnessed numerous children in Coney Island rolling around in the newly sprayed weeds and riding their bicycles and scooters through them. They tracked the chemicals into their classrooms and parents' apartments. So the first of our suggestions (apart from our main demand that all spraying be stopped at once) was that the Parks Department at least inform teachers when spraying would occur near schools and parks, and have them warn students to stay away from those areas. Spiegel agreed with the coalition's proposal to contact all school principals in advance of any spraying by the city, who would then inform their young students to stay away from those areas.

New York City's biologist didn't agree. He defended the use of Roundup. Glyphosate, he said, was "perfectly safe—unless you are a weed." Commissioner Spiegel, however, took note of the city biologist's arrogance and overruled him. He agreed to implement the proposal and thought favorably of a second one: to set up a pilot project in which volunteers would take responsibility for hand-weeding unsprayed sections along Eastern Parkway. This too was opposed by the city's biologist, who now appeared to be taking personally our criticisms of the herbicide. Officials, we learned, are frequently wary of establishing a prototype relying on the mobilizing of local volunteers—risky business for those in positions of authority, whenever people get together around *any* project and begin to discover their own power.

The city's biologist sneered: "What do you propose for areas where you can't get volunteers?" My reply came as a shock: Goats. Sheep. I had recently met Lani Malmberg, a sheep and goat herder from Wyoming, at a Beyond Pesticides conference in Washington, DC. Beyond Pesticides was one of the organizations (along with Disabled in Action, Save Organic Standards, and the Brooklyn Greens) that had joined with the No Spray Coalition in suing New York City's government to stop its aerial and truck-spraying of chemical pesticides. Beyond Pesticides–formerly the National Coalition Against Misuse of Pesticides–sponsored annual gatherings of activists, scientists, and occasionally government officials, who would strategize ways to vastly reduce, if not outright ban, the use of chemical pesticides. Lani explained that she travels around the western United States bringing hundreds of animals to "weed" urban areas. This approach seemed to me both beautiful artistically and functional—a way to solve problems in line with nature rather than going to war with it, a change from the *modus operandi* of the pesticide industry. The city's biologist ridiculed the idea. Even Spiegel was taken aback by it.

Over the years, more and more people have come to understand the importance of the anti-pesticides work. But activists kept meeting the same intransigence from government officials, even after a federal judge ruled in favor of the No Spray Coalition and against New York City in a lawsuit the Coalition brought under the Clean Water Act,[39] and even after the City administration signed a settlement agreement with the No Spray Coalition that it soon ignored. When New York City's new mayor, Bill de Blasio, assigned two assistants in 2015 to meet with the Coalition, it became clear that the City had learned nothing in the sixteen years following the filing of the lawsuit; there was no institutional memory. They asked the Coalition's Cathryn Swan and me the same question: "Which less toxic chemical sprays should we use?" And we answered with the same response we'd given Spiegel eight years earlier: "None! Bring in goats, dragonflies, bats!" The use of natural predators of the so-called pests in question was to them so far outside the boundaries of the dominant chemical mindset as to seem absurd.

In the years since the No Spray Coalition first made that proposal, Rev. Billy Talen and Savitri D., the choreographer and director of the Church of Stop Shopping's performances, led New Yorkers in the fight to block NYC officials from applying Roundup to city parks. Weekly performances of their show, "Monsanto is the Devil," sold out fashionable Joe's Pub in the Public Theater in SoHo and mobilized audiences for protests. At last, in the summer of 2016, goats were finally brought into the city from an upstate New York farm to feed on weeds in a fenced-off section of Prospect Park. The goats chomped happily on the wild plants. Children got to see them "at work," and appreciated their new connection to nature. That section of the park was safely "weeded" in record time. But instead of expanding the use of goats or assigning workers to weed by hand, the city government continued to spray glyphosate, which bioaccumulates in bone marrow.[40] It also "applied"—and continues to apply—other cancer-causing and endocrine-disrupting adulticides throughout the five boroughs, despite the well-documented harmful effects on people and the environment.

As a result of the work of the No Spray Coalition and sister groups such as East Bay Pesticide Alert (Don't Spray California), Beyond Pesticides, and Organic Consumers Association, California has now listed glyphosate as a likely carcinogen. Meanwhile, an important study (2018) of glyphosate's affect on pregnant women confirms that the widely used herbicide ends up in women's bodies and that prenatal glyphosate exposure may be linked to shorter pregnancies and lifelong adverse health consequences for children.[41] In New York City, Paula Rogovin, a teacher at Public School 290-Manhattan New School on the Upper East Side, and jazz pianist Jill McManus are currently mobilizing students, teachers, and the community to press the City Council and the Mayor to ban all synthetic pesticides from city parks.

Despite the increasing consciousness over the dangers of pesticides, environmental and health activists are at every turn confronted by corporate power, corporate profiteering, and corporate ways of thinking that Monsanto and others, in conjunction with governments, propagate at the expense of future biodiversity and children's health. What we do today makes all the difference in whether there will even be a future in which to argue the finer points of our philosophies, theories, and proposals. There's a saying from the 1960s: *"The future will be what we the people struggle to make it."* While micro-organisms will no doubt continue to thrive no matter what, humans won't. The fate of complex life on this planet, including human beings, is in our hands.

2

Better Active Today
than Radioactive Tomorrow

The United States emerged from its military victory over Nazi Germany, Japan, and Italy in World War II awash in technological innovations and, in a few years, bristling with nuclear weapons. The U.S. government's decision to drop atomic bombs for the first time in world history on the Japanese citizenry of Hiroshima and Nagasaki (August 1945), launched a cold war with the Soviet Union and a never-ending manufacture of military weapons. The country enjoyed a burgeoning economy kick-started by profiteering from wars whose widespread devastation never reached its own shores.

America's newly powerful agribusiness sector and its mass application of pesticides were direct outgrowths of war production.[1] Along with advertising, heavy industry, and pharmaceutical/chemical infrastructure, agribusiness and pesticides held out the promise of securing "the good life" for all, although that application of democracy in real life would require powerful social justice movements to overturn the country's legacy of white supremacy and Jim Crow laws. Millions of people were excluded from partaking in what was portrayed as the American dream and that remained, for many, an American nightmare.

From the perspective of an expansive capitalism, nature stood in the way. The environment, to the extent that nature was noticed at all in the 1950s, was seen by many as an obstacle to overcome and control. The mass application of pesticides would become the way to both protect America's citizenry from the weedy monsters of the remaining wilderness and to secure food for all. In her book *Silent Spring* (1962), Rachel Carson challenged the chemicalization of agriculture and in so doing undermined the ideological imperative of the suburban fantasy for what constitutes "the good life"—a Hollywood propagandized fantasy serving as the glue, the big idea—uniting an aspiring middle class around the illusion of a singular domestic national identity, and thrusting American capitalism forward into the post–World War II era.

More than half a century has now passed since Carson meticulously exposed some of the insidious consequences of what the quest for "Progress" and "The Good Life" meant in terms of the environment, government and corporate poisoning of the planet with synthetic pesticides and radiation from nuclear bomb tests. Serialized in the *New Yorker* in weekly installments, *Silent Spring* was read by thousands before it was officially published as a book. In the throes of the lingering and officially orchestrated anti-Communist hysteria of the 1950s where women were pushed back into their role as housewives in the nuclear family, *Silent Spring* was read by many of these women in installments as they were published. Rachel Carson not only exposed the prevalence of chemical pesticides but also the fact that radioactive Strontium 90, a byproduct of the above-ground nuclear bomb tests, had tainted the nation's milk supply. This was shocking information indeed, and it inspired an army of young women in particular, who anguished over the threats to the health of their children. Many men became alarmed and active as well, but as the Cold War between the United States and the Soviet Union heated dramatically, it was the mobilization of women, primarily, that challenged the government on the nuclear bomb tests and the possibility of nuclear annihilation.

Their challenge extended beyond the issue of nuclear radiation, as horrific as that was. Women working in blue-collar jobs had participated by the tens of thousands in support of the anti-Nazi effort, and they brought their newly experienced independence into organizing the new mass movement, shaking up what we think of as the typical or traditional family structure—a structure that was never as typical or traditional as we're led to believe.

There has never been a shortage of brilliant and courageous people who work against all odds to bring about a different world. Hollywood is replete with stories of heroic individuals standing up to and defeating powerful villains. America's fantasy machine has, by now, ushered Rachel Carson into that panoply of exalted but isolated figures—the better to cast our hero as a romantic lone wolf against the system that audiences crave. In fact, those individual heroes are almost never isolated, and the social changes they inspire succeed only when large numbers of people forge new ways of looking at their own place in the world and join together as part of something that becomes larger than them and that transcends their own immediate self-interest.

Today's revising of history has forgotten the role these women played in the movement to ban the bomb. Carson, bravely unconventional, was part of a revolt in the 1950s and 1960s of scientists, mothers, and leftist thinkers alarmed not only by chemical pollution and pesticides but also equally by the above-ground nuclear bomb tests and the harmful effects of radiation. These life-and-death matters in the early 1960s "spurred scientists, emanating primarily from the left, to raise searching questions about the destructiveness of our civilization. From this work the modern ecology movement emerged," wrote John Bellamy Foster and Brett Clark in a revelatory 2017 article in *Monthly Review*.[2] Rachel Carson inspired thousands to join an environmental movement that grew many tentacles; they did not separate their condemnation of chemical pesticides from support for banning the bomb. The issues were organically related and indivisible, and the joining of both efforts helped to spur an emerging ecological sensibility.

A decade after her death from cancer in 1964, that movement succeeded in forcing the banning of domestic applications of DDT, an organochlorine pesticide that had been developed to replace the arsenic-based insecticides previously in use. New national efforts were mobilized to preserve rivers, oceans, parks, and forests. Public opposition to pest-spray pollution was overwhelming. By the 1970s, President Richard Nixon, one of the most reviled and hostile politicians of the era, nonetheless responded by outlawing DDT and instituting the Environmental Protection Agency, Clean Water Act, Safe Water Drinking Act, Department of Natural Resources, National Oceanic and Atmospheric Administration, Clean Air Act, Pesticide Control Act, Endangered Species Act, and Water Quality Improvement Act—all of them currently under attack. Whatever Nixon's private sentiments concerning the environment (environmental aide John Whitaker years later told Nixon he would be remembered for his domestic-policy successes, especially on the environment, to which Nixon responded, "For God's sake, John, I hope that's not true"),[3] Nixon's policies were purely expedient. He believed that by sponsoring environmental legislation at home, he could make the environment a Republican issue[4] and break off some of the liberal support for antiwar radicals who were in the streets and clogging the jails in opposition to U.S. military bombardment of North

Vietnam's cities of Hanoi and Haiphong, the mining of their ports, invasions of Laos and Cambodia, and authorization of massive spraying of chemical pesticides over forests throughout Southeast Asia.[5]

The pesticide made for that purpose by Monsanto, Dow, and DuPont—Agent Orange—was used as a defoliant to strip Vietnamese forces of ground cover and food, while American helicopters strafed overhead. Agent Orange stays in the soil and sediment at the bottom of lakes and rivers for generations. It can enter the food supply through the fat of fish and other animals and has been found at alarmingly high levels in breast milk that dioxin-contaminated mothers have fed their children.[6] Agent Orange poisoned millions of Vietnamese people and hundreds of thousands of U.S. soldiers, with chromosome-damaging and cancer-causing dioxin. An estimated five million Vietnamese were killed in more than a decade of direct U.S. military bombardment and warfare, and by the lingering effects of the U.S. government's intentional mass-spraying of pesticides over Southeast Asia's forests, rice paddies, and drinking water.

The U.S. chemical companies that made Agent Orange—and the government and military authorities who ordered it sprayed on Vietnam—were fully aware of the terrible toll it would take, and they were thus complicit in enacting massive crimes against humanity and nature.

Such are the revelations compiled by journalist Jon Dillingham from the records Monsanto was forced to release in 2017, but which have escaped the kind of media attention that, in another era and in another country, would have sent government officials and corporate executives to the guillotine. Decades after the U.S. military was forced to withdraw from Vietnam, spokespersons for Monsanto were still concealing evidence of Agent Orange's devastating effects[7]—and the corporation continues to do the same today, lying to federal authorities and falsifying studies in defense of (and to win approval for) the company's signature weed-killer Roundup.

Why didn't U.S. government agencies ban Monsanto's products, instead of fast-tracking them and enabling Monsanto and others to market them as quickly and as widely as possible? Writing in *Thanh Nien,* the flagship publication of the Vietnam National Youth Federation, Jon Dillingham explained just how insidious the pesticide industry was and continues to be.

The massive spraying of Agent Orange was a war crime and crime against humanity under international law, which government officials knew well before approving the defoliant's use.[8] In fact, Dillingham discovered a declassified letter by V. K. Rowe at Dow's Biochemical Research Library to Bioproducts Manager Ross Milholland, dated June 24, 1965, which reveals that the company knew in advance that the dioxin in its products, including Agent Orange, would be toxic to people—and was proud of it! Rowe stated: "This material is exceptionally toxic; it has a tremendous potential for producing chloracne and systemic injury." Dillingham notes that Rowe worried the company would suffer if word got out: "The whole 2,4,5-T industry would be hard hit and I would expect restrictive legislation, either barring the material or putting very rigid controls upon it," Rowe wrote.[9]

Rowe advised the company to keep quiet about the toxicity: "There is no reason why we cannot get this problem under strict control and thereby hopefully avoid restrictive legislation. . . . I trust you will be very judicious in your use of this information. It could be quite embarrassing if it were misinterpreted or misused. . . . P.S. Under no circumstances may this letter be reproduced, shown, or sent to anyone outside of Dow," Rowe concluded. Dillingham summed it up: "Dow played its cards right, never getting in serious trouble. The spraying of Agent Orange in Vietnam went on for another six years."[10]

Included among the information concealed by Dow and Monsanto was that several factory workers had fallen sick after exposure to dioxin. It turns out, Dillingham wrote, that the chemical companies could have manufactured their pesticides without dioxin, as others had done, but the process was slower and more expensive. So they chose a more dangerous method that enabled the companies to maximize their profits. According to plaintiffs in a lawsuit that ended up before the U.S. Supreme Court in 2009, the companies "secretly tested their products for dioxin and *hid its extreme toxicity* from the military."[11]

Are we really to believe that the U.S. military and government officials were clueless? Just as R. J. Reynolds and other cigarette manufacturers defended Big Tobacco against citizen lawsuits[12] (the tobacco corporations eventually lost, but not until many people lost their lives to cancers caused by their products), and just as the giant wireless cellphone corporations have been allowed to suppress the results of numerous studies indicating the dangers of wireless technology,[13] the pesticide corporations have provided (and continue to provide) government officials with "credible denial," enabling government officials to pretend to look the other way and feign not to know, while contracting for and administering the chemicals that are poisoning the planet.

Suppression of research establishing that pesticides, highly pesticided cigarettes, and cellphone technologies cause cancer and neurological illnesses, has become the standard operating procedure for giant multinational corporations. Dow and Monsanto, for example, funded "studies" by one "expert," Dr. Alvin L. Young, whose role was to trivialize health concerns of U.S. ground forces poisoned by Agent Orange. Young—a doctor who was frequently paid to testify at the behest of the pesticide industry—testified that dioxins are *not* harmful, and then, when challenged, was forced to concede that pesticides are indeed harmful, after all.[14]

Was the U.S. military aware of the pesticide industry's hiding of data showing the harmful effects of dioxin on the human body? Of course it was. Did it conspire with Dow and others to cover up those intentional gaps in the data that its hirelings presented? Yes.

"Though reports point to the fact that chemical companies like Dow and Monsanto knowingly hid evidence of dioxin-related medical problems from the government, the declassified 1990 Zumwalt Report suggests that U.S. military experts knew that Agent Orange was harmful at the time of its use," Jon Dillingham wrote. He continued: "The report quotes a 1988 letter from Dr. James R. Clary, a former government scientist with the Chemical Weapons Branch. Dr. Clary was involved in designing tanks that sprayed herbicides and defoliants in Vietnam, according to the report." Clary told Daschle the following:

> When we (military scientists) initiated the herbicide program in the 1960's, we were aware of the potential for damage due to dioxin contamination in the herbicide. We were even aware that the 'military' formulation had a higher dioxin concentration than the 'civilian' version due to the lower cost and speed of manufacture. However, because the material was to be used on the 'enemy,' none of us were overly concerned. We never considered a scenario in which our own personnel would become contaminated with the herbicide. And, if we had, we would have expected our own government to give assistance to veterans so contaminated.[15]

Supporters of the U.S. government's Agent Orange campaign tried to mask their complicity in the destruction of millions of lives by rationalizing the government's environmental toxification of Vietnam as an "herbicide program" to clear away forests–bad enough!–rather than what it actually was: chemical and biological warfare, and genocide.

Dillingham points to a U.S. Senate's Congressional Record dated August 11, 1969. There, "a table presented to senators showed that congress clearly classified 2,4-D and 2,4,5-T (main components of Agent Orange) in the Chemical and Biological Warfare category. The table also includes cacodylic acid, a main component of Agent Blue, another chemical sprayed on Vietnam"[16] to destroy the rice paddies and force the villagers to leave their homes in search of food, as well as to destroy bamboo, which provided cover.[17] The table describes it as "an arsenic-base compound . . . heavy concentrations will cause arsenical poisoning in humans. Widely used in Vietnam. It is composed of 54.29 percent arsenic." As Vietnam War scholar and U.S. veteran W. D. Ehrhart put it concisely in a *Thanh Nien Daily* interview in 2009, "It would be hard to describe Agent Orange as anything other than a chemical weapon. Dioxin is a chemical." Dillingham says: "So is arsenic."[18]

* * *

Arsenic is both a naturally occurring rock as well as a synthetic poison. Lead and calcium arsenate were widely used as pesticides in the nineteenth and first half of the twentieth centuries on apples, cotton, and rice, and arsenic was used to treat lumber, such as telephone poles and wood flooring. In the 1944 film *Arsenic and Old Lace,* arsenic was made infamous as the murder agent of preference discovered by Cary Grant's character. While the maximum allowable arsenic in drinking water was set in the U.S. at 50 parts per billion in 1942, the United Nations and other bodies, concerned with arsenic's culpability in causing cancer of the skin, lungs, bladder, and prostate and its record as an endocrine disruptor, had long recommended that the levels around the world be lowered to 10 parts per billion.

President Bill Clinton and Vice President Al Gore did nothing to address arsenic during their eight years in office. Environmentalists advocated that the maximum level of arsenic in drinking water be lowered not to 10 ppb, but to 3 parts per billion. It was not until three days before leaving office that Clinton finally adopted a tougher standard. On becoming president in 2001, George W. Bush was informed that much of the drinking water in the United States was contaminated with unexpectedly high levels of arsenic. Nevertheless, the Bush administration suspended Clinton's newly lowered arsenic maximum, estimating that it would cost at least $200 million for local communities to enact the tougher standard. Bush also questioned the scientific basis behind the arsenic studies that had been cited to justify lowering the amount of arsenic allowed in drinking water.

But the National Academy of Sciences (NAS) responded with a report[19] in 2001 claiming the Environmental Protection Agency had greatly *underestimated* the cancer risks of arsenic in drinking water. CBS News summed up the report: "The risks [from arsenic] are much higher than the agency had acknowledged under the Clinton administration as well as the current Bush administration, even for low levels of arsenic in tap water."[20]

"Even very low concentrations of arsenic in drinking water appear to be associated with a higher incidence of cancer," said Robert Goyer, chair of the committee that wrote the report and professor emeritus of pathology, University of Western Ontario, now living in Chapel Hill, North Carolina "We estimated the risk of developing cancer at various arsenic concentrations, and now it is up to the federal government to determine an acceptable level to allow in drinking-water supplies."[21]

Doctors should certainly be routinely ordering the testing of urine for heavy metals (usually arsenic, lead, mercury and sometimes chromium), toxicologist Robert Simon maintains. In addition to drinking water, mercury poisoning can come from eating tuna; arsenic poisoning can come from rat poison as well as contaminated drinking water; and lead poisoning can come from children eating chips of paint and breathing paint particles in dust. "These types of tests should become routine, because even if we can convince the government agencies to lower the allowable arsenic levels in drinking water down to less than 3 parts per billion, there will still be people who will have contaminated water or face other sources of harmful chemical exposures. Good medical practice must be included in protecting people from heavy metal poisoning."[22] Even at that lower level of three parts per billion, the risk of bladder and lung cancer from arsenic exposure is still between four and ten deaths per ten thousand people, according to the aforementioned NAS report. The EPA's maximum acceptable level of risk for the past two decades for all drinking water contaminants has been one death in ten thousand. At each level of arsenic in drinking water, investigators found the cancer risks to be much higher than the EPA had estimated.

Arsenic at various levels is also associated with heart disease, high blood pressure, and diabetes, among other diseases. The report rejected arguments by industry that there is a clear, safe threshold below which arsenic could not cause cancer.

A 2014 overview by David Heath for the Center for Public Integrity[23] covered the political gyrations and jockeying over the arsenic issue. Updated in 2015, the story is astounding to read—made more so by the fact that so few in the United States are aware of anything at all about arsenic contamination of drinking water, and even less about how a relative handful of legislators, working hand in hand with pesticide manufacturers, managed to prevent a disinterested Obama administration from taking critical measures needed to protect the country's drinking water.[24]

Just as the broader public is now learning that much of the U.S. drinking water is contaminated with arsenic, a shocking study published in *Toxicology Reports* shows that the current regulatory assessments of the world's most used herbicides are wrong, and that arsenic is being regularly found in Roundup and other glyphosate-based pesticides at toxic levels.[25]

Professor Gilles-Eric Séralini from the University of Caen Normandy, France, and Drs. Nicolas Defarge and Joël Spiroux, discovered that so-called "inert" ingredients in glyphosate-based pesticides like Roundup add even more to the toxicity of the pesticide, and that glyphosate-based herbicides contain endocrine-disrupting heavy metals such as arsenic, which would otherwise be banned due to their toxicity.[26] (Large amounts of arsenic were found in Roundup previously sold in Sri Lanka before that country banned such herbicides due to concerns about chronic kidney disease.) "Inert" ingredients (or "formulants") in Roundup and all pesticides need to be listed, labeled, researched, and regulated—not just the primary ingredient. And, says Séralini, glyphosate-based herbicides should be banned.

Henry Rowlands, a project director for *Sustainable Pulse*, concluded: "These results show that the difference between 'active ingredient' and 'inert compound' is a regulatory assertion with no demonstrated toxicological basis. . . . An immediate ban on glyphosate-based herbicides is the only way to protect public health."[27]

Could the decades of mass spraying of corn and soy fields with Monsanto's Roundup have poisoned the nation's waters with arsenic, thereby contributing to the skyrocketing rates of many types of cancer across the United States? What did Monsanto's corporate officers (and U.S. and European government officials) know about *these other* components in Roundup, and when did they know it?

3

The Future Ain't What It used to Be . . . and What's More, It Never Was

President Nixon's hopes for splitting support for the antiwar movement and co-opting a significant fraction of it were not without precedent. At the turn of the twentieth century, reforms such as workers' compensation were enacted despite those issues not having been raised as part of any radical or trade-union campaign. Workers' compensation was solely a corporate innovation at the time, enacted not as a result of mass protests but when leading industrial capitalists decided they preferred federal regulations that supplied a predictable stream of workers over being subjected to patchwork, disruptive state measures and costly lawsuits that were of less concern to smaller businesses.[1]

When mass movements, such as the environmental movement, are not strong enough to project issues to the fore by dint of their own actions, reforms can sometimes be won by finding ways to exploit the interests of one section of capitalists against others. Especially as giant corporations gain more and more control over the state apparatus, popular movements not strong enough to achieve victories on their own should take the current structural realities into consideration when developing strategies for moving forward and not assume that moral suasion or the power of one's ideas alone will be sufficient to achieve victories.

Nixon, in line with a sector of U.S. corporations in the 1970s (as was Franklin Delano Roosevelt forty years earlier), believed it to be in the capitalist system's interest to support some regulation of industrial pollution. "By the time Nixon was elected, the nation was pumping out 200 million tons of air pollutants and throwing out 100 million automobile tires and 30 billion glass bottles annually, with much of the refuse piled in mountainous open dumps."[2] Many large corporations, and politicians looking out for their long-term interests, happily recognized that compliance with Nixon's legislation would be more exacting on and costly for their smaller corporate competitors than for themselves. Up to a point, they were willing to risk the rising of environmental movements in the expectation that they could eventually de-fang them before they did much damage to the corporations' core interests, while also turning a portion of those movements to their own advantage against the other sectors. Large industries looked to their successes in co-opting a sector of the labor movement in the 1910s and again in the 1940s as a sort of blueprint for how to fight it out with their competitors, without too much disruption to their long-term profits.

Over a few short years in the 1970s, and especially culminating with Ronald Reagan's election to the presidency in 1980, many of the environmental regulations enacted under the Nixon administration were rolled back. Even under Nixon's pro-environment policies, industry-sponsored bills were passed when it mattered, exempting, for example, the Alaskan pipeline from the review requirements of the National Environmental Policy Act (NEPA)

> and allowing the temporary licensing of nuclear power plants without environmental impact statements. The influence of big corporations on federal policy was evident, for example, in an agreement Nixon signed with Canada to improve water quality in the Great Lakes. He agreed to address the dumping of dredging spoils and phosphates from detergents that had fouled the water and caused a giant algae bloom, but pressure from detergent manufacturers weakened the water-quality standards. . . .
>
> The administration likewise maintained a piecemeal approach to DDT restrictions, banning the [DDT] treatment of fruits, vegetables, and forest trees but not cotton crops, which accounted for almost 75% of the pesticide used. The federal government also failed to address the effects of strip mining for coal in Kentucky and West Virginia, which, writes [historian J. Brooks] Flippen, "turned lush landscapes into permanent pits, polluted the air with sulfuric acid from oxidized coal, and contaminated streams with chemical-laden overburden."[3]

Rachel Carson and the new environmental movement succeeded in winning the demand to ban DDT; but that pesticide was replaced by organophosphates, which themselves were soon replaced by pyrethroids and by glyphosate-based chemicals. The official mindset hadn't changed, only the particular chemicals had. When the Clinton/Gore administration promoted genetic engineering (as had former President Jimmy Carter) and encouraged farmers to plant expansive acres of Roundup Ready corn and soy, the national pesticide regulations were already so weakened—with the encouragement of Democrats as well as Republicans—that there was no control over the massive amounts of pesticides used on genetically engineered (GMO) crops. The supposed regulators welcomed bioengineered crops, whose success depended on highly profitable carcinogenic pesticides like Roundup, as their gift to the industry from which many of their own numbers were spawned and to which they continued to cast their loyalty, as opposed to watching out for the public good.

Monsanto and a few other giant chemical companies were thus able to circumvent those regulations and continue—even accelerate—their destructive practices. In fact, they succeeded not only in developing ever-more dangerous and widely used pesticides but also in consolidating and centralizing their private ownership and control of agricultural *seeds*. Bayer's $66 billion purchase of Monsanto[4] threatens to put the vast majority of the planet's original seed stocks into the hands of one gigantic super corporation.

Many involved in today's environmental movements, including some of its leaders, lack Carson's visionary framework and the radical analysis needed to make sense of the vast swamps of information now available on the internet with the touch of a finger. Despite the outpouring of information, where anything of interest is tweeted from cellphones and posted within seconds to Facebook and Instagram (and Facebook now owns Instagram), all sorts of significant findings have been buried or lost. Much of the technology so useful for acquiring and sharing information and for connecting us to much larger educational, social, and activist networks than existed in the 1960s, have become instruments of surveillance. Obsessive postings let our "friends" (and the government) know exactly what we think, where we've been, who we love, with whom we congregate, what we purchase, and what we are doing at any given moment, providing the authorities not only with data concerning our individual consumer preferences but with pictures of our social and political networks—information that Facebook, Google, and others provide to Monsanto and other entities and advertisers for profit. As *the New York Times* reported, concerning Facebook and the Cambridge Analytica scandal, "The number of people whose data was harvested reached about 50 million. Most of those people had no idea that their data had been siphoned off (after all, they hadn't installed the app themselves), let alone that the data would be used to shape voter targeting and messaging for Mr. Trump's presidential campaign."[5]

We've already seen how Monsanto and the U.S. Department of Defense intentionally hid studies proving the dangers to humans of many pesticides, including Agent Orange and Roundup. So too with some of Monsanto's other products, such as "white phosphorous," which was deployed illegally as a weapon in both Iraq and in Israel's ongoing invasions of Gaza. The use of white phosphorous is a war crime under U.N. conventions. A report of Monsanto's involvement in the manufacture of white phosphorous first surfaced in 2013,[6] yet Monsanto's ties to this horror and the protection afforded the company by the U.S. government remain largely unknown.

A heavily redacted U.S. Army document, signed in 2012, has come to light, linking the Monsanto Company to the manufacture of white phosphorus for the U.S. military and authorizing the U.S. government to provide Monsanto with special protection from competition by other corporations.[7] Kurtis Bright, writing for *Natural Blaze*, explains:

> If you've never heard of white phosphorus, there's a good reason for that: the U.S. government and the government of Israel would rather you didn't know about it, nor about their use of the horrific substance against civilian populations in conflicts in Fallujah and Gaza, respectively.
>
> White phosphorus ignites spontaneously at 30°C and is purportedly used to illuminate enemy positions or to create a smoke screen to cover troop movements. However, although it is prohibited [under the 1993 *Convention on the Prohibition of the Development, Production, Stockpiling and use of Chemical Weapons and on their Destruction*[8]] it is also sometimes used to firebomb enemy positions. It has been widely reported [to have] been used against civilians in Iraq and Gaza.

Bright concludes his description of the horror of white phosphorous this way: "When it comes into contact with a person, white phosphorous ignites, and once it does it continues to burn until it runs out of fuel or is deprived of oxygen. It sticks to clothing or skin, causing horrific, often fatal chemical burns. It is said that it 'burns to the bone' according to witnesses who have seen white phosphorous in action, as its victims often have no way to stop its progression until it burns itself out."

For many years, Monsanto was the sole company in the United States to manufacture the stuff, most likely because its complex production required highly specialized techniques and skills. So the government's support of this domestic capability, in Bright's words, was critically important:

> And the next time you hear some defender of GMOs talk about how companies like Monsanto are just benevolent corporations looking to feed the hungry people of the world, remind them that Monsanto is a chemical company first, and a seed company second. And that they have only one motive: profit.
>
> And if the death and maiming of children and civilians is the path to profit, they will not hesitate to take it.[9]

* * *

The new waves of pesticides, genetic engineering of crops, private patenting of DNA sequences in the living biological cell, and centralized ownership and control of seeds have all dramatically reshaped the environment in the decades since Rachel Carson wrote *Silent Spring*. Most of us experience that "brave new world" viscerally, if not intellectually, as those in positions of authority ignore our protests and enact decisions that severely impact on ourselves, our society, and nature.

Opposing the mass application of *all* chemical pesticides enables organizers to reexamine how they see our relationship to nature. While some encourage us to tackle the problems presented by pesticides one chemical or one corporation at a time (thus constraining our efforts to repeatedly beat back their individual chemical products as though each was the result of an isolated problem), it was Henry David Thoreau more than a century-and-a-half ago[10] and Rachel Carson, 108 years later in her classic *Silent Spring*, who engaged in a deeper, more comprehensive critique, and did so with furious eloquence and unrelenting focus. Carson did not shrink from naming names, and was reviled by the American corporate structure, which libeled her mercilessly for challenging the corporatocracy and U.S. government officials beholden to it. The then-dominant ideology fostered by leaders of American liberalism presented human activity and nature as necessarily non-reconcilable adversaries in a never-ending war for "progress." *Silent Spring*, as a radical manifesto, challenged all that, and the modern ecology movement was launched on the worthy foundations of both Thoreau's and Carson's searing systemic critiques.

4

Monsanto: Origins of an Agribusiness Behemoth

By Brian Tokar

Many years ago, in the spring of 1998, I was invited to be part of a publication that would make history, but not for the reasons its editors and publisher anticipated. The item in question was a special issue of the UK-based magazine, *The Ecologist*, profiling the Monsanto corporation and its expanding push to genetically engineer common food crops. Genetically engineered foods, or GMOs (i.e., foods made from genetically modified organisms) had only been grown commercially for two short years but had already inspired massive worldwide opposition. The first Global Day of Action against GMOs featured demonstrations and public gatherings in 19 U.S. cities, as well as 17 European countries, India, the Philippines, Malaysia, Japan, Canada, Australia, New Zealand, Brazil, and Ethiopia.[1] Greenpeace painted a 100-foot biodegradable "X" on a field of Roundup-resistant soybeans in Iowa and later blocked a ship containing genetically engineered soybeans from leaving Cargill's grain facility on the Mississippi River, outside of New Orleans.[2] European activists fought to block imports of GMO corn and soybeans from the United States, uprooted experimental plots of engineered crop varieties in broad daylight, and would soon succeed in their push for labeling requirements for products of genetic engineering throughout the European Union.

Monsanto, which was already the most aggressive promoter of GMO agriculture on both sides of the Atlantic, fought back with as much hype and political clout as they could muster. They had already derailed congressional efforts to regulate GMOs in the United States, forcing public officials to rely on food and plant safety laws that long predated the new technology. The company tried to brand itself literally as a savior of humanity, uniquely able to feed the world's nutritionally insecure masses. A pervasive advertising campaign across the United Kingdom aimed to convince readers that the future of global agriculture depended on Monsanto's ability to enhance future harvests.

In those days, *The Ecologist* was probably the leading popular journal of environmental research and activism in the English-speaking world. Its readers were active on every continent, and its reputation was unsurpassed, even after a team of internationally acclaimed editors departed following a dispute with the magazine's publisher. A special issue of *The Ecologist* was planned focusing on Monsanto and its history. With a foreword by Prince Charles and articles by some of the leading United States and United Kingdom critics of the global biotechnology and pesticide industries, this special edition was bound to have a major impact.

Apparently, someone on the other side felt the same way. Just as subscribers were anticipating receipt of their September/October 1998 issue of *The Ecologist*, word got out that the printer, Penwells of Saltash Cornwall, had physically destroyed all 14,000 copies.[3] The printer refused to comment on the incident, and Monsanto denied any involvement, but commentators widely agreed that fear of a libel suit was the most likely motive. The "Monsanto Files" special issue did not appear for several more months, but it was eventually reprinted several times and *The Ecologist* later reported that they had distributed over 100,000 copies.

Today Monsanto is merely a division of the German pharmaceutical and agrochemical giant Bayer, which has long been the world's largest manufacturer of insecticides. Bayer's purchase of Monsanto was finalized in 2018, and the new parent company has taken steps to remove the Monsanto name from its products. While it is true that the Monsanto of the early 2000s to 2010s focused almost entirely on biotechnology, seeds, and a limited range of agrochemicals, especially its Roundup brand herbicides, it was once a broadly diversified chemical manufacturer, one of only four to be listed among the top ten U.S. chemical companies in every decade from the 1940s to 1990s.[4] In 1997, Monsanto spun off various industrial chemical divisions into a new company called Solutia, seeking to limit its liability from a vast array of lawsuits. These included a successful suit by Vietnam veterans contaminated by the company's notorious Agent Orange herbicide mixture, which the U.S. Army used to destroy the rainforests of Vietnam, and others brought by communities in the U.S. South where Monsanto manufactured PCBs (polychlorinated biphenyls) for numerous industrial uses. A merger and subsequent divestment just a few years later led to the acquisition of Monsanto's profitable pharmaceutical division, G. D. Searle, by the transnational drug company Pharmacia. Monsanto lost half its share value during those complex transactions and saw its fortunes ebb and flow quite dramatically as its focus narrowed to the highly volatile agribusiness sector. To better understand the company's emergence as the leading developer and promoter of genetically manipulated agriculture and Roundup herbicides, it is important to first examine its history.

Origins

The Monsanto Chemical Company was founded in St. Louis in 1901 by John Francis Queeny. A self-educated chemist, Queeny brought technology from Germany to the United States to manufacture saccharin, the first artificial sweetener. In the 1920s, Monsanto became a leading manufacturer of sulfuric acid and other basic industrial chemicals, and by the 1940s, plastics and synthetic fabrics were the centerpiece of its business.

In 1947, a French freighter carrying ammonium nitrate fertilizer blew up at a dock 270 feet from Monsanto's plastics plant outside Galveston, Texas. More than 500 people died in one of the U.S. chemical industry's first major disasters.[5] The plant was manufacturing styrene and polystyrene plastics, which are still important constituents of food packaging and other products; in the 1980s the U.S. Environmental Protection Agency (EPA) listed polystyrene as fifth in its ranking of the chemicals whose production generates the most total hazardous waste.[6]

In 1929, the Swann Chemical Company, soon to be purchased by Monsanto, developed polychlorinated biphenyls (PCBs), which were widely praised for their nonflammability and chemical stability. The most widespread uses were in the electrical equipment industry, which adopted PCBs as a nonflammable coolant for a new generation of transformers. By the 1960s, Monsanto's growing family of PCBs were also widely used as lubricants, hydraulic fluids, cutting oils, waterproof coatings, and liquid sealants. Evidence of the toxic effects of PCBs appeared as early as the 1930s, and Swedish scientists studying the biological effects of DDT also began finding significant concentrations of PCBs in the blood, hair, and fatty tissue of wildlife in the 1960s.[7]

Research in the 1960s and 1970s revealed PCBs and other aromatic organochlorines to be potent carcinogens and also traced them to a wide array of reproductive, developmental, and immune system disorders.[8] Their high chemical affinity for organic matter, particularly fat tissue, was responsible for their dramatic rates of bioaccumulation and their wide dispersal throughout the North's aquatic food web: Arctic cod, for example, once carried PCB concentrations forty-eight million times that of their surrounding waters, and predatory mammals such as polar bears harbored tissue concentrations of PCBs more than fifty times greater than that. Though the manufacture of PCBs was banned in the United States in 1976, its toxic and endocrine disruptive effects persist worldwide.[9]

The world's center of PCB manufacturing was Monsanto's plant on the outskirts of East St. Louis, Illinois. East St. Louis is a chronically economically depressed suburb, across the Mississippi River from St. Louis, bordered by two large metal processing plants in addition to the Monsanto facility. "East St. Louis," reported education writer Jonathan Kozol, "has some of the sickest children in America." Kozol reported that the city had the highest rate of fetal death and immature births in the state, the third highest rate of infant death, and one of the highest childhood asthma rates in the United States.[10]

Dioxin: A Legacy of Contamination

The people of East St. Louis continued to face the horrors of high level chemical exposure, poverty, a deteriorating urban infrastructure, and the collapse of even the most basic city services, but the nearby town of Times Beach, Missouri, was found to be so thoroughly contaminated with dioxin— a byproduct of Monsanto's herbicide manufacture—that the U.S. government ordered the whole town evacuated. Hundreds of horses, pets, and wild birds had died, and children born to mothers exposed to the dioxin-contaminated oil demonstrated evidence of immune-system abnormalities and significant brain dysfunction.[11] While Monsanto consistently denied any connection to the incident, the St. Louis–based Times Beach Action Group (TBAG) uncovered laboratory reports documenting the presence of high concentrations of PCBs manufactured by Monsanto in dioxin-contaminated soil samples from the town.[12] "From our point of view, Monsanto is at the heart of the problem here in Missouri," explained TBAG's Steve Taylor in a 1998 interview.

The contamination and attempted cover-up at Times Beach reached the highest levels of the Reagan administration in Washington and was a factor in the resignation of Reagan's EPA administrator, Anne Gorsuch Burford, mother of current Supreme Court Justice Neil Gorsuch. In one widely reported incident, the Reagan White House ordered Burford to withhold documents on Times Beach and other contaminated sites in Missouri and Arkansas, and her special assistant, Rita Lavelle, was cited for shredding important documents; eventually Lavelle was jailed for perjury and obstruction of justice, and Burford was forced to resign.[13] An investigative reporter for the *Philadelphia Inquirer* identified Monsanto as one of the chemical companies whose executives frequently hosted luncheon and dinner meetings with Lavelle.[14] The evacuation sought by residents of Times Beach was delayed until 1982, eleven years after the contamination was first discovered, and eight years after the primary cause was identified as dioxin.

Monsanto's association with dioxin can be traced back to its manufacture of the potent herbicide 2,4,5-T, beginning in the late 1940s. "Almost immediately, its workers started getting sick with skin rashes; inexplicable pains in the limbs, joints, and other parts of the body; weakness; irritability; nervousness; and loss of libido," explains Peter Sills, author of *Toxic War* and a leading expert on dioxin and Agent Orange. "Internal memos show that the company knew these men were actually as sick as they claimed, but it kept all that evidence hidden."[15] An explosion at Monsanto's Nitro, West Virginia, herbicide plant in 1949 drew further attention to these complaints. The contaminant responsible for these conditions was not identified as dioxin until 1957, but the U.S. Army Chemical Corps was already interested in this substance as a possible chemical warfare agent. A request filed by the *St. Louis Journalism Review* under the U.S. Freedom of Information Act revealed nearly 600 pages of reports and correspondence between Monsanto and the Army Chemical Corps on the subject of dioxin, going as far back as 1952.[16]

The herbicide mixture known as "Agent Orange," which was used by U.S. military forces to defoliate the rainforest ecosystems of Vietnam during the 1960s, contained 2,4,5-T and 2,4-D herbicides from several sources, but Monsanto's formulation had concentrations of dioxin many times higher than that produced by Dow Chemical, the defoliant's other leading manu-facturer.[17] This made Monsanto the key defendant in the lawsuit brought by Vietnam War veterans in the United States, who faced an array of debilitat-ing symptoms attributable to Agent Orange exposure. When a $180 million settlement was reached in 1984 between seven chemical companies and the lawyers for the veterans, the judge ordered Monsanto to pay 45.5 percent of the total.

In the 1980s, Monsanto undertook a series of studies designed to mini-mize its liability in the Agent Orange suit and also in continuing instances of employee contamination at its West Virginia manufacturing plant. A three-and-a-half-year court case brought by railroad workers exposed to dioxin following a train derailment revealed a pattern of manipulated data and misleading experimental design. An official of the U.S. EPA concluded that the studies were manipulated to support Monsanto's claim that dioxin's effects were limited to the skin disease chloracne.[18]

The court case, in which the jury granted a $16 million punitive damage award against Monsanto, revealed that many of Monsanto's products, from household herbicides to the Santophen germicide once used in Lysol brand disinfectant, were knowingly contaminated with dioxin. "The evidence of Monsanto executives at the trial portrayed a corporate culture where sales and profits were given a higher priority than the safety of products and its workers," reported the Toronto's *Globe and Mail* after the close of the trial.[19] "They just didn't care about the health and safety of their workers," explains Peter Sills. "Instead of trying to make things safer, they relied on intimida-tion and threatened layoffs to keep their employees working."

A subsequent review by Dr. Cate Jenkins of the EPA's Regulatory Development Branch documented an even more systematic record of fraud-ulent science. "Monsanto has in fact submitted false information to EPA which directly resulted in weakened regulations under RCRA [Resources Conservation and Recovery Act] and FIFRA [Federal Insecticide, Fun-gicide, and Rodenticide Act]," reported Dr. Jenkins in a 1990 memoran-dum urging the agency to launch a criminal investigation of the company. Jenkins cited internal Monsanto documents revealing that the company "doctored" samples of herbicides that were submitted to the U.S. Depart-ment of Agriculture, manipulated technical arguments to deflect attempts to regulate 2,4-D and various chlorophenols, hid evidence regarding the contamination of Lysol, and excluded several hundred of its sickest former employees from its comparative health studies: "Monsanto covered-up the dioxin contamination of a wide range of its products. Monsanto either failed to report contamination, substituted false information purporting to show no contamination or submitted samples to the government for analysis which had been specially prepared so that dioxin contamination did not exist."[20]

New Generation Herbicides

The use of chemical pesticides increased manifold during the post–World War II era. For example, total revenues from insecticide production rose from $10 million in 1940 to $100 million in 1950 to over $1 billion in the early 2000s. With this expansion, agrochemical companies in the United States and Europe established a profound degree of control over agricultural practices. The agrochemical industry set the agenda for changing farm practices, came to dominate agricultural policymaking and the information available to farmers, and forged strategic alliances with the emerging global grain trading companies, such as Cargill, Conagra, and Archer Daniels Midland. Insecticide and herbicide use continued to grow, even as DDT and many other widely used chemicals were banned in the 1970s due to their extreme toxicity.

With sales of PCBs and 2,4,5-T eroding by the 1970s, Monsanto turned its focus to glyphosate-based herbicides, relying on the company's Idaho rock phosphate mine for the extraction of raw ingredients. The mine was reportedly a source of radioactive waste and chemical contamination in the region, leading to its designation as an EPA Superfund site.[21] By the late 1990s, a new generation of glyphosate-based weed killers such as Roundup accounted for at least one sixth of Monsanto's annual sales and half of the company's operating income.[22] The importance of herbicide sales increased further after the company spun off its industrial chemicals and synthetic fabrics divisions in 1997. Monsanto has promoted Roundup as a safe, general-purpose herbicide for use on everything from lawns and orchards to large coniferous forest holdings, where aerial spraying of the herbicide has been used to suppress the growth of deciduous seedlings and shrubs and encourage the growth of profitable fir and spruce trees.[23] The

Oregon-based Northwest Coalition for Alternatives to Pesticides (NCAP) reviewed over forty scientific studies on the effects of glyphosate, and of the polyoxyethylene amines (POEA) used as a surfactant in Roundup, and concluded in 1991 that the herbicide was far less benign than Monsanto's advertising suggests:

> Symptoms of acute poisoning in humans following ingestion of Roundup include gastrointestinal pain, vomiting, swelling of the lungs, pneumonia, clouding of consciousness, and destruction of red blood cells. Eye and skin irritation has been reported by workers mixing, loading and applying glyphosate. EPA's Pesticide Incident Monitoring System had 109 reports of health effects associated with exposure to glyphosate between 1966 and October, 1980. These included eye or skin irritation, nausea, dizziness, headaches, diarrhea, blurred vision, fever and weakness.[24]

Roundup was far less widely used in 1966–1980 than it is today.

A series of suicides and attempted suicides in Japan during the 1980s using Roundup herbicide allowed scientists to calculate a lethal dose of six ounces. The herbicide is 100 times more toxic to fish than to people and is toxic to earthworms, soil bacteria, and beneficial fungi. Scientists have measured a number of direct physiological effects of Roundup in fish and other wildlife, in addition to secondary effects attributable to defoliation of forests.[25] Breakdown of glyphosate into N-nitroso glyphosate and other related compounds heightened concerns about the possible carcinogenicity of Roundup products, long before glyphosate was declared a likely carcinogen by the World Health Organization in 2015.

A 1993 study at the University of California at Berkeley's School of Public Health found that glyphosate was the most common cause of pesticide-related illness among landscape maintenance workers in California, and the third leading cause among agricultural workers.[26] A 1996 review of the scientific literature by members of the Vermont Citizens' Forest Roundtable— a group that successfully lobbied the Vermont Legislature for a statewide ban on the use of herbicides in forestry—revealed evidence of lung damage, heart palpitations, nausea, reproductive problems, chromosome aberrations, and numerous other effects of exposure to Roundup herbicides.[27] In 1997,

Monsanto responded to five years of complaints by the New York State Attorney General that its advertisements for Roundup were misleading; the company altered its ads to delete claims that the herbicide is "biodegradable" and "environmentally friendly," and paid $50,000 toward the state's legal expenses.[28]

In March of 1998, Monsanto agreed to pay a fine of $225,000 for mislabeling containers of Roundup on 75 separate occasions. The penalty was the largest settlement ever paid for violation of the Worker Protection Standards of the Federal Insecticide, Fungicide and Rodenticide Act (FIFRA).[29] This was only one of several major fines and rulings against Monsanto during the 1980s–90s, including a $108 million liability finding in the case of the leukemia death of a Texas employee in 1986, a $648,000 settlement for allegedly failing to report required health data to the EPA in 1990, a $1 million fine by the state Attorney General of Massachusetts in 1991 in the case of a 200,000 gallon acid wastewater spill, a $39 million settlement in Houston, Texas, in 1992 involving the deposition of hazardous chemicals into unlined pits, and numerous others.[30] In 1995, Monsanto ranked fifth among U.S. corporations in EPA's Toxic Release Inventory, having discharged 37 million pounds of toxic chemicals into the air, land, water, and underground.[31]

In the early 2000s, Roundup came to play a central role in the U.S. "drug war" due to its widespread use to eradicate coca and poppy plants in Colombia and other countries.[32] Colombian agronomists uncovered the use of an additive that reportedly increased herbicide exposures to more than 100 times Monsanto's recommended dosage for conventional agricultural applications, and coca and poppy plants were not the only casualties. U.S. aerial spraying of tons of Roundup over the Colombian countryside destroyed local subsistence crops, such as manioc, bananas, palms, sugarcane, and corn; poisoned creeks, rivers and lakes; and destroyed indigenous fish populations.

GMOs Enter the Food Chain

Monsanto's aggressive promotion of its biotechnology products, from recombinant Bovine Growth Hormone (rBGH), to Roundup Ready–branded soybeans and other crops, to its insecticide-producing varieties of corn and cotton, is a substantial continuation of its many decades of ethically questionable practices. "Corporations have personalities, and Monsanto is one of the most malicious," explains author Peter Sills. "From Monsanto's herbicides to Santophen disinfectant to BGH, they seem to go out of their way to hurt their workers and hurt kids."

In the late 1980s, Monsanto was one of four chemical companies seeking to market a synthetic Bovine Growth Hormone to artificially boost milk production, using genetically engineered *E. coli* bacteria to manufacture the protein hormone. As Jennifer Ferrara described in *The Ecologist's* "Monsanto Files" issue, the company's fourteen-year effort to gain approval from the U.S. Food and Drug Administration (FDA) to bring recombinant BGH to market was fraught with controversy, including allegations of a concerted effort to suppress information about the hormone's ill effects.[33] One FDA veterinarian, Richard Burroughs, was fired after he accused both the company and the agency of suppressing and manipulating data to hide the effects of rBGH injections on the health of dairy cows.[34]

In 1990, when FDA approval of rBGH appeared imminent, a veterinary pathologist at the University of Vermont's agricultural research facility released previously suppressed data to two state legislators documenting increased rates of udder infection in cows that had been injected with the then-experimental Monsanto hormone, as well as an unusual incidence of severely deforming birth defects in offspring of rBGH-treated cows.[35] An independent review of the University data by a Vermont farm advocacy group documented additional cow health problems associated with rBGH, including high incidences of foot and leg injuries, metabolic and reproductive difficulties, and uterine infections. Congress member Bernie Sanders urged Congress's General Accounting Office (GAO) to investigate, but the agency was unable to obtain the necessary records from Monsanto and the University, particularly with respect to suspected teratogenic and embryotoxic effects. The GAO auditors still concluded that cows injected with rBGH had mastitis (udder infection) rates one third higher than untreated cows, and recommended further research on the risk of elevated antibiotic levels in milk produced using rBGH.[36]

Monsanto's rBGH was eventually approved by the FDA for commercial sale beginning in 1994. The following year, Mark Kastel of the Wisconsin Farmers Union released a study of Wisconsin farmers' experiences with the drug. His findings exceeded the 21 potential health problems that Monsanto was required to list on the warning label. Kastel found widespread reports of spontaneous deaths among rBGH-treated cows, high incidences of udder infections, severe metabolic difficulties and calving problems, and in some cases an inability to successfully wean treated cows off the drug.[37] Many experienced dairy farmers who experimented with rBGH found they needed to replace large portions of their herd. Instead of addressing the causes of farmers' complaints about rBGH, Monsanto went on the offensive, threatening to sue small dairy companies that advertised their products as free of the artificial hormone, and joining with several dairy industry trade associations to sue the state of Vermont after it passed the first and only mandatory labeling law for rBGH in the United States.[38] Evidence for the damaging effects of rBGH on the health of both cows and people continued to mount, until most leading dairy manufacturers in the United States eventually prohibited their suppliers from using the drug, and Monsanto sold the rights to rBGH to Eli Lilly's veterinary drug division in 2008.[39]

Monsanto's efforts to prevent labeling of genetically engineered soybean and corn exports from the United States continued the same strategy that sought to squelch complaints against the dairy hormone. While Monsanto still argues that its Roundup Ready soybeans reduce herbicide use, crop varieties genetically engineered to withstand chemical weed killers are far more likely to increase farmers' dependence on these chemicals. Weeds that emerge after the original herbicide has dispersed or broken down are often treated with further applications of herbicides.[40] "It will promote the overuse of the herbicide," Missouri soybean farmer Bill Christison told Kenny Bruno of Greenpeace International. "If there is a selling point for RRS, it's the fact that you can till an area with a lot of weeds and use surplus chemicals to combat your problem, which is not what anyone should be doing."[41] Christison refuted Monsanto's claim that herbicide-resistant seeds are necessary to reduce soil erosion from excess tillage, and explained that midwestern farmers have developed numerous methods of their own to reduce herbicide use.

Defying these concerns, Monsanto stepped up its production of Roundup. After Monsanto's U.S. patent on Roundup expired in 2000, the company's strategy to compete against generic glyphosate products came to increasingly depend upon the packaging of Roundup herbicide together with Roundup Ready seeds capable of withstanding repeated herbicide treatments.[42] By the mid-2000s, Roundup use on soybeans was increasing by 9 percent every year, and glyphosate-resistant varieties of common weed species began to appear in California and the Midwest. By 2012, more than 20 resistant weed varieties had been identified, reducing some farmers' crop yields by as much as 50 percent.[43] Monsanto's response was to develop new GMO varieties of soybeans and corn with combined resistance to several additional herbicides, including highly volatile, drift-prone dicamba—which has become such a nuisance to neighboring farms that several states have instated partial bans on its use[44]—and the highly toxic 2,4-D, formerly best known as a component of Agent Orange.

In addition to its herbicide-tolerant varieties, Monsanto has also been at the forefront of genetically engineering varieties of corn, cotton, and other crops to a class of bacterial toxins in order to reduce damage from crop pests. The insecticidal toxins, derived from *Bacillus thuringiensis* (Bt), have been utilized by organic growers in the form of a freeze-dried bacterial spray since the early 1970s. But while Bt bacteria are relatively short-lived and secrete their toxin in a form that only becomes activated in the alkaline digestive systems of particular worms and caterpillars, genetically engineered Bt crops produce an active form of the toxin throughout the plant's life cycle.[45] Today, many common genetically engineered crops contain a combination of several different herbicide tolerances and Bt toxin-producing traits, a technology known as *gene stacking*.

As predicted more than 20 years ago, the presence of Bt toxins throughout a plant's life cycle has encouraged the development of resistant strains of common crop pests. Early on, the EPA determined that widespread resistance to Bt could render natural applications of Bt bacteria ineffective and required growers to plant refuges of up to 40 percent non-Bt corn or cotton in an attempt to forestall this effect, a mandate that large growers almost universally ignored. The active toxin secreted by these plants was also found to harm beneficial insects, moths, and butterflies.[46]

With all these problems, how did GMOs become so pervasive in U.S. commodity crop production, as well as in several other countries such as Brazil and Argentina? Two reasons stand out in particular. The first is that growers of commodity crops like corn and soybeans are under continual pressure to increase production amidst highly volatile crop prices. The ability to spray crops with chemical weed killers throughout the growing season has made it easier for farmers to expand their acreage without additional labor costs. Second, and probably more important, is the increasing concentration of ownership of the seed supply by Monsanto and a handful of other GMO producers. By 2015, only seven companies had come to control 71 percent of the global commercial seed supply, including GMO producers Monsanto, DuPont and Dow (now merged), Syngenta, and Bayer. All these companies originally came to the forefront as producers of agricultural chemicals, and many had their origins as producers of nerve gases and other weapons of war.[47] Monsanto has leveraged its increasing control over seed production to limit farmers' ability to choose whether or not to grow GMOs. As public-sector research in traditional plant breeding has declined precipitously, Monsanto has become a dominant player in numerous areas of agronomic research. New conventionally bred traits that are widely sought by growers—affecting crop yields, oil content, and nutrition—are now often only available from seeds that also carry genetically engineered traits for herbicide tolerance and Bt insecticide production.

Monsanto started its climb toward becoming the world's largest seed producer in the late 1990s, when it began to acquire ownership of many of the largest, most established seed companies in the United States. First it bought Holden Foundation Seeds, supplier of germplasm used on 25–35 percent of U.S. maize acreage, and Asgrow Agronomics, which Monsanto described as "the leading soybean breeder, developer and distributor in the United States."[48] In India, Monsanto took control of the country's flagship seed supplier Mahyco, originally the Maharashtra Hybrid Seeds Company. Monsanto also bought De Kalb Genetics, once the second largest seed company in the United States and the ninth largest in the world, and spent more than two years trying to acquire Delta and Pine Land Company, the largest U.S. cotton seed company and the codeveloper (with the U.S. Department of Agriculture) of the notorious Terminator seed technology, a genetic intervention that induces plants to produce sterile seeds that could never be successfully replanted.[49] For Monsanto and other GMO producers, this was once perceived as the key to maintain control over GMO patents, but farmer activists and critical scientists saw it as an existential threat to the integrity of the seed supply.

In October 1999, amidst international public uproar, Monsanto made world headlines with the announcement that it would not seek to market

Terminator seeds.[50] For Monsanto, this was a small price for salvaging the future of genetically engineered crops. It was almost a textbook case of modern corporate public relations, in which companies are urged to admit mistakes and seek wider credibility by appearing to involve activist groups in corporate decision-making.[51] However, Monsanto didn't stop trying to prevent farmers from using their patented GMOs unless they agreed to sign restrictive contracts and pay steep royalties. Instead of engineering sterility into their GMO crops, Monsanto took to the courts and to various extra-legal forms of harassment. As of 2012, the Center for Food Safety had documented 142 lawsuits against 410 farmers in 27 U.S. states, and an additional 4,500 investigations of farmers the company accused of "seed piracy," i.e., saving or reselling seeds harvested from GMO crops.[52] Most famously, Saskatchewan canola grower Percy Schmeiser faced a Monsanto lawsuit for having replanted his own seed after it apparently cross-pollinated with a neighbor's Roundup-tolerant canola. The Canadian Supreme Court ultimately ruled that Monsanto had the right to sue Schmeiser to protect the company's patent rights, but overturned all the monetary damages that had been imposed by a lower court; Schmeiser didn't use Roundup, and thus had gained no tangible benefit from the herbicide-tolerance trait.[53]

Monsanto further cemented its dominance in global seed production in 2005 when it purchased Seminis Seeds, a Mexican company that had become the premier supplier of vegetable seeds in the Western Hemisphere.[54] By 2015, Monsanto controlled more than a quarter of the world's production and sale of agricultural seeds.[55]

Monsanto's Greenwash

Given this history, it is easy to understand why people around the world refused to trust Monsanto with the future of our food and our health. But Monsanto has gone to great lengths to appear unperturbed by this opposition. Through efforts such as the British advertising campaign that inspired "The Monsanto Files" in response, their involvement with the prestigious Missouri Botanic Garden, and their sponsorship of a state-of-the-art biodiversity exhibit at the American Museum of Natural History in New York in the early 2000s, they sought to appear greener, more righteous, and more forward looking than their opponents.

In the United States, Monsanto and other biotech companies have cultivated close ties to people at the highest levels of every recent presidential administration. In May 1997, Mickey Kantor, an architect of Bill Clinton's 1992 election campaign and United States trade representative during Clinton's first term, was elected to a seat on Monsanto's Board of Directors. Marcia Hale, once a personal assistant to Clinton, served as Monsanto's public affairs officer in Britain.[56] Vice President Al Gore's chief domestic policy advisor in the late 1990s, David W. Beier, was formerly the senior director of government affairs at the pioneering medical biotech company Genentech.[57] In the George W. Bush administration, cabinet secretaries Donald Rumsfeld (Defense), John Ashcroft (Attorney General), Tommy Thompson (Health and Human Services), and Anne Veneman (Agriculture) all had historic ties, either as officials of companies absorbed by Monsanto or as recipients of large campaign contributions. Rumsfeld, the notorious Iraq War architect, had been the CEO of the G. D. Searle pharmaceutical company prior to its purchase by Monsanto in 1985. A few years earlier, he exercised his influence as a leading member of Ronald Reagan's presidential transition team to urge the Food and Drug Administration to approve Searle's artificial sweetener, aspartame (a.k.a. NutraSweet), despite its known neurotoxicity.[58]

Barack Obama's Secretary of Agriculture, former Iowa Governor Tom Vilsack, had once been honored as Governor of the Year by the Biotechnology Industry Organization.[59] His term as governor was also noted for the widespread expansion of large-scale factory farms (concentrated animal feeding operations, or CAFOs) and expanding unsustainable ethanol production from corn. Initially, Obama was praised for his early campaign statements in support of GMO labeling, Michelle Obama's famous organic vegetable plot on the White House lawn, and an early antitrust initiative in support of small farmers. However, the administration soon succumbed to an intensive industry lobbying campaign, stepping back from antitrust enforcement and largely abandoning its rhetorical challenges to the corporate food giants.[60] In Obama's second term, Vilsack's USDA initiated an effort to "modernize" the assessment of new GMO crop varieties, which eventually resulted in a proposed rule to limit the scope of new crop trials that fall under the agency's jurisdiction.[61] When Donald Trump moved into the Oval Office in 2017, he appointed Sonny Purdue, a former Georgia governor and agribusiness operator—mainly involved in fertilizers, grain, and trucking—as his Agriculture Secretary. Most likely, this will further exacerbate the problem of agribusiness dominance over the mammoth department, which is second only to the Pentagon in its size and scope.

For much of its recent history, however, Monsanto was on the defensive. Several attempts to merge with larger companies—mainly to help pay for the billions spent on seed company acquisitions—had fallen through.[62] Monsanto engaged in merger talks with American Home Products, DuPont, Novartis, and several other companies and eventually merged for two years with the pharmaceutical giant Pharmacia, which spun off a much more agriculturally focused Monsanto in 2001. Still, the company's aggressive promotion of genetically engineered corn and soybeans had made the name Monsanto synonymous with everything threatening and out of control about genetic engineering. In 1999, Germany's Deutsche Bank declared genetically modified crops an economic "liability to farmers," advising its investors to stop buying agricultural biotech stocks, and the *Wall Street Journal* announced that Monsanto, once the seemingly invincible world leader in biotechnology, would be worth significantly more to investors if it were to be broken up.[63]

Even the smaller and more focused Monsanto that emerged from the Pharmacia merger remained vulnerable. In 2003, when the *New York Times* investigated the growing problem of Roundup resistant weeds, the paper quoted Idaho agronomist Charles Benbrook saying he had been approached to consult with investment bankers considering a purchase of Monsanto in order to break it up.[64] Then, in 2016, following an unsuccessful attempt to purchase the Swiss agrochemical and GMO producer Syngenta, Monsanto agreed to sell itself to the pharmaceutical and agribusiness giant Bayer. The merger has further increased monopolization in the already highly concentrated seed and agrochemical sectors, controlling 34 percent of the global herbicide market, 23 percent of insecticides, 29 percent of all seed sales, and close to 60 percent of the world's cotton seeds.[65]

Monsanto also made headlines in the mid-2010s as its efforts to manipulate research results in its favor began to attract more scrutiny. In 2015, the *New York Times* highlighted Monsanto's funding support for horticulturist Kevin Folta of the University of Florida, who travels the United States advocating for GMOs.[66] The *Times* reported that many of Folta's talking points were drafted by the Ketchum public relations firm on behalf of the biotech industry. Then in 2017, two decades worth of documents uncovered by attorneys representing cancer patients who had been exposed to Roundup revealed a consistent pattern of payments to scientists on Monsanto's behalf. The documents, stated one of the attorneys, "show that Monsanto has deliberately been stopping studies that look bad for them, ghostwriting literature, and engaging in a whole host of corporate malfeasance. They (Monsanto) have been telling everybody that these products are safe because regulators have said they are safe, but it turns out that Monsanto has been in bed with U.S. regulators while misleading European regulators."[67]

Today, with surrogates for the fossil fuel, chemical, agribusiness, banking, and pharmaceutical industries firmly in charge of federal regulatory agencies, the public is increasingly reliant on unofficial sources to reveal those agencies' inner workings.[68] The current deregulatory fervor in Washington may be a boon for megacorporations like Bayer/Monsanto, but it is clearly at the expense of those corporations' victims and the public at large.

5

Poisoning the Big Apple—Forgotten History in the Lead-Up to 9/11

Two years before what we now know as 9/11, a terrorist attack hit New York—or so we were told. U.S. government officials falsely announced that Iraq's president, Saddam Hussein, had sent to New York City some arcane virus that was killing birds, mostly crows, and that it could be transmitted to people by mosquitoes. Panic ensued. The mysterious disease threatened to infect and kill people throughout the metropolitan area. New Yorkers were told to prepare for emergency measures around West Nile virus and that without such measures thousands of people were likely to die.

"September 4th, 1999, was an extremely important day in the history of New York City,"[1] one analyst noted. Indeed, the future history of our city, and the country as well, was about to change. I was strolling through Prospect Park in Brooklyn on that warm day near the end of summer. Hundreds of people were out in the park sunbathing, reading, kissing, walking their dogs. Kids were everywhere playing baseball and soccer. Suddenly, helicopters buzzed just above the tree line spraying a substance we later learned to be malathion—one of a class of organophosphate pesticides invented as a nerve gas[2] by the Nazis in World War II—spewing out in substantial bursts.

The Mayor of New York City at that time, Rudy Giuliani, ordered the toxic pesticide malathion to be sprayed from helicopters, airplanes, and trucks to kill mosquitoes, poisoning the city's population, wildlife, soil, parks, and waterways. They drenched 526-acre Prospect Park that afternoon, spraying the malathion over and onto hundreds of children. There were a few police cars patrolling, but none of them warned people to get out of the park and off the streets. I ran like a lunatic trying to get the kids away from the spray. And then I held my breath for as long as I could and ran out of the park.

Over the next few days, the City sprayed the subways, food markets, sewer system, schools, religious institutions, daycare centers, and restaurants. Spraying also occurred over or near open waterways. A scientist working with the newly formed No Spray Coalition, Jonathan Logan, and videographer Roy Doremus, followed the trucks and filmed the City spraying pregnant women on 125th Street in Harlem early in the evening. The Coalition presented the video as evidence in a lawsuit filed a few months later.[3] (The No Spray Coalition won its lawsuit after seven years, achieving in 2007 a historic settlement with the City administration.) Reporters covering the suit saw the video in court, and that night every TV station broadcast the alarming footage.

The Mayor's sprayers also hit Boro Park as Jews walked to shul (synagogue) on Yom Kippur, creating panic, and triggering memories in them of the Holocaust—and outrage. They sprayed over or near lakes and rivers, in reckless disregard for the warnings required by the Environmental Protection Agency against spraying over or near bodies of water,[4] poisoning fish and marine life and wreaking havoc with delicate ecosystems. The spraying killed mosquitoes' natural predators (dragonflies, frogs, and bats—a single bat can eat more than a thousand mosquitoes in one hour[5]). The neurotoxins drifted out over the lakes and bays and into the Atlantic Ocean, killing bees, butterflies, fish, birds, and, in the Long Island Sound, lobsters.

Ecologists and bird-watchers had tried to call attention to clusters of dead crows *months* earlier in 1999, mostly around Fort Totten on the northern shore of Queens, where large amounts of pesticides were suspected to have been used. But it was not until August that animal pathologist Tracey McNamara and her team at the Bronx Zoo disclosed that exotic birds had begun dying from an unknown disease. They feared it would spread to other birds, attack human beings, and present a serious health emergency. McNamara—who had also been associated with the top security U.S. government animal disease and biowarfare lab researching dangerous pathogens on Plum Island (off the eastern tip of Long Island, New York, around 150 miles from Fort Totten as the crow flies)—contacted the NYC Department of Health and the federal Centers for Disease Control. Health and government officials released incorrect and ever-changing information about the nature of the illness, which by now had affected several people as well as birds. At first the birds were said to have died from St. Louis encephalitis, and then West Nile-like virus. The Centers for Disease Control announced, finally—after competing diagnoses with the U.S. Army laboratories—that a rare mosquito-borne virus, West Nile, was causing the disease.[6]

In actuality, far fewer people died from the virus in this "emergency" than from diseases caused or exacerbated by the massive aerial and truck-based pesticide spraying that ensued.[7] Similarly, the Parks Department's applications of Roundup put thousands of people at risk for serious illness.[8] Four people in New York City died from what was said to be West Nile encephalitis that first year, a disease previously unknown in the United States. (Encephalitis is an inflammation of the brain that could be caused by exposure to pesticides and other toxins, vaccines, allergic reactions, and certain viruses and bacteria.) Strangely—and this information had to be pried a year later from Department of Health officials with a crowbar—not a trace of West Nile virus was found in the brain tissue of those who'd died,[9] even though studies at Yale showed that "people who develop full-blown cases of West Nile encephalitis . . . are those in whom the virus has penetrated the blood-brain barrier."[10] If the deaths were truly due to the virus, fragments should have been found in the brain of anyone suffering or having died from West Nile.[11]

Mayor Giuliani held daily press conferences at his bunker in World Trade Center #7 surrounded by maps, graphs, and armed guards, whipping up hysteria, which enabled officials to circumvent civil liberties and manipulate the media and public. Giuliani and others falsely pinned the upsurge of mosquito-borne ailments and the handful of cases of West Nile encephalitis on Saddam Hussein, just as officials would falsely claim that Iraq held secret troves of weapons of mass destruction, and later insisted—similarly, with nary a shred of evidence—that Saddam was behind the attacks on the World Trade Center in 2001 and the Anthrax attacks that followed. It took a very long two weeks before the Centers for Disease Control publicly denied there was a connection between Saddam Hussein and the New York West Nile illnesses.[12]

Meanwhile, thousands of people were being sickened *by the spraying*. But that didn't stop officials from the City and the Centers for Disease Control from ordering the City to repeatedly spray a toxic barrage of Fyfanon ULV (96.5 percent malathion). The ULV stands for "ultra low volume," which officials spun as "safer—it's low volume." In actuality, ULV is much worse—its extremely fine droplets (low volume) hang in the air longer and penetrate deeper into the lungs. They also sprayed the synthetic pyrethroids Scourge (resmethrin) and Anvil 10+10 (sumithrin), which in humans as well as in insects impair the endocrine system, mimic hormones such as estrogen, and may cause breast cancer, prostate cancer, erectile dysfunction, miscarriages, asthma, and drastically lower sperm counts.[13]

Instead of a pesticides hotline for people to call who were sickened by the spraying, the city's Department of Health set up a West Nile virus telephone answering service, where phone calls were outsourced to ill-informed non-unionized operators 200 miles away in Pennsylvania, and who were only allowed to take information from those suspected of having contracted West Nile virus; they refused to take information from callers who were sickened by the spraying itself and had no advice to offer them. The No Spray Coalition taped some of those calls, and submitted transcripts as part of its lawsuit.[14]

The City did not warn people with asthma, compromised immune systems, cancer, or allergies nor did it warn those facing repeated exposure (homeless people, subway workers, spray truck drivers), let alone the general public, about the dangers of the pesticides. Instead, Mayor Giuliani vilified those opposing the spraying as "environmental terrorists" who "like to get you angry because it gets them on television."[15] Office of Emergency Management Coordinator Jerome Hauer, who had been appointed by Giuliani at the behest of the Manhattan Institute, a neo-conservative think tank, dismissed concerns over the pesticide spraying as "irresponsible environmental hysteria and stupidity."[16] Giuliani, Hauer, and Health Commissioner Neal Cohen, a psychiatrist, repeatedly assured the citizenry that pesticide spraying was harmless. "There is absolutely no danger to anyone from this spraying. . . . There are some people who are engaged in the business of wanting to frighten people out of their minds," the mayor charged.[17]

The mayor rejected protests over the hazards of spraying, saying, "There's no point in not spraying, because there's no harm in spraying. So even if we're overdoing it, there's no risk to anyone in overdoing it."[18] Giuliani's irresponsible and misleading statements so angered then–New York State Attorney General Eliot Spitzer that, responding to numerous complaints by anti-pesticide activists, Spitzer and others in his office admonished city officials over their claims that the pesticides were safe. Peter Lehner, the top environmental lawyer in the New York State Attorney General's office, told the *New York Times* that it was important not to gloss over the fact that malathion is a chemical that was designed to kill things, that it had sickened people in the past, and that the Federal Environmental Protection Agency prohibits those who sell any pesticide from describing it as harmless. "The EPA clearly says don't call these things harmless, because they are not." Lehner encouraged city officials to change the way they described malathion, and warned the mayor that private companies making such claims would be in violation of federal and state law.[19]

But Giuliani persisted. "I've been sprayed seven times, and I'm perfectly healthy," Giuliani said in October 1999. Over the next few years, Mayor Giuliani, Police Chief Howard Safir, and a half-dozen other City officials were diagnosed with prostate cancer. Eight members of the No Spray Coalition died from cancers and other disorders caused or exacerbated by the spraying. Immune-compromised illnesses such as common colds, flu, and asthma increased dramatically across the city over the spraying months, compared to the same period in previous years. Longer-term consequences of the spray campaign will emerge slowly over the shortened lifetimes of those exposed.[20]

In years that followed, the City stopped using massive doses of malathion and substituted what officials termed "a very low rate" of Anvil 10+10 containing pyrethroid, phenothrin and piperonyl butoxide, for its adult mosquito control efforts. Piperonyl butoxide is a so-called "synergist" that slows the body from breaking down and excreting the pyrethroids, dramatically increasing their toxicity.[21] The U.S. Environmental Protection Agency lists piperonyl butoxide as a possible cancer-causing agent. Also in the toxic mix are unlabeled "inert" ingredients, likely carcinogens, mostly petroleum and benzene compounds.[22,23]

Nine years earlier, the manufacturer of U.S. military uniforms had soaked them in permethrin, a similar pyrethroid. The pesticide is thought to be one of the factors contributing to Gulf War Syndrome in tens of thousands of U.S. soldiers. Brain scans of veterans who became ill after serving in the Gulf War clearly showed evidence of significant brain-cell loss.[24] Researchers have now succeeded in linking Gulf War soldiers' exposure to pesticides and nerve gas to debilitating brain damage by using magnetic resonance spectroscopy scanning techniques, which detect changes to the brain at the chemical and molecular levels. Sick Gulf War veterans had 20 percent fewer brain cells in the brain stem than healthy veterans.[25] The sick veterans also showed a 12 percent loss in the right basal ganglia and 5 percent loss in the left basal ganglia. The basal ganglia are associated with control of motor functions.[26] The British government considers Gulf War syndrome a result of organophosphate (i.e., malathion) poisoning.[27]

Even after 11 years of mass pesticide spraying, New York City officials continued to claim that thorough environmental review and epidemiologic analyses conducted subsequent to spray events have shown that the public in general "is not expected to experience symptoms given the low level of exposure that may occur during the spraying events."[28] However, the City provided no evidence whatsoever to justify its trivializing of the spraying's effects. It ignored numerous reports and testimony from people who were sickened by the chemicals in the mosquito sprays. In 2003, the Centers for Disease Control reviewed reports of poisonings due to West Nile virus spraying from nine states in the country that collected such data. The CDC found 262 cases of pesticide-related illnesses *from the spraying*. The majority of cases resulted in respiratory (66 percent) and neurological (61 percent) reactions.[29]

Beyond Pesticides—a national organization headquartered in Washington, DC (formerly known as the National Coalition Against the Misuse of Pesticides) and a co-plaintiff in the *NoSpray, et al.* lawsuit against the NYC Department of Health and mayor's office—reviewed a federal General Accounting Office report that examined claims relating to the reported incidence of illnesses due to pesticide exposure and that concluded the following:

> Pesticide poisonings in the U.S. are not well tracked and are commonly misdiagnosed, unreported, and severely underestimated. Physicians receive little training on identifying poisonings and even when correctly diagnosed, rarely are they reported to authorities. EPA recognizes that poisonings are underreported and that the lack of national data on the extent of pesticide illnesses is a problem. It is therefore wholly imprudent for public officials to dismiss the hazards of broadcast spraying and the need for safer practices simply because pesticide poisonings are not making headlines.[30]

Beyond Pesticides and a Long Island–based network of breast-cancer survivors exposed the fact that, unfortunately, the EPA was not assessing the endocrine disruption potential of chemicals, although required to do so by law. It reported that endocrine disruptors, even in very small doses (such as those in ultra-low volume sprays), can cause neurological, developmental, and reproductive health problems in both humans and animals.[31] Beyond Pesticides also cited studies that linked endocrine disrupting pyrethroids to illnesses such as cancer and birth defects, refuting the City's "dose makes the poison" argument for the safety of ULV pesticides. It argued that far from being less toxic than malathion and organophosphates, pyrethroids are just as dangerous but work in a different way and warrant greater precautionary approaches. But City officials ignored this data and made sweeping and false claims about the pesticides' safety.

Clearly, the Giuliani administration was not being truthful when it claimed that it performed "thorough environmental review and epidemiologic analyses . . . [that showed] that the public in general is not expected to experience symptoms given the low level of exposure that may occur during the spraying events." Jay Feldman, executive director of Beyond Pesticides, announced that the group had "asked the EPA for the data on pesticide product effectiveness (efficacy) for public health mosquito control" and was told that there is none.[32] Cornell University Professor David Pimentel argued that "ground spraying in general is a waste of money. Most ground spraying is political and has very little to do with effective mosquito control." Members of the Brooklyn Greens—the founding local of the New York State Green Party—were joined by other Greens from around the City, New York State, and Long Island who were outraged at exposure to the toxic spraying and the lies broadcast in their communities, as exemplified in Mayor Giuliani's statements, and began organizing a campaign in New York against pesticides.

In New York, more than 70 people filled a classroom in Harlem to participate in the No Spray Coalition's first public forum organized by cofounder Valerie Sheppard, and provided the funds needed to get the Coalition off the ground. The No Spray Coalition organizers quickly educated each other about the issues. They discovered and read the City's comprehensive mosquito plan and met local activists at community board meetings and hearings in every borough, where together they challenged the mosquito plan's conclusions point by point, and gathered everyone's separate points into an initial program that called on the City to do the following:

- Stop the indiscriminate pesticide spraying
- Establish a pesticide-exposure hotline
- Develop environmental impact statements that study the effects of aerial spraying of pesticides and implement non-toxic alternatives for controlling mosquitoes
- Test sprayed areas for toxic pesticide residues
- Notify schools and daycare centers to carefully wash children's play areas after spraying
- Avoid spraying areas designated as "cancer clusters" as well as the homes of those choosing to opt out of being sprayed

City officials scorned all of those suggestions, and the mayor publicly castigated their proponents.

Meanwhile, tens of thousands of fish turned belly-up in Clove Lake in Staten Island and Alley Pond Park in Queens. Samples tested confirmed the fish were killed by malathion poisoning.[33] Cancer survivors in Nassau County organized forums outlining the dangers of pesticides;[34] they pointed out that since President Richard Nixon declared "war on cancer" in 1971, childhood cancers had, by 1999, *increased* overall by 26 percent. Rates of some specific cancers increased even more dramatically: acute lymphocyte leukemia by 62 percent, brain cancer by 50 percent, and bone cancer by 40 percent.[35] Increased exposure to pesticides—*not* faulty genes—is the main reason for this cancer explosion in children in the United States.[36]

I interviewed dozens of doctors and health care researchers about the pesticide spraying. Most medical practitioners in New York were cowed into silence. They were willing to talk, but only off the record. Each one agreed that pesticides were especially dangerous to children, elderly people, and those who are immune-compromised and that mass spraying of pesticides for any reason was a bad idea. But despite the consensus of health care professionals in New York City, few would come forward and publicly oppose the indiscriminate spraying of pesticides, unlike, for instance, health providers in places like Stamford, Connecticut, where scores of medical professionals signed petitions and took public stands against spraying. "We're afraid our funds will be cut off," one high-level researcher told me—a common complaint.

The Studies

Numerous studies have shown terrible developmental consequences, especially to children who have been exposed to pesticides, and reveal the City's recklessness and disregard for the science that runs counter to the drumbeat for its spray campaign. These included studies that show the following:

- All residents of the United States now carry dangerously high levels of pesticides and their residues in their bodies, which may have onerous effects on health.[37]
- A large portion of waterways and streams throughout the United States has been found to contain environmentally destructive pesticides that may severely affect animal and aquatic life.[38]
- Pesticides are both a trigger for asthma attacks and a root cause of asthma, which is epidemic throughout New York City, and even worse in poorer areas that are often racially segregated.[39]
- Pesticides killed off the natural predators of mosquitoes, and mosquitoes came back much stronger after the spraying, because many of their natural predators (which have longer reproductive cycles) were dead. These studies were done in New York State for mosquitoes carrying Eastern equine encephalitis. They found a fifteen-fold increase in mosquitoes after repeated spraying, and virtually all of the new generations of mosquitoes were pesticide-resistant.[40]
- Pesticides have cumulative, multigenerational, degenerative impacts on human health, especially on the development of children, which may not be evident immediately and may only appear years or even decades later.[41]
- Pesticides make it easier for mosquitoes and other organisms to get and transmit West Nile virus due to damage pesticides cause to their stomach lining.[42]
- Pyrethroid spraying is ineffective in reducing the number of the next generation of mosquitoes.[43]

Federal Judge George B. Daniels ruled in 2006 in *No Spray Coalition, et al., v. New York City*, in the case first filed in 2000, that the City of New

York had violated the Clean Water Act by spraying pesticides over navigable waterways. Suddenly, City officials decided they'd better negotiate.[44] A year later, the City signed a settlement spelling out the problems with the pesticide spray program. The admissions rebuked the hype that the City's own public relations onslaught had typically propounded in promoting the spraying, while minimizing the risks. Pesticides, the City admitted:

- May remain in the environment beyond their intended purpose
- Cause adverse health effects
- Kill mosquitoes' natural predators (such as dragonflies, bats, frogs, and birds)
- Increase mosquitoes' resistance to the sprays
- Are not presently approved for direct application to waterways

Despite New York City's admissions that were now enshrined in a legal settlement agreement, New York continued to spray pesticides to kill mosquitoes. It also began to apply glyphosate to parks and sidewalks to kill weeds, adding to our toxic burden. Grassroots activists quickly learned that relying primarily upon legal maneuvering and the presentation of scientific research, without taking action to shut down the pesticide spray programs altogether, ends up in a cycle of forever refuting officials' double-speak. The science against exposure to pesticides is overwhelming, but in "emergency conditions," whether real or contrived, truth rarely is able to meaningfully speak to power. Here, as in all movements, activists had to wrestle over how compliant they were willing to become in modulating their opposition to pesticides in order to achieve the slightest advances.

When the No Spray Coalition sued New York City's government over its massive spraying of toxic pesticides in 2000, it found that almost all the laws that once provided standing for citizen lawsuits had been eviscerated. Attorneys Joel Kupferman and Karl Coplan of the PACE University Law Clinic, who had joined the lawsuit, were not permitted to present a case based on the City's toxic spraying per se, but were forced to use the last remaining relevant law that allowed citizens to sue government—the Clean Water Act. Under the CWA, one had to show that the pesticide spraying constituted a "toxic pollutant" from a point source emitted over or near navigable waters (which the No Spray Coalition eventually succeeded in showing in federal court). Every year, corporations fund candidates and then lobby them to remove provisions protecting citizens' right to sue, and every year activists have an uphill fight to hold onto them. Legislation expanding corporate and governmental secrecy—that is, the protection of their profits through malignant wrong-doing and pollution of the environment—has been for decades a main objective of corporations and the elected officials they've bought.

Under the Clinton/Gore administration, federal funds for combating West Nile virus were coming not from the Department of Health but through bio-terrorism budgets, via the Centers for Disease Control and quasi-military agencies. "We have no choice," health providers told me. "To get funded, we have to frame our applications in terms of bio-terrorism, even when we know terrorism has nothing to do with it," since over the years Public Health had been defunded. The researchers repeated that mantra. They said nothing as City and Federal officials reiterated incorrect statements that the West Nile virus had been sent by Saddam Hussein, even though they knew it to be untrue. Senator Chuck Schumer promoted this ideological framework and was, as a consequence, able to pry loose $16 million for New York City to combat West Nile—but only as a potential bioterrorism threat.

The ruse had unanticipated negative consequences. In 2006, the U.S. government's Department of Homeland Security contracted Halliburton—the infamous corporation formerly headed by Bush administration Vice President Dick Cheney—to construct "temporary detention and processing facilities" or internment camps,[45] a conception soon expanded to include "quarantine" camps in the United States;[46] top officials again falsely pinned West Nile disease on Saddam Hussein. Framing West Nile virus as a bio-terror attack to obtain needed funds provided an early rationalization for the torture and rendition that was to follow. Their devil's bargain regarding West Nile allowed the demonization of Saddam Hussein to go unchecked, which in turn created an acceptance for attacks on civil liberties, which would be given their ultimate rationalization two years later.

A few No Spray Coalition activists sneaked onto the lawn surrounding Gracie Mansion (the official residence of the mayor of New York City) and took their own soil samples. No Spray Coalition members Kimberly Flynn and Stephanie Snow collected filters from air conditioners around the city. They shipped all the samples off to toxicologist, Dr. Robert Simon, in Virginia, whose lab found that they contained extremely high levels of pesticide residue *six months after the spray had ended*, contrary to official statements that predicted they would dissipate. The Coalition asked the NYC Department of Health to recommend that residents change their air conditioner filters before starting up their machines as summer approached. The DOH officials privately admitted the danger of pesticides blasting into peoples' apartments when they turned on the air conditioners for the first time that season, but they refused to notify the public.

The attorney for the No Spray Coalition, Joel Kupferman of the New York Environmental Law and Justice Project, attended the Mayor's press conference and tried to question him about the pesticide residues. Under the direction of the mayor, armed marshals converged on Kupferman and threatened to arrest him if he did not leave the press conference immediately. At packed New York City Council hearings, and at congressional hearings conducted by Representative Gary Ackerman, Kupferman, toxicologist Dr. Robert Simon, a handful of courageous health professionals such as Dr. Adrienne Buffaloe,[47] and numerous environmental, disability, and health advocates testified that the City officials were wrong and that far from breaking down into harmless ingredients as officials had assured, the malathion broke down into malaoxon and isomalathion—chemicals even more toxic than the original malathion itself. The City stored metal drums of malathion in warehouses baking in the sun in Calverton, Long Island; the pesticides had already broken down into isomalathion and malaoxin in their containers before the spraying even began!

The mass spraying was run as a military operation, not a public health program. The spray program sought to integrate federal, state and local emergency management teams—the same teams that were shortly established in response to anti-globalization protests in Seattle, Boston, Washington, Philadelphia, and Los Angeles—and was coordinated not by the Department of Health as one might have expected but by the new Office of Emergency Management (OEM), headed at the time by Jerome Hauer and headquartered in Mayor Giuliani's "bunker" on the 23rd floor of World Trade Center #7–a building that, unknown to most of us at the time, also housed, a few floors above, the largest CIA offices in the country outside of Langley, Virginia. It also housed Securities and Exchange Commission files on Enron and other corporate scandals, scheduled to be heard shortly after 9/11. Most of those files were destroyed in the attacks on the World Trade Center.)

Attorney Joel Kupferman and No Spray Coalition cofounder Robert Lederman, who had been repeatedly arrested for his artistic portrayals of Giuliani as Hitler (after having voted for Giuliani a few years earlier), managed to bypass security and enter the "bunker" in order to serve Freedom of Information papers on Hauer and the OEM, which stonewalled the activists. When he returned to his office, Kupferman received a message threatening legal action against Lederman. Apparently "someone" had glued stickers attacking the mayor to the undersides of furniture and equipment in the "impenetrable" bunker, and the OEM was none too pleased. Hauer and the OEM refused to comply with requests for information made under the Freedom of Information Act. A judge ordered them to comply. They claimed to have no records pertaining to the purchase, storage, or use of the pesticides.

Lederman obtained and publicized a secret memo issued by the NYC Police Department amplifying the Coalition's concerns. The memo quoted the mayor's *Chem-Bio Handbook*, which had been distributed to every police station and firehouse in the City. It warned police officers accompanying spray trucks to stay at least 25 feet away from the spray, wear protective clothing, keep their patrol car windows tightly shut, and avoid all contact with the pesticides—while Giuliani and other city officials continued to tell the public that the spraying was "perfectly safe."

Neither the memo nor the book's accurate and scary warnings of the dangers of pesticides exposure had been publicized by the government or appeared in the media. The *Handbook* described in detail the dangerous nerve gases that were sprayed over the entire City in violation of the manufacturers' labeling instructions, which read in part: "This product is toxic to fish. Keep out of lakes, streams, ponds, tidal marshes and estuaries. Do not apply when weather conditions favor drift from areas treated. This pesticide is highly toxic to bees exposed to direct treatment or to residues remaining on the treated area." But the government failed to present this information to the public. It was left to the small No Spray Coalition to do so.

When the spray truck drivers heard Kupferman and Valerie Sheppard interviewed on an AM radio program, they contacted the Coalition. Some were very sick, and they suspected their illnesses were related to the spraying.

The drivers came from poor communities and had been hired as independent contractors at low wages and no health benefits. The mayor trivialized their health issues—until Kupferman arranged for doctors at Mount Sinai Hospital to examine them. They diagnosed the workers as suffering from pesticide poisoning.[48]

Foreshadowing the mistreatment of rescue personnel following 9/11, spray truck drivers like Kent Smith testified—at great risk to their jobs, which they desperately needed—that neither the company nor the government had provided them with training, safety equipment, respirators, or health care. *New York Daily News* columnist Juan Gonzalez wrote several front-page stories exposing the workers' plight and the dangers of pesticides. Attorney Kupferman, his New York Environmental Law and Justice Project, Kimberly Flynn, and the No Spray Coalition pressed New York's agencies to hold Clarke Environmental Mosquito Control, Inc., of Illinois, accountable. Much to the activists' surprise, the New York State Department of Environmental Conservation fined Clarke $1 million, and they also succeeded in blocking a $267 million contract bid by Clarke to run the City's spray program for the next few years. A week after blowing the whistle on the company's sweetheart contract with the City, one of the workers whose testimony had been key to exposing the company, Corey Gregory, was found murdered in an east Brooklyn elevator with seventeen bullets in his body. His wallet was still in his pocket.

Federal, state, and local authorities were responding, so they claimed, to a "health emergency" in New York. Yet, no health emergency was officially declared. Only 3–5 percent of the birds autopsied that first year were found to contain antibodies to the West Nile virus. What did thousands of birds and animals die from?

State-certified animal rehabilitator Bob Zink, in Staten Island, treated animals injured by the pesticides, and meticulously tracked animal deaths, providing copies of his logs to the No Spray Coalition. Researcher Jim West superimposed EPA maps of petrochemical air pollution emissions over the counties where most of the dead animals and birds were found (the city claimed they had died from West Nile virus). Those maps showed strong correlation between the illnesses alleged to have been caused by West Nile virus and the hours and drift currents of heavy petrochemical-related air pollution. Could petroleum emissions from New Jersey's oil refineries have caused the deaths attributed to West Nile? West proposed the possibility that MTBE (a common fuel additive prior to being banned on January 1, 2004) or some other chemical pollutant emitted in the refining of oil had been the actual cause of death. WNV antibodies, West hypothesized, were biomarkers for petrochemical-related air pollution illness, but not indications that West Nile virus by itself caused the deaths.[49]

How could we find out? Across the country, localities did not determine for themselves the cause of death in suspected West Nile cases; they sent blood samples to the Centers for Disease Control, whose central lab in Fort Collins, Colorado, determined that the presence of antibodies to the virus in the affected tissues meant that the virus was the cause of illness or death. But the chief wildlife pathologist for New York State, Ward Stone, revealed that most of the dead birds he autopsied had been killed not by the West Nile virus, but by pesticides and "perhaps" by air pollutants. New York State, however, denied Stone the funds needed to perform full toxicological screenings that would determine the specific causes of death in birds sent to his lab, including those said to have died from West Nile. So, once Stone found antibodies to West Nile, that was that; Stone's investigation went no further, and the government made the assumption that the presence of WNV antibodies was sufficient to attribute mortality to West Nile, rather than being a sign of a healthy immune system exposed at some time to the virus and producing antibodies in overcoming it.

Officials echoed the CDC's decision that presence of the West Nile virus constituted a "health emergency." This mantra was trumpeted incessantly in the press and by the cowed medical establishment. However, the necessary declaration of "emergency" was never legally invoked. To do so would have required a higher standard of scientific evidence and review, not simply repetition in the media.

With the founding of the No Spray Coalition in 1999, New Yorkers became part of and helped inspire nationwide grassroots movements against pesticides that were emerging across the country. A number of them established their own No Spray Coalitions. These were not actual chapters of a larger organization. There was no national structure, registered trademark, or charter imposed. Each group acted autonomously and did what its local members decided, and shared each other's work. In New York, the Coalition structured its lawsuit and demands with the purpose of contributing to the creation of a nationwide movement against pesticides.

In the Bay Area of California, East Bay Pesticide Alert (also known as Don't Spray California) built a coalition that pressured then-Governor Arnold Schwarzenegger to suspend plans to spray the entire Bay area with synthetic pheromone pesticides[50] to eradicate the harmless light brown apple moth (LBAM). Environmental-health activists Max Ventura and Isis Feral had to battle not only the government of California but also large, well-funded corporatized environmental groups—a similar problem that existed in New York as well—when those groups decided it would be impolitic to fight for an outright *ban* on all forms of chemical pesticides. Grassroots activists repeatedly pointed out that it was not only the blanket *spraying* that was dangerous, but so too were the pheromone pesticides and "inert" ingredients in "twist-ties" and traps that California was using, which were specifically designed to release chemicals into the air and linger, drift and "saturate" large areas. That "alternative" to spraying nevertheless chronically exposed entire neighborhoods for indefinite periods.[51] Don't Spray California challenged attempts to apply synthetic pheromone pesticides in any form; the group won a partial victory when the LBAM spraying was shut down in urban areas where the resistance to it was particularly strong. But in rural areas and in cities in Big Ag territory, they could not overcome the established environmental corporations' decision to accept twist ties and non-spraying forms of pesticide applications.

The case of the light brown apple moth differed somewhat from the West Nile situation in New York. In California, no serious claims were made that the moth was a danger to public health. Nevertheless, pesticide applications went on regardless. Max Ventura and Isis Feral explain the hidden government purpose for doing so:

"What the government agencies are defending here is not our food supply nor our ecosystems but capitalist interests in international trade. The LBAM is no threat to us, but it is a threat to a complex system of agribusiness trade agreements, formed not to safeguard human or environmental health, but rather to guarantee supremacy in the marketplace for the U.S., specifically to crowd out competition from their countries. The LBAM quarantine is a tool of big agribusiness to achieve this supremacy," particularly with regard to the U.S. economic competition with China.

Other areas of the country presented similarly hidden rationales, and local activists had to try to disentangle and expose them. Often, it came down to trying to figure out why governments were spraying *at all*, a question that became into focus when, in 2017, No Spray Nashville (Tennessee) issued a report analyzing data based on statistics provided by the health departments of fourteen major cities that spray pesticides, including Dallas, Nashville, and Baltimore. The report also listed cities that don't spray—Washington, DC; Charlotte, North Carolina; Cincinnati, Ohio; and Fort Worth, Texas, to name a few. The final report compared the rate of West Nile disease in locations that sprayed pesticides to the rates in areas with no spraying. The Nashville group determined that there was *no significant difference in West Nile virus rates between communities that sprayed and those that didn't.*[52]

"When we find West Nile present in mosquito pools here in Washington, D.C.," said Peggy Keller, chief of the Bureau of Community Hygiene and Animal Disease Prevention in the DC Department of Health, "we don't spray. We've learned that the best way to protect the public from both the virus and the pesticides is to intensify our larval program and distribute outreach and education information that emphasizes prevention and protection techniques to the public in the surrounding area."[53]

The City of Lyndhurst, Ohio, a suburb of Cleveland, carefully evaluated the reported effectiveness of reducing the number of mosquitoes by spraying, the actual risks of contracting the virus, and exposure to pesticides. The city determined it would not spray.

In May 2017, the residents of Lincoln County, Oregon, became the first county in the United States that, by popular vote, banned aerial pesticide spraying. "Back in 1976, folks here put Lincoln County on the map by winning a huge landmark case against the United States government, stopping federal spraying of Agent Orange on our forests and homes and waterways," said Susan Parker Swift. "Now Lincoln County has done it again. I couldn't be prouder to share this repeat victory!"[54]

Some areas use predators of mosquitoes, such as dragonflies and bats, to control mosquito infestations instead of chemical pesticides. Besides the aforementioned bat's appetite of more than one thousand five hundred mosquitoes each night, a single dragonfly can eat thirty to hundreds of mosquitoes per day. But even to this day, New York City has failed to shift away from reliance on chemical pesticides. The Department of Health did issue some warnings about eliminating standing water, which serve as mosquito breeding grounds. But administrators failed to consider the unanticipated consequences of what they were proposing, and they did the same for Zika fifteen years later. They warned to "make sure backyard pools are properly maintained and chlorinated,"[55] but made no consideration for toxic combinations of chlorine with malathion and other pesticides.[56] As so often happens when one does not look holistically at complex problems, the "answer" has unanticipated consequences. A number of studies now show that mosquitoes come back after spraying in larger numbers and that they are resistant to the pesticides.[57] And so, in 2017, after eighteen years of spraying with first malathion and then Anvil 10+10[58] with piperonyl butoxide to kill mosquitoes, City decided to spray an enhanced toxic formulation called "Duet," disregarding the projected increases in cancers, endocrine disruption, and greatly reduced sperm counts. Mayor Bill de Blasio actually *increased* the application of pesticides.

How did the City get away with circumventing the law? For all those years, the Department of Health took advantage of a loophole in the law that allowed it to issue to itself "emergency" waivers from New York's environmental regulations. Each year the same agency has sprayed the City with toxic pesticides to kill mosquitoes, as though the federally certified settlement agreement with the No Spray Coalition and Local Law 37 were merely paper tigers to be ignored at will. City officials also approved the application of glyphosate in public parks and on sidewalks across the five boroughs. They ignored Local Law 37's requirements for issuance of waivers,[59] as the No Spray Coalition had warned would happen, enabling the Department of Health to grant itself waiver after waiver from the prohibitions against pesticide spraying. That same circumvention of environmental laws also occurred in Oakland, California, and in other cities throughout the country. What at first was intended to be a rarely used measure in case of an actual emergency quickly became a major loophole through which governments have regularly driven whole fleets of pesticide-spraying trucks.

In July 2017, the New York City Department of Health resumed its annual toxic pesticide spraying for mosquitoes said to be carrying West Nile virus in neighborhoods throughout all five boroughs. The city presented no evidence that its seasonal spraying prevents West Nile encephalitis, its alleged goal. The risk of contracting West Nile disease is very low to begin with.[60] The previous year the City had sprayed the same poisons, ostensibly to combat a nonexistent Zika "epidemic" in New York. The *pesticides*, however, presented significant risk and harm to human health, wildlife, pets, birds, frogs and insects, including natural predators of mosquitoes.

Max Schmid, the long-time host of WBAI radio's Sunday night show *The Golden Age of Radio*, was heading home from work at two o'clock in the morning and was in the streets on a night that NYC was spraying the new combination of pesticides. He got out of the subway at Broadway and 46th Street in Astoria, Queens, his regular station. "After the long trip up the stairs I had to stop to catch my breath. I heard a loudspeaker coming down the block, coming down Broadway: 'Clear the streets immediately, seek shelter, go inside until the truck passes. New York City is spraying for West Nile virus, so get off the street.' Of course, Broadway is totally closed; there's nowhere to go except back down into the subway. I opted to jaywalk across the busy street against the light, and hightailed it up 45th Street. I got around thirty feet up and turned to watch as the truck came by spraying. They went right down Broadway, y'know, about ten feet behind the announcement truck. I hope my friends in the Boss of Tacos truck got their flaps closed in time, 'cause they would have been directly hit by this thing. There's nowhere to go. There's not enough warning, and I was sprayed."[61, 62]

The NYC Department of Health took over direction of the spraying in the early 2000s from the Office of Emergency Management. At the prompting of anti-pesticide activists, in 2005 the New York City Council responded to the growing disquiet regarding its pesticide policies by announcing its intention to reduce the amount of pesticides used on public land by City agencies. It passed Local Law 37, which provided new requirements for pesticide applicators, posting of warning notices prior to applications, new recordkeeping provisions, and burdens to be met by city agencies seeking emergency waivers of those stipulations. It also prohibited certain pesticides.

Nevertheless, the Council refused to hear informed testimony on the final version of the bill as to its shortcomings. I was "escorted" out of those hearings in City Hall by two guards when I questioned from the audience the City Council's failure to consider how easily circumvented, in practice, those safeguards against the granting of waivers would become. (The No Spray Coalition had predicted that the issuing of exemptions to the law would become *pro forma*.) Pesticide applicator Steve Tvedten, whose essay "Why I Stopped using Pesticide Poisons" appears in this book, was another of those prevented from testifying at that hearing. He'd recently retired from his job as a professional pesticide applicator in Michigan and had written a book on alternative means for safely repelling pests without killing them and without endangering one's family, and was offering his nontoxic services for free to the City, to no avail.[63]

The Instructive Case of Dr. Omar Shafey

Dr. Marcelle Layton contributed to the pesticide drumbeat with an article on the spray campaign in the CDC's journal, *Morbidity and Mortality Weekly Report*. But even here, in the supposedly objective halls of science, political intrigue abounds: Layton's report was chosen at the last-minute to replace a very critical article by Florida epidemiologist Dr. Omar Shafey, who headed the Florida Department of Health's Division of Environmental Hazards and Health Effects about spraying malathion. Shafey's study, "Surveillance for Acute Pesticide-Related Illness During the Medfly Eradication Program—Florida, 1998," detailed 138 reported cases of pesticide-related illnesses among Florida residents following intensive malathion spraying there.

Dr. Shafey's findings were slated to appear in the October 22, 1999 issue of MMWR. But on October 19, two days before publication, Dr. Steven Ostroff, then acting deputy director for science and public health at the CDC, called the editor of the MMWR, and convinced him to bump Shafey's article critical of spraying and replace it with Layton's. The CDC had come under pressure from New York City Hall, which did not want to contend with an anti-pesticide article just as their experimental spray program was going into full gear. At the very last minute, Dr. Shafey was asked to alter statistics and omit key passages. Shafey refused. Layton's pro-spraying article was published instead. Shafey's very important article did not ran until November 12, 1999, after the spray campaign in New York had ended for the season and too late to have any influence on breaking events.

Shafey was clear about why his article had been pulled: "It is a simple cover-up and I have little doubt that we are being censored to protect Giuliani's Senate bid. The CDC director should be ashamed of his complicity in suppressing a scientific report for political reasons."[64] Shafey explained that the NYC spraying began after the outbreak *had already peaked*, indicating that the City's spray program had little to do with protecting the people of New York City and everything to do with politics and appearances.

Ostroff's complicity with New York City officials in their pressure to delay publication of Shafey's scientific article was a deliberate attempt to postpone New Yorkers from learning the truth behind their frequent dosing of malathion. Ostroff himself was on-site as CDC WNV coordinator during much of the early decision-making in New York, and he testified the following year as the City's expert witness in the lawsuit brought by the No Spray Coalition. One hand washes the other, and Ostroff in turn received cooperation from the City in conducting the CDC's door-to-door serosurvey in parts of Queens and Staten Island. Ostroff's involvement revealed that the hierarchy at the highest echelons of public health in this country was willing to suppress scientific reports that exposed the health dangers of pesticides, for political purposes.

Dr. Shafey continued his research on workers exposed to pesticides on the job, but he was harassed and ultimately sacked for resisting pressure from his supervisors to present results more pleasing to powerful agriculture interests.[65] Instead of leading to collective outrage and action against the firing of Dr. Shafey, many researchers and health providers retreated into silence, for fear that the same might happen to them.

Pesticides and Biowarfare in the Lead-Up to 9/11

All told, that first year, seven people (four in New York City) died from what officials said was West Nile encephalitis. And over the next twelve years, a grand total of twenty-six people died in New York City from

West Nile viral encephalitis, out of a total of 198 people who contracted the disease. Yet in the year 2000 alone, 2,680 people died in New York City from the common flu.[66] Each year since, the flu has averaged close to three thousand deaths in the City. Many others died from heart disease, chronic fatigue, immune disorders, asthma, cancer, and infections associated with AIDS—conditions that are also caused or exacerbated by *pesticide spraying*. Those diseases involved far more deaths than West Nile, but no "emergency" had been declared for them. West Nile became the first in an annual train of announced emergencies concerning potentially pandemic diseases. Next was anthrax. That one dropped out of the news when the powdered anthrax—sent to key officials in government and media—turned out to have been "weapons grade" and traced to the U.S. Army's top-secret biowarfare laboratory in Fort Detrick, Maryland.[67] Next came the great smallpox scare and mandatory vaccination edicts. And then came bubonic plague, SARS, Avian Flu, Swine Flu, and finally Zika. Officials said each would require emergency measures, bypassing normal civil liberties and environmental protections. Each time government and media spread panic, pharmaceutical corporations sold hundreds of millions of dollars of worthless drugs.

Beginning with West Nile in 1999, each time a new health scare was announced, so too were more facets of a massive surveillance and repressive apparatus nailed into place under the rubric of a "health emergency."[68] The USA Patriot Act, for example, had already been written and was waiting for the right moment to be introduced as legislation by the time 9/11 occurred. Proponents of the repressive legislation (with the help of a fearful populace) pocketed millions of dollars in drug company campaign donations while chipping away at laws protecting freedom of speech and assembly, hammering nail after nail into the Bill of Rights' coffin.

In 2013 and again in 2017, Avian Flu became the disease of the moment. Avian Flu is a genetic mutation of a virus induced in birds, particularly chickens, and quickly spread by the horrible, cramped and unsanitary conditions of factory farming in China and Vietnam.[69] One would hope that Congress would pass laws banning the sale of chickens raised and slaughtered under such conditions wherever they might occur, including here in the U.S. But that never happened. (A 2011 proposition in California was one of the few to address the horrible conditions under which animals are raised.) Former U.S. Senator Bill Frist (R-Tennessee) was also the only medical doctor in the Senate, which gave his opinions added weight. He gave the following advice about what to do if Avian Flu were to hit the United States: Dose yourself and your kids with the pharmaceutical, TamiFlu. Neither Frist nor any other U.S. government official dared point out that TamiFlu may lessen the severity of some flu symptoms but does nothing to block or weaken the viruses said to cause Avian Flu, Swine Flu, or West Nile. They also neglected to mention that the patent was owned by Gilead, Inc., of which the CEO and major shareholder was former Secretary of Defense Donald Rumsfeld, the architect of "shock & awe" in Iraq. As late as 2005, Rumsfeld was known to profit from every box of TamiFlu sold.[70] (Rumsfeld was also CEO of Searle, Inc. when the company first began manufacturing NutraSweet [aspartame], ignoring studies that showed the sugar substitute caused lesions in the brain.) The panic was needed to sell TamiFlu and gain acceptance for the onerous conditions in the Model Emergency Health Powers Act on both federal and state levels. (See below)

In the panic generated by yearly health-related "emergencies," the pharmaceutical companies played much the same insidious role in setting the political and economic agendas of this country as have the giant oil companies and banks, with many of the same corrupt individuals determining policies to augment profits and undermine the fightback against them. Donald Rumsfeld is an important player here, but he is not the only one. Fresh from his stint as director of the CIA and before becoming Vice President and then President of the U.S., from 1977 to 1979 George H.W. Bush presided over the Executive Board of Eli Lilly & Co.—one of the pharmaceutical corporate giants. The company is controlled by the Quayle family of Indiana, and Dan Quayle became Bush's Vice-President. The Bush administration's Budget Director, Mitch Daniels, was also a Lilly senior executive. It was Lilly who bankrolled a Rudy Giuliani speaking tour during Giuliani's 2008 Presidential campaign.

Here's a good example of the interplay between profits and policy: Lilly boosted the sales of Sen. Bill Frist's book on bio-terrorism, published in the blink of an eye immediately following 9/11. The corporate behemoth bought 5,000 copies right off the press and distributed them to doctors along with its drug samples. Meanwhile, Sen. Frist, coincidentally, was busy submitting last-minute amendments to the Homeland Security Act that would limit Eli Lilly's liability from lawsuits by people suffering negative reactions to the company's vaccines and by families whose children were harmed by the mercury-based preservative thimerosal and other vaccine additives.[71]

Frist's amendments sought to lock corporate records away from public view and shield corporations in advance from exposure of any prior knowledge they might have had. Doctoring data in their drug trials? No problem. Endangering people from so-called "side effects" of their products? Leave it to Frist.

The pharmaceutical protection legislation served as a blueprint, a few years later, for "protecting the citizens" from learning the truth behind the huge financial scandals and bank bailouts. When did Bank of America and its subsidiaries decide to purchase billions in so-called "derivatives" and seize ownership and control of hundreds of thousands of homes and farms in the U.S., and evict the rightful owners from them? What did Morgan-Chase know and when did they know it?

In the case of the pharmaceutical corporations, Frist submitted last minute "technical corrections" into the hundreds of mostly unread pages of the USA Patriot Act just hours before it was set to go to a vote. The Act also exempted foreign security companies from lawsuits that would have otherwise forced them to provide documents related to the events of 9/11, using the justification that 9/11 itself impelled secrecy. Thus, ICTS-International, for example, and its subsidiary—Huntleigh USA Corporation, the corporation running Security at key airports on 9/11—could no longer be compelled by a U.S. court of law to produce the airports' video surveillance tapes from 9/11, nor answer questions about any other aspect of its security measures on that fateful morning.

In the weeks leading up to 9/11—all but forgotten today!—the U.S. government's secret development of biological weapons at Ft. Detrick under the aegis of preparation for a bio-terrorist attack was finally coming to light. The *New York Times* reported that "Over the past several years the United States has embarked on a program of secret research on biological weapons that, some officials say, tests the limits of the global treaty banning such weapons." The *Times* continued:

> The 1972 treaty forbids nations from developing or acquiring weapons that spread disease, but it allows work on vaccines and

other protective measures. Government officials said the secret research, which mimicked the major steps a state or terrorist would take to create a biological arsenal, was aimed at better understanding the threat.

The projects, which have not been previously disclosed, were begun under President Clinton and have been embraced by the Bush administration, which intends to expand them.

Earlier this year, administration officials said, the Pentagon drew up plans to engineer genetically a potentially more potent variant of the bacterium that causes anthrax, a deadly disease ideal for germ warfare.[72]

Genetically engineered bacteria *and* untested antidote/vaccines? New bio-weapons *and* resistance to them? Chemical and biological warfare? Mass vaccinations of military personnel with untested vaccines—"vaccines" that, we now learn, the U.S. government believed could be aerosolized and sprayed over large populated areas, and which at the very least should have required extensive testing?

Government-orchestrated health scares begat biowarfare "drills." Drills begat the release of toxic substances into the air, water and soil, which led to a stripping away of freedoms, establishing an Orwellian surveillance infra-structure, in the name of "health" and "fighting bio-terrorism."

NYC's Office of Emergency Management—which conducted the West Nile spraying in 1999 and 2000—became the model for the Department of Homeland Security. Federal agencies pumped millions of taxpayer dollars into the pesticides industry and giant pharmaceutical corporations. They submitted legislation that would require the mass-inoculation of the entire population for smallpox to "fight terrorism." After the attacks on September 11, 2001, DHS and other federal agencies urged cities across the country to spray toxic pesticides over their populations and ecosystems to prevent West Nile. These included heavily populated areas that had not had any indica-tion of West Nile virus. Under the plan, DHS would send survey-takers door-to-door in targeted communities to take blood and DNA samples.

The author of those policies, Jerome Hauer, had, by the time of the 9/11 attacks, left the OEM and was coordinating security with Kroll Associates, Inc. for the World Trade Center.[73] Hauer has close ties to the

U.S. military's secret biological warfare development programs. In 1998, he introduced Col. Thomas Monath—a virologist with a long history with U.S. government secret forces and genetically engineered vaccines— to President Clinton. Monath, Hauer, Rockefeller University president Joshua Lederberg[74] and Dr. J. Craig Venter (president of The Institute for Genomic Research and co-owner of the Human Genome Project) had been pressuring President Clinton to spend billions on a nationwide vaccine and pesticides program under the guise of opposing germ warfare. With Clinton's support, they stumped for funds for Monath's company, Oravax (now Accambis), for production of a West Nile vaccine—this was before that disease hit New York City, and before it became widely known in the United States.[75] "Coincidentally," Monath's company announced it was "almost ready" in developing a vaccine for West Nile encephalitis at just the moment the disease was first detected in birds at the Bronx Zoo, and as Hauer and Giuliani were about to begin their mass spray campaign. Oravax was awarded a $3 million grant from the National Institutes of Health the following summer to create a live viral vaccine for West Nile, using the yellow fever vaccine as its base. Oravax, according to Monath, was also working on a vaccine for Dengue fever at the time that Cuba filed a complaint that the U.S. was engaging in bio-terrorism against it by spreading mosquito-borne illnesses there.[76] A mysterious outbreak of Dengue fever also hit Hawaii at that time.

Some have suspected that the discovery of West Nile disease on the heels of the announcement of the vaccine was not coincidental, that political and economic motives intersected and reinforced each other, and that the West Nile virus was released to create a market for the vaccine as well as to engage in bio-terrorism preparedness drills and experiments. "The theory I and a number of other activists have been suggesting," writes researcher Patricia Doyle, "is that the entire WNV panic has been created specifically in order to justify the mass distribution of this very vaccine throughout the entire U.S. Oravax was granted a license by the U.S. Army bio-warfare lab in Ft. Detrick in 1996 (Monath was previously a researcher at the same lab) to manufacture a Japanese encephalitis vaccine derived from a genetically-altered virus the Army itself created. WNV is a variation of Japanese encephalitis. Numerous medical and scientific institutions, many run by the Federal government, have been quietly experimenting with WNV in NYC and the surrounding area for decades. Most of this research involved bio-warfare applications."[77]

It was Hauer who, more than anyone else, persisted in raising the "for-eign terrorism" diversion for West Nile virus (claiming that Saddam Hussein was behind it) long after that disinformation and attempt at public manipu-lation had been discredited, as his friend Col. Monath pushed for financial subsidies to his company to manufacture the vaccine. Hauer was next hired as Assistant Secretary of Health and Human Services on—please note the date—Sept. 10, 2001. Together with Tommy Thompson (the secretary of the agency), it was Hauer who told the White House staff and President Bush to begin taking the drug Cipro, made by Bayer, ostensibly for Anthrax exposure.[78] Why? The date of that recommendation? The morning of Sep-tember 11, 2001, prior to the anthrax attacks.

Hauer next went on to head the federal Office of Public Health Emer-gency Preparedness,[79] created in June 2002. He used his position as a bully pulpit to stump for emergency mandatory vaccination programs and ham-mer them into place. He was the main force behind legislation known as the Model Emergency Health Powers Act (MEHPA), which would estab-lish a bio-terror preparedness "czar"—Hauer clearly envisioned himself in that role. The "czar" would have authority to override state health laws and declare a "health emergency" in response to the emergency disease of the moment, and order mandatory vaccinations of health care workers and first responders (police, fire-fighters, soldiers, medical personnel). He could also order mass relocation of those who fall victim to smallpox or other "emergency" diseases to "quarantine" facilities, which would also hold those rounded up for refusing to take the experimental, genetically engineered smallpox vaccine. This bears repeating: Under the legislation that Hauer fought for, *those refusing mandatory vaccinations would be incarcerated in the same facilities as people who contracted the highly contagious smallpox.* By 2009, Halliburton had already completed construction of the "camps."

Bits and pieces of MEHPA were implemented here and there. But growing popular movements against corporate pollution, genetic engineer-ing of crops and vaccines, mandatory vaccinations, corporate farming, the mass spraying of pesticides, and globalization and trade treaties stood in the way of what would amount to a massive accumulation of billions in profits by a few corporations assisted by the government. Those different threads of resistance were woven together during the anti-"free trade" protests in Seattle in 1999 and over the next two years, as campaigns against privatization fueled people's distrust of government and the corporations it all too clearly represented. Those movements had to be suppressed in order for the government to fully implement its package of repressive legislation in the wake of 9/11, which provided the rationalization needed to crush the organized political resistance.

On the federal Centers for Disease Control's website, one lone footnote cites four different vaccines to be administered for smallpox, not just the one that many older folks got in the 1950s—and even *that* was more dangerous than we were led to believe. Three of the four kinds of vaccines would be *genetically engineered* and administered to unsuspecting people, despite having never been tested. Basically, the CDC proposed a massive experiment on a population driven to hysteria by the government and media. "At the end of the day, the numbers [of first responders receiving mandatory smallpox vaccinations] could be significantly greater than 500,000," Hauer crowed.[80]

At first, a panel of outside experts—the Advisory Committee on Immunization Practices—rejected the proposal to mandate smallpox vaccinations to the general public. But Hauer and several other officials overruled that moderating recommendation and produced a firestorm within the Bush administration over the question of mandatory emergency vaccinations. Hauer recommended "that a phased approach be used, starting with 500,000 and then moving in steps to 10 million."[81] But in a decision of unheralded courage and historical importance, the California Nurses Association—now National Nurses United—heroically refused to allow themselves or their patients to be vaccinated with unnecessary, untested, and genetically engineered experimental vaccines. The nurses ignited a resistance movement across the political spectrum that threw a wrench into the gears. In mobilizing against the government's "emergency" forced vaccination program for smallpox—and then again for flu—the nurses saved tens of thousands of lives and pointed the way for new forms of resistance in the post-9/11 era.

The Need to Think Holistically

Health officials in the U.S. work in a system influenced by private business interests out to maximize their profits. They fail to think holistically in terms of entire ecosystems and the interrelationship of species, which has repeatedly led to disasters, on large as well as smaller scales. During a malaria outbreak in Borneo in the 1950s, the World Health Organization sprayed DDT to kill mosquitoes. But the DDT also killed parasitic wasps which were the natural predators of thatch-eating caterpillars. As a result of the spraying, the caterpillars' numbers increased and the thatched roofs of many homes collapsed. Meanwhile, the DDT-poisoned insects were eaten by geckos, which, according to some versions of this story, were in turn eaten by cats. The cats perished—either from eating the geckoes or from direct DDT poisoning—which led to an explosion of the rat population. As a consequence, there were outbreaks of sylvatic plague and typhus. To put an end to this destructive chain of events, the World Health Organization had to parachute live cats into the area to control the rats, with the help of the British Royal Air Force.[82, 83]

A 1962 article in *The New York Times* reported a similar chain of events, this time from Vietnam: "American DDT spray killed the cats that ate the rats that devoured the crops that were the main props against Communist agitation in the central lowlands."[84] The politics once buried in the science are suddenly brought out front and center. No matter how "scientific" a project may seem, its political context and implications are not far from the minds of even the most "objective" health experts.

It is no coincidence that dozens of military and police agencies in Seattle, 1999–operating domestically under a new joint command structure for the first time–sprayed mixtures of malathion-like cholinesterase inhibitor chemicals as part of the tear gas, defining a new phase in the application of U.S. biological and chemical warfare techniques as applied not only to mosquitoes but against the domestic civilian population.[85] During those two years between 1999 (West Nile, Seattle) and 9/11, what began with concern over West Nile virus was manipulated to gain public acceptance and acquiescence for the repeated spraying of malathion and pyrethroid pesticides against protesters in Seattle and then over the largest urban population centers and ecosystems in the country. Following the attacks on the World Trade Center and Pentagon on September 11, 2001, the government was able to orchestrate the ensuing panic, enabling it to divide, demobilize and repress the emerging confluence of civil rights, labor, anti-war, and radical ecological movements.

The events of 9/11 continue to cast twin shadows over the fight against militarization of public health in the service of the pharmaceutical and pesticide companies and extension of government control of the populace. Its umbra gives renewed cover to government and corporate attacks on civil liberties, while the umbrella of social and ecological justice movements has been unable to protect participants from the rains of repression, manufactured "health threats," and calls to "fight bio-terror." New movements are beaten back, only to reemerge—today's militant movements against hydrofracking, genetic engineering, and tar sands pipelines are good examples of the ongoing resistance—like waves of consciousness and action, action and consciousness. If, as Milán Kundera has written, "the struggle of human beings against power is, in some important sense, the struggle of memory against forgetting," awareness of how fears were (and continue to be) marshaled to beat back the environmental and other movements in the wake of annual pandemic hysteria, in the shadow of 9/11. Such a holistic understanding is a prerequisite for reclaiming the hopeful worlds of "the possible" from the Abu Ghraibs and Guantánamos to which they've been renditioned.

6

Children & Pesticides
By Patricia Wood

In the mid-1990s, when my two children were in grade school, I became actively involved in improving access to healthy food by creating organic school gardens in my community. I learned about the routine use of pesticides on school grounds and playing fields, and began to discover the scientific studies linking pesticide exposure with negative impacts on children's health.

Over a period of several weeks, I had meetings with parents and decision-makers. I was able to convince our district to pass a school-board policy prohibiting the use of pesticides. After a few more school districts followed suit, several years later, New York State passed the most comprehensive pesticide law for schools in the country—a ban on the use of pesticides at schools, K through 12, including daycare centers. As of this writing, only Connecticut has a similar law regarding the use of pesticides at schools, prohibiting their use in grades K through 8 and daycare centers.

A robust and growing body of scientific knowledge is now linking pesticide exposures to a wide array of health problems in children, including asthma, neurological harm, endocrine disruption, birth defects, and certain types of cancers. This compelling research linking exposure to pesticides with serious health outcomes in our children necessitates that we use every opportunity to help reduce those exposures.

Young children are uniquely vulnerable to toxic exposures due to their immature and rapidly developing bodies and typical childlike behaviors. Pound for pound, children breathe more air, eat more food, and drink more water than adults, so the impact of any chemical contaminants in their environment is magnified.

During critical windows of vulnerability during a child's development, the rapid growth of different organs and body functions can be disrupted, in some cases by even extremely low-level exposures to pesticides and other chemicals. These windows of vulnerability occur during fetal growth, infancy, early childhood, and puberty.

Another critical factor raising the risk level for children is their natural curiosity about the world and the instincts they have for discovery. Young children play close to the ground, engage regularly in hand-to-mouth behavior, and are instinctively focused on their faces, where they learn about their world using their senses to see, smell, touch and taste. Normal human behavior in an increasingly toxic world is putting our children at risk!

A 2012 report by the American Academy of Pediatrics (AAP) stated that "Children encounter pesticides daily and have unique susceptibilities to their potential toxicity. Acute poisoning risks are clear, and understanding of chronic health implications from both acute and chronic exposures are emerging."[1]

How Children Are Exposed

Some children are more at risk from involuntary exposures to pesticides than others. Those who live in rural agricultural areas are often exposed through pesticide drift off fields or from aerial spraying. A study of 210,723 live births in Minnesota farming communities found that children of pesticide applicators had significantly higher rates of birth defects than unexposed populations.[2]

Those living in cities are likely to be exposed through regular extermination services to control insects and rodents in their apartments, daycare centers, and other facilities. Accidental poisonings of children involving pesticides are not uncommon, as many brightly colored packages intended to sell rodent and insect poisons are also attractive to small children.

Suburban neighborhoods with chemically maintained lawns also present significant pesticide exposure risks for children. Over the past sixty years, the use of pesticides on residential properties in the United States has skyrocketed. Today, approximately 70 million pounds of pesticides are spread on suburban lawns each year. Scientists say that lawn pesticides are designed to break down in sunlight, and with rainfall and soil microbial activity. That does occur when all climate factors needed for pesticide decomposition are ideally present, but that ideal is rarely the case. In addition, pesticides travel. Studies show that in homes where there are children and pets, lawn pesticides are often carried indoors on the soles of shoes, bicycles, and wheelchairs, where they take much longer to break down, extending the potential exposure for children.

One pesticide exposure that all children share is through the consumption of commercially produced food. Pesticide residues contaminate the majority of our conventionally grown fruits and vegetables, as well as processed foods containing genetically modified ingredients.

Most GM crops are Roundup Ready, which means that glyphosate and other chemicals contained in the popular pesticide formulation can be sprayed heavily during the growing season to prevent weeds without harming the primary crop. Residues of glyphosate have now been found in small but not insignificant quantities in most processed foods. The Food and Drug Administration (FDA), under public pressure after the World Health Organization determined glyphosate to be a probable human carcinogen, has started testing samples of food produced in the United States. In a recent FDA examination of honey samples from around the country, glyphosate was found in all of them.

Fetuses and breastfed infants are exposed via their mother's diet, which is likely to contain GM foods. The three largest manufacturers of baby formula, representing 90 percent of the supply, use GM corn, sugar beets, and soy in their products.

Children also eat a greater percentage of corn and soy in their diets than adults. It is estimated that more than 80 percent of all processed foods, including institutional and restaurant foods, contain GM ingredients, and our children are the biggest consumers of processed foods.

Scientists are still developing evidence regarding the role of glyphosate in human disease. Although Monsanto, the manufacturer of Roundup, has claimed that it is almost harmless to humans, the fact is that it kills certain kinds of bacteria and has the potential to disrupt the normal functioning of the human gut, where billions of bacteria exist.[3] One of the most interesting (and controversial) theories of the impact of Roundup involves the role of glyphosate in altering the behavior of children by interfering with their normal gut-brain communication.

Today, with many children eating only limited diets, it is important that the few foods they do eat be free of pesticide residues. Under current USDA definitions, the "organic" label means "without pesticides or genetically modified organisms." Providing organic food for children is a small but important step in reducing pesticide exposure.

Brain Development

A growing body of peer-reviewed studies shows damage to developing brains occurs from pesticide exposures, especially organophosphate and synthetic pyrethroid pesticides.[4] The human brain begins developing in the womb and continues into early adulthood. During this long and vulnerable developmental period, many complex processes take place that can be easily impacted by neurotoxic pesticides. Damage to the brain during this period cannot be reversed, and may manifest itself as ADHD (attention-deficit/hyperactivity disorder), a decline in cognitive abilities, behavioral problems, or even autism. Dr. Theodore Slotkin, a scientist at Duke University Medical Center, has published dozens of studies on rats exposed to the insecticide chlorpyrifos (Dursban) and concluded, "There doesn't appear to be any period of brain development that is safe from its effects."[5] Leaders in the field of children's environmental health from Mt. Sinai School of Medicine and Harvard call the damage to children's developing brains from chemical exposures a "silent pandemic."[6]

Chlorpyrifos, a proven neurotoxin, was banned by the EPA in 2000 from use in home, lawn, and garden bug killers. It is no longer sold to the public because of its unacceptable health risks to children. Its use was still permitted on agricultural crops, but the new EPA rules sharply limited chlorpyrifos on grapes, apples, and tomatoes, all common foods in children's diets. After additional study, the EPA announced in 2016 that new research determined that the pesticide posed an unacceptable risk to everyone and recommended a ban for all uses. This was scheduled to happen in 2017, but new leadership at the EPA under the Trump administration overruled its own scientists and declared that there was actually no risk after all, reversing its prior decision. No ban!

Asthma

Asthma is a serious lung disease characterized by recurring attacks of bronchial constriction, which causes breathlessness, wheezing, and coughing. It's a dangerous, life-threatening illness that has become almost epidemic in the United States, especially among children.

A child's lungs and airways are still developing, making them more vulnerable to the effects of pesticides and other pollutants. Almost 10 percent of all boys under the age of eighteen have been diagnosed with asthma, and 7 percent of all girls. It's the third leading cause of hospitalization among children.[7]

The timing of pesticide exposure seems to play a vital role in the potential development of asthma. A 2004 peer-reviewed study found that young infants and toddlers exposed to herbicides within their first year of life were four-and-a-half times more likely to develop asthma by the age of five, and almost two-and-a-half times more likely when exposed to insecticides. [8]

Cancer

An American Academy of Pediatrics literature search identified sixteen studies that examined the possible association between residential pesticide use and childhood cancers. They found that indoor use of pesticides, especially frequent application of insecticides, was associated with an increased risk of leukemia and lymphoma. They also found that outdoor use of herbicides was associated with a slightly higher risk of childhood cancers in general.[9]

Researchers have also found that children born to mothers living in households with pesticide use during pregnancy had over twice as much risk of getting cancer, specifically acute leukemia and non-Hodgkin's lymphoma.[10] It cannot be stressed enough that the risk of early life exposures must be taken into consideration when making decisions about the use of pesticides of any kind. Nontoxic alternatives to pest control need to be more widely available, and education of parents-to-be should be part of prenatal care. We have an epidemic on our hands; it is incumbent on us to advocate for our children and eliminate pesticides where they live, learn, and play.

7

It's Not That Anyone *Wants* to Kill Butterflies

By Cathryn Swan

> *In all things of nature there is something of the marvelous.*
> —Aristotle
> *The difference between a flower and a weed is a judgment.*
> —Unknown

It's not that anyone *wants* to kill the butterflies. Or the bees. Or the hawks. Or the owls. Or the ladybugs. They are collateral damage in the war against "weeds" or "pests," deemed unwanted interlopers in our society's quest for perfectly manicured, pristine surroundings.

Modern farmers are more and more abandoning time-honored methods, in order to prevent nature from "getting in the way" of their goal of efficient crop production. In fact, all organisms that inhabit the earth may become casualties in day-to-day decisions being made by farmers, landowners, parks officials, golf course CEOs, and perhaps your next door neighbor, aspiring to control nature and achieve a more sanitized world. The poisons they employ, designed to banish these "interlopers," put all living beings at risk.

And worse, we are up against government policies heavily influenced by powerful corporations, the chemical companies and their lobbyists. They may not intend to kill, but their actions, most often motivated by financial profit, can and do cause deadly harm. They are dismissive not solely of scientific research but also of centuries-old wisdom, and of environmentalists and activists who use that knowledge in their prolonged and varied fights to save complex life on earth.

The weeds targeted for destruction often serve as food sources for birds and animals. Weeds also provide food and nectar for insects, which in turn feed birds.[1]

Who determines what is a "weed" or a "pest"? These terms are bandied about as if we understand the full complexity of how the larger ecosystems work. Branded with what are meant to be derogatory terms, these so-called weeds or pests have lives of their own and often contribute to this complexity in ways we cannot see. Do they need to contribute some larger benefit to humans for us to allow them to inhabit this earth? The blanket killing of organisms resulting from this mindset is now evincing repercussions well beyond the eradication of an immediate short-term target.

Rachel Carson described it in *Silent Spring* in 1962:

> Our attitude toward plants is a singularly narrow one. If we see any immediate utility in a plant we foster it. If for any reason we find its presence undesirable or merely a matter of indifference, we may condemn it to destruction forthwith. . . .
>
> The earth's vegetation is part of a web of life in which there are intimate and essential relations between plants and the earth, between plants and other plants, between plants and animals. Sometimes we have no choice but to disturb these relationships, but we should do so thoughtfully, with full awareness that what we do may have consequences remote in time and place.

For example, the endangered monarch butterfly's life cycle "is exquisitely synchronized to the seasonal growth of milkweed, the only plant its larvae will eat."[2] Many butterflies, and their survival as larvae and caterpillars, are dependent on milkweed, which Monsanto's Roundup is designed to kill.

"In a game of hopscotch," Warren Cornwall at *Slate* Magazine writes, "successive generations of monarchs follow the springtime emergence of milkweed from Mexico as far north as Canada. The hardy plant once flourished in grasslands, roadsides, abandoned lots, and cornfields across much of the continent. It fueled a mass migration that ended each winter with more than 60 million butterflies converging on pine forests in the Sierra Madres [in Mexico]. Then came Roundup."[3]

Milkweed is necessary to the very existence of monarchs and other butterflies, and their crucial pollination of flowers and plants. Roundup's poisoning of milkweed eliminates these butterflies' food source, and then the butterflies themselves, depriving us of the delicate beauty that butterflies bring to the world.

These colorful, beloved species are not the only victims of glyphosate—a wide variety of insects, as well as birds, fish, amphibians, and mammals are among its victims. Pesticides overall are causing grievous harm to many species as diverse as bees, dragonflies, frogs, owls, hawks, and robins. Humans are not exempt.

The environmental organization Beyond Pesticides notes that glyphosate applications directly affect a variety of non-target insects such as earthworms,[4] ladybugs, lacewings, and parasitoid wasps.[5] Glyphosate also kills fish,[6] and the food sources for birds and small mammals.[7]

It's a Catastrophe

"The biomass of flying insects in Germany has dropped by three quarters since 1989, threatening an 'ecological Armageddon.'"[8] Indeed, it is a catastrophe, and those in positions to reverse that road to Armageddon are trapped in what I call an assembly-line mentality. Speed of production is a primary consideration for success in a profit-driven society. Toxic sprays may increase the frequency of the crops' growth cycles versus more careful, life-preserving methods. The methods may be slower and possibly more costly in the short-run, and may be dismissed as unnecessarily burdensome. It is no matter that the toxic approach ends in destroying vital members of our ecosystems: plants, flowers, bees, butterflies, insects, and other beings. Humans await their turn in this ecocidal merry-go-round. This is the price that corporate owners, backed by their government partners, are willing to pay for their immediate financial gain, and apparently they consider themselves, their families, and fellow elites to be somehow immune to the coming dire consequences.

The assembly lines of chemicals used in these corporations' toxic arsenal provide an efficient killing machine. Any organism suspected of hindering the singular goal of increased production has to go. *Slate*'s Cornwall explains that the herbicides used to douse Roundup Ready corn and soy—both genetically modified to withstand the poisonous spray—also kill the milkweed.

By the turn of the twenty-first century, Monsanto had induced many farmers to plant genetically modified seeds. Production of GMO crops skyrocketed, along with their dependence on Roundup, which caused a tremendous decline in the monarch butterfly population. The amount of milkweed in farm fields fell by more than 80 percent, according to Karen Obenhauser, a conservation biologist at the University of Minnesota. Obenhauser determined that "the loss of milkweed almost exactly mirrored the decline in monarch egg production," and that "before Roundup, patches of milkweed grew among the corn and along the edges of fields. After the herbicide—nothing but corn."[9]

Milkweed belongs to the genus *Asclepias,* named after Asclepius, the Greek god of medicine. The genus is comprised of herbaceous perennials, dicotyledonous plants that encompass over 140 known species.[10]

It is important to understand milkweed's significance in pollination. The flowers produced in the *Asclepias* genus are essential to it. Butterflies, as well as bees and wasps, carry the pollen from those flowers.

Pollination in this genus is accomplished in an unusual manner. The pollen is grouped into complex structures called pollinia (or "pollen sacs"), rather than being individual grains or tetrads, as is typical for most plants. The feet or mouthparts of flower-visiting insects such as bees, wasps and butterflies slip into one of the five slits in each flower formed by adjacent anthers. The bases of the pollinia then mechanically attach to the insect, so that a pair of pollen sacs can be pulled free when the pollinator flies off, assuming the insect is large enough to produce the necessary pulling force (if not, the insect may become trapped and die). Pollination is effected by the reverse procedure, in which one of the pollinia becomes trapped within the anther slit.[11]

The female butterfly lays her eggs on the underside of the milkweed plant, which is poisonous to predators and even to horses.[12] The caterpillar emerges from the egg situated on the milkweed, then eats the egg and feeds on the milkweed leaves in its early life. This plant is essential to the life cycle of the butterfly and the species' survival. After the caterpillar makes its dramatic metamorphosis into a butterfly, the milkweed consumed from youth remains in its body; it acts as a deterrent, and often poisons or sickens animals that eat the plant. This is another way the butterfly survives.

Since butterflies rely on milkweed to survive, and glyphosate is implicated in the wide decline of milkweed growth, glyphosate use is a key (if not defining) factor causing the current decline and potential future eradication of butterfly populations.

Humanity's *own* survival depends on monarch butterflies and other pollinators. The pollinator helps the plant or flower continue its evolution, and that same plant or flower provides energy and is a food source for the pollinator. Pollinators are typically birds, bats, and insects. (The wind and sometimes the plant itself also act as pollinators.) "In the economy of nature, the pollinators provide an important service to flowering plants, while the plants pay with food for the pollinators and their offspring."[13] Yet, the food sources in our ecosystem are intricately linked to pollinators, and pollinators are most at risk from pesticides and herbicides. "Every third bite of food we eat comes to our table courtesy of a pollinator. Monarchs, bees and many other pollinators share much of the same habitat—so what happens to monarchs, happens to other pollinators. Monarchs are an indicator of the damage done to our environment—we can count them as they gather by the millions in Mexico. They are an indicator of what we cannot fully quantify—the loss of our pollinators and their habitat."[14]

Though many of us are unaware of or may have forgotten this important interrelationship as we become further disassociated from the sources of our food, we are all intricately linked: the survival of our own species and our planet depends on us—all of us—remembering those interconnections, celebrating them, and acting always to protect them.

Other consequences of glyphosate and related pesticides are equally alarming. Don M. Huber, professor emeritus at Purdue University, and American Phytopathological Society (APS) coordinator of the U.S. Department of Agriculture (USDA) National Plant Disease Recovery System, has studied pathogens for more than fifty years. In 2011, Huber wrote to then–US Secretary of Agriculture Tom Vilsack and reported that his team of plant and animal scientists had serious concerns about Roundup. They were alarmed by its ability to "significantly impact the health of plants, animals, and probably human beings." They were also seeing reports of infertility rates in dairy heifers "of over 20%, and spontaneous abortions in cattle as high as 45%. . . . For example 450 of 1,000 pregnant heifers fed wheatlege experienced spontaneous abortions. Over the same period, another 1,000 heifers from the same herd that were raised on hay had no problematic births. High concentrations of the pathogen glyphosate were confirmed on the wheatlege."[15]

I reached out to Professor Huber in May 2018 for an update. What kind of response had he received to his letter to the head of the USDA in 2011?

Prof. Huber noted that there was "one official response"—a group of scientists were given an opportunity to meet with USDA and EPA administrators to share results from about 130 published, peer-reviewed scientific papers documenting the concerns expressed in his letter and requesting permission for USDA and EPA scientists to follow up on their concerns. "We were treated cordially," Huber said, "but were told to let them know when we get more information—so no action or acknowledgment resulted."

Huber reported:

> There is a wealth of information on the damage to the soil, crops, environment, animals and man, which demonstrates that the damage being done is much greater than even anticipated in 2011. I receive three to five phone calls and emails a week from other scientists (soil scientists, veterinarians, doctors, and research consultants) as well as growers/producers asking for help and direction with health problems. . . .
>
> Glyphosate and its formulated product, Roundup, are probably the most chronically toxic compounds ever indiscriminately released into the environment! The exponential increase in over twenty human diseases can be understood only as the physiological disruptions caused by the essential mineral chelation and antibiotic effects of glyphosate are recognized. Similar damage to animal, crop, and environmental entities can also be documented scientifically. I know of *no* peer-reviewed scientific toxicological studies which have documented the safety of either the GMO product or the glyphosate residues in our food, water, and environment.
>
> There are many scientists who have sacrificed their jobs and reputation in an attempt to share the damaging effects their science has demonstrated! Scientific censorship is intense in these areas.

And what of the affected cattle? I asked Huber if anything was being done for these cows and whether the alarming damage to these animals is continuing. It appears this is slowly becoming more well known beyond the researchers, but not fast enough. Huber responded:

> The problem persists and has intensified as the exposure to glyphosate (and forthcoming dicamba and 2,4-D) increases as a result of Roundup resistant weeds. More veterinarians and medical doctors are becoming aware of the problems. It is a revelation to many. When veterinarians have their clients change to non-GMO, glyphosate-free feed (and bedding straws) many of the health issues resolve in two to four weeks. The contamination of our food and water with glyphosate is so extensive that it isn't quite as simple a restoration process for humans, but there is a dramatic difference in both behavior and general health (autism, gut issues, allergies, etc.) when [humans' diets are] changed to organic [food sources].

Again, we see how cattle certainly are not the species being targeted by Roundup, and yet their reproduction and very life is threatened (while also suffering from the many other abusive and inhumane conditions known to occur widely under this country's factory-farm system).

Another class of pesticide, neonicotinoids, has been shown to harm other pollinators such as honeybees. Imidacloprid, manufactured by Germany's Bayer, is the flagship neonicotinoid, and has long been considered toxic to bees,[16] as have other pesticides in this class. Imidacloprid has been found to have "a negative effect on honeybee colonies," and neonicotinoids overall are "accused of crippling insects' nervous systems and decimating bee colonies."[17] The Environmental Protection Agency has finally and definitively assessed the blame for the bee colony catastrophe on neonicotinoids.[18] The European Union, at this writing, has been considering a ban on three specific chemicals in the neonicotinoid class, including imidacloprid. The way these pesticides work is so aggressive that we need to stop to consider what it means to "cripple" the nervous systems of countless billions of honey bees.

Anti-Pesticide Activism

I was handing out flyers one summer day at Brooklyn's Prospect Park, home to many birds and wildlife, including swans and Canada geese, as well as raccoons, fish, squirrels, and many (so-called "beneficial") insects, as the 526-acre green space was set to again be sprayed with pesticides. I encountered a woman who had taken an early morning walk through the park after pesticides were sprayed by truck the previous evening. She told me she saw pathways strewn with multiple dying ladybugs. Ladybugs, which are properly classified as "beetles," are not the target of this misguided spray program, mosquitoes are. Yet the ladybugs, which eat insects such as aphids, scale bugs, and mealybugs (they have huge appetites: a single ladybug can eat five thousand aphids across its lifetime[19]) and control pests naturally, are also being indiscriminately killed by pesticides. And ladybugs are not the only unintended victims of these pyrethroid pesticides in New York City's spray program: dragonflies, bats, bees, fish, and more are also killed.

We hear news of the problems with neonicotinoids or Roundup around Earth Day in the spring when bees and butterflies appear. The pesticide companies and their lobbyists, who appeal to government to ease regulations around these chemicals, work hard to suppress this information and make what regulations that do exist difficult to enforce.

In 1974, Alan Watts wrote, "I—and others—have been saying for years that destruction of the environment is based on contempt for everything outside the human skin, failure to see that as a field flowers, the planet peoples, and ignorance of the fact that the oceans, the air, and even the solar system are as much our vital organs as heart and stomach. We are not *in* nature; we *are* nature. But as masters of technical weapons we are fighting the environment as if we still believed ourselves to be strangers on the earth, sent down into this world from a purely abstract, ideational, and spiritual heaven."[20]

Mass-scale decisions are being made by governments and large corporations that end up killing living beings in large, unthinkable numbers, simply because they are "in the way" of "progress" and efficiency. But each of us makes smaller-scale decisions on whether to accept or counter those policies and that mindset, as the system perpetrates the destruction of countless beings, plants, insects, animals, and wildlife. Can we transform our ways of thinking, being, acting, living, and interacting with these other species in our day-to-day lives? And can we do so in the short amount of time left to us? To do so requires a shift in our ways of thinking that have led to the ecological disaster we now face.

8

Pesticides and U.S. cigarettes

By John Jonik

Sadly, there seems to be little interest in the matter of pesticide residues on tobacco, no matter the number of illnesses and deaths related to contaminated products, and no matter what the rap sheets are for the complicit pesticide manufacturers.

One hundred and forty thousand people are killed annually in the U.S.A. by "smoking" or "tobacco-related disease." *But there's hardly any tobacco in U.S. cigarettes.* What if what really is killing them is pesticide-contaminated dioxin-delivering smoking products and other exposures to those industrial substances?

We need to stop ignoring pesticides in U.S. cigarettes.

Certain radiation-contaminated fertilizers have been "legal" on tobacco for a long time . . . and, despite the radiation being carcinogenic, this mass public endangerment is allowed to continue. And, of course, no rad symbol is required on contaminated products.

Most typical cigarettes contain residues of chlorine pesticides that are and have been legal on tobacco. Also, most cigarettes, with rare exceptions, are wrapped in chlorine-bleached paper. Both those pesticide residues and the paper expose unwitting, unprotected, guinea-pigged smokers to dioxin, as if our "regulators" never heard of Agent Orange or that dioxin causes immune system damage, learning disabilities, fertility and birth problems, and cancer. To say that undefined "smoking" or "smoke" or tobacco are causes of such maladies is to serve the interests of cigarette makers, adulterant suppliers, all their investors (incl. even health insurers), and the sold-out officials who psychopathically allowed this for so many decades.

The "Family Smoking Prevention and Tobacco Control Act"—a federal statute in the United States that was signed into law by President Barack Obama in 2009—gave the Food and Drug Administration the power to regulate the tobacco industry. *But it forbids the FDA from doing anything about all the 450 or so tobacco pesticides and their residues,* though most "smoking-related" diseases are identical to symptoms of exposures to pesticides, especially the dioxin-emitting chlorine chemicals.

It also ignores certain phosphate tobacco fertilizers that leave carcinogenic levels of radiation in typical cigarettes.

The mainstream-promoted crusade against "tobacco" and "smoking" (terms never qualified) pretends concern for health that plays on a public perception of tobacco as being "sinful," and smoking, a "dirty habit." This trickery, playing further with everyone's natural concern for health, just

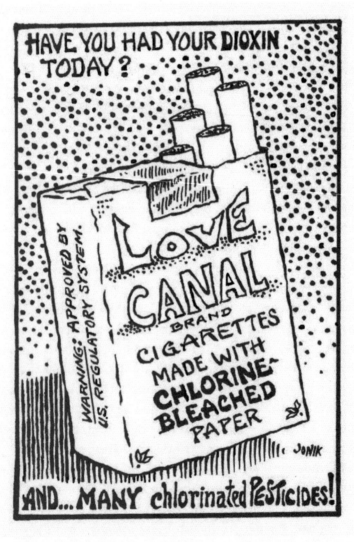

happens to relieve the broad cigarette cartel, including ingredient suppliers and industry investors, of astronomical liabilities, penalties, and profit losses that would occur if public and legal attention was rightly refocused on the toxic/carcinogenic/addiction-enhancing nature of government-approved cigarette additives and adulterants. Exposing this charade might do wonders to, at last, seriously expose, and perhaps eliminate corporate economic linkages with any public officials in regulatory positions. This relates to the safety of all industrial products and processes. After all, the same agencies responsible for typical cigarettes (aka "pesticide pegs," "radiation rods" and "dioxin dowels") are the same agencies that regulate our food and drug safety.

Warning: Trusting this U.S. Regulatory System may be dangerous to your health.

Since the worst possible exposure route for pesticides and dioxin is via inhalation, we need to make the "smoking" and inhalation argument, and the government's failure to ban pesticides from cigarettes, front and center in the anti-Pesticide Activism movement.

In April 2003, the U.S. General Accounting Office (GAO) condemned lax government oversight of pesticide residues in cigarettes. http://www .gao.gov/atext/d03485.txt

9

Why I Stopped using Pesticide Poisons

By Steve Tvedten

I had been a pest-control operator for over fifty years, licensed at various times to apply economic (pesticide) poisons in twenty-one states in the United States. I, along with my firm, Get Set, Inc., routinely made many thousands of pesticide poison applications every year to homes, yards, farms, orchards, groves, commercial buildings, offices, stores, bars, and restaurants, none of which were ever inspected by government agencies to see if the poisons had been correctly applied. The only time a state "regulator" ever came, it was to see if "enough" poison had been applied. I was scheduled to go to court for refusing to use the maximum amount of chlordane the label allowed, just before all applications of this cancer-causing chemical was banned in the United States.

No animal, plant, or insect is automatically or naturally a pest. Only the way we feel about it in a particular location determines whether we welcome, or are repelled by, its life and consider it to be a pest, so the terminology pest is really only an issue if one considers its damage or annoyance intolerable. The economic injury level is the level of pests at which the cost to manage the pest is equal to the losses that pest causes. The action threshold is the pest density at which action must be taken to prevent the pest from reaching the economic injury level and/or tolerance level in any specific area.

The first step in controlling unwanted pests is to prevent them from entering in the first place. The second step is not to feed and water the pests or they will become "pets."

Let me quickly mention some of the many reasons I stopped using any synthetic pesticide poisons and will never go back:

First, many of my friends and family were getting poisoned and dying, including my Uncle Joe and our infant son.

Second, I had been repeatedly told that I could bathe in the various pesticide poison mixes safely. Of course I did not believe these "regulators." My men and I wore protective masks, gloves, and other safety gear to protect ourselves. We were told by the same set of "regulators" that we were unnecessarily scaring the public and that the Michigan Department of Agriculture would appreciate it if we stopped. Even after using all the various safety/protective equipment, I found I could no longer count pocket change and remember dates, names, or addresses. My moods would shift from suicidally depressed to murderously angry in the space of a few seconds, so I sought medical advice and was told I was very intoxicated, not with alcohol but by pesticide toxins, and I needed to detox.

Third, after my detoxing in a sauna I then had to find a safer way to control all of the various pests without killing myself. I invented and patented the perfect pesticide.[1] Then I researched and found hundreds of safer and far more effective natural pest-control alternatives. After using these safe and far more effective and inexpensive alternatives to remove *all* pest problems inside and outside in over 280 schools, I wrote *The Best Control II* and gave it away for free.[2]

Fourth, man-made synthetic pesticide poisons are known to be very persistent and dangerous contaminants! Today, *all* life has become contaminated with these man-made pesticide poisons. "Medically incurable" chronic diseases and conditions, such as heart disease, stroke, Alzheimer's, cancer, type 2 diabetes, obesity, and arthritis, are *now* among the most common, costly, and preventable of all health problems. As of 2012, approximately half of all U.S. adults—117 million people—had one or more of these chronic health conditions. One in four adults had two or more chronic health conditions. Seven of the top ten causes of death in 2014 were chronic diseases caused or exacerbated by exposure to pesticides. Two of these chronic diseases—heart disease and cancer—together accounted for nearly 46 percent of all deaths. I and many other researchers now believe one needs to personally detox *all* of the various man-made toxins in one's body to truly heal.

Fifth, man-made, synthetic pesticide poisons never did and still do not control all of the pests! The advent of synthetic insecticide poisons after World War II launched a new era of pest "control." The number of registered pesticide poisons rose from thirty in 1936 to more than nine hundred in 1972. No one has ever checked the dangers from multiple exposures or measured interaction of different pesticides on the human body.

For several years, poison sprays were used on a routine, preventive basis, regardless of whether or not a pest was even present in damaging numbers. Soon, insect pests began to develop resistance to these insecticide poisons. Some insects that had previously been kept in check by their natural enemies now reached pest status, because their natural enemies were killed off by the spraying, creating new problems. In the absence of natural enemies, growers often misapplied even more toxic products in an effort to control these secondary pests, in addition to resurging populations of the insecticide-resistant target pest. Growers became trapped in a cycle of using more and more insecticide poisons to try to "cure" one pest problem, which resulted in the worsening of another pest problem. The cost of "control" increased while the degree of actual pest control often declined, and the harmful effects on our environment greatly escalated. History has clearly proven that after over seventy years of waging continuous chemical warfare on about one thousand kinds of pests we have now contaminated the entire earth and every living thing. But we have not controlled much less eliminated a single pest species. In fact, there are *now* many more pests causing much more damage than when we first began to spray pesticides after World War II!

Sixth and most important, I have never found any pest that could not be controlled more effectively and more safely using alternative controls.

True IPM (or Intelligent Pest Management)—also known as Ecologically Based Pest Management (EBPM), Situational Pest Management (SPM), or Get Set Pest Management (GSPM)—is an environmentally sane, innovative approach to managing weeds, insects, vertebrates, and other pest organisms. IPM strives to provide economical, long-term protection from pest damage. Simply stated, it means preventing or controlling infestations in ways least toxic to people, pets, and the environment. True IPM, when properly practiced, removes the cause rather than treating the symptom with poison! Knowledge of the pest's species, local breeding areas, feeding, and territorial habits (entire native environments) is essential to controlling pest populations.

IPM involves safely controlling, preventing, reducing, or eliminating unwanted pests using common sense, enzymes, and science, and Pestisafes® rather than synthetic pesticide poisons. This is done by limiting their access to food, water, and shelter; changing the conditions conducive to their growth and entry; encouraging their natural enemies; creating structural modifications; reducing habitat; and reducing the carrying capacity of the site. To properly accomplish these tasks, one must know the structure, habits, and life cycles of the many pests and beneficial organisms, and understand the natural processes and conditions conducive to each pest population. Then buildings and pests must be inspected and monitored to determine if controls are needed.

Managing pests through prevention is usually less expensive, more effective, and safer than trying to control an established pest population with volatile poisons. Pest prevention also reduces the chance for substantial economic loss or damage, and it avoids disruption of people and the many dangers associated with synthetic pesticide poison control efforts that may be used after pests become established.

10

Where and How Is It *Still* Possible to Eat *Relatively* Safely?

By Carolina Cositore

It is disturbing what Monsanto, now Bayer, has been perpetrating on our present and our future; not only regarding enormous damage to plant and animal diversity and our planet's climate, but also the appalling direct effects on our health; especially recognizable since the advent of Covid-19.

Globally, environmental and health organizations have long been lobbying, protesting, and organizing to ban glyphosate, among other dangerous chemicals. The danger made international headlines in 2015 when the UN World Health Organization found glyphosate to be a probable carcinogen for humans and animals. Monsanto then began the fight to overturn that decision and it became increasingly difficult to separate science from economic and political interests. Despite the corporate pr claims to the contrary, that the accusations of cancerous tumors as well as severe liver and kidney damage and hormonal have credence can be surmised by the huge court payouts demanded and made both before and since Bayer AG bought out Monsanto and the glyphosate headache for $63 billion in 2018 when the company ranked 199th on the 2018 Fortune 500 of the largest United States corporations by revenue.

Around 2017–2018 the possibilities for healthy eating, i.e., non-GMO/ glyphosate foods, certainly seemed to be looking up as concern about the pesticides spread and more consumers were raising questions and more and more nations were responding by partially or completely taking steps to ban the cultivation of GMO crops. Nine countries (Algeria, Bhutan, Cuba, Kyrgyzstan, Madagascar, Peru, Russia, Venezuela, and Zimbabwe) had also even banned the importing of any GMO crops.

Positive expectations increased both by the benchmark Bayer payout in June 2020 of $8.8 to $9.6 billion for claims and, while no admission of liability or wrongdoing was part of the deal hence Bayer could continue marketing Roundup without warning labels, there was a hopeful $1.25 billion cushion to establish an independent expert panel to resolve whether glyphosate causes cancer, and if so, what is the minimum danger exposure level.

So what happened? Is it now safe to eat everywhere or anywhere?

Not surprisingly, a company that makes money by dirtying our soil, water, plants, and animals, also plays dirty when it comes to advertising, pr and lobbying. A European Parliament report issued in January 2019 found that EU regulators based their decision to re-license glyphosate on an assessment that was plagiarized from a coalition of pesticide companies, including Monsanto. The scandal has caused a number of countries in the bloc to introduce individual legislation banning or restricting the use of the substance.

A superficial Google search would indicate erroneously that our food safety was still looking up. While the United States continues to allow GMO and pesticides in our food, glyphosate has been or will be banned in ten jurisdictions, including Germany, Saudi Arabia, and Vietnam, and at least fifteen additional countries restrict its use. The World Health Organization's International Agency for Research on Cancer formally classified glyphosate as a "probable carcinogen" in July 2020.

Moreover, non-GMO crops have the potential to be extremely profitable for farmers and investors. As people grew more concerned about what was is in their food, the market for non-GMOs and sustainably raised food products continued to grow.[1]

A search of the internet would feed those positives and give us sanguine feelings with long lists of countries seemingly "banning GMOs"; hence their built-in pesticides.

Unfortunately, a closer look greatly dims that perception.

- Not all countries claiming to ban GMO crops have a total ban on GMO cultivation.
- To date, only France, Germany, Austria, Greece, Hungary, the Netherlands, Latvia, Lithuania, Luxembourg, Bulgaria, Poland, Denmark, Malta, Slovenia, Italy and Croatia have chosen a total ban.
- But don't plan on a gourmet vacation in those "total ban" countries just yet.
- This is a total ban on all crops grown within their borders, but not on

imports. Imports focus on, but are not limited to, livestock feed. Over 80 percent of GMO crops grown in the world have been engineered for protection by the use of toxic herbicides, with concomitant negative impact on the environment and human health. The vast majority of North American crops are genetically modified, again including but not limited to packaged foods containing sugar, corn, soy, and canola. Livestock, agriculture, and aquaculture products are also considered to be high-risk for GMOs and all of these products are exported by the United States.

• The nine countries that had previously stood up to Monsanto and banned both the use of glyphosate and the importing of products grown with it [Algeria, Bhutan, Cuba, Kyrgyzstan, Madagascar, Peru, Russia, Venezuela and Zimbabwe] have been seriously damaged both by Covid-19's constraints on travel, which most seriously afflicted tourist-dependent countries, and/or by political changes, especially increased sanctions. These recent developments have added to food limitations, often with outright shortages and hunger. Food shortages, in turn, have frightened some governments into relaxing GMO restrictions believing the hype that GMO is the answer to a hungry maiden's prayer although such seeds do not increase yield or reduce water use. These GMOs are crops engineered to withstand, work in partnership with, and self-generate pesticides. They are not engineered to increase yield or face climate-related challenges to growth, such as drought tolerance. What minimal increases in yield there have been have come with major problems including water pollution, pollinator loss, and soil degradation that put future food security at risk. After decades of attempts, Big Biotech has not been successful in breeding GE seeds that increase yield or reduce water use. Conventional farming is far superior when it comes to nitrogen use efficiency (the ability of crops to pull nitrogen out of soil, developing a more efficient use of fertilizer, ultimately decreasing the demand for fertilizers) and water use efficiency. But the propaganda has been pervasive and has had a real effect.

Taking the effects on the nine one at a time:

Algeria continues to outlaw both the cultivation and import of GMO products, however, it is only recently that there has been any attempt to oversee this law. A recent survey of maize-derived foods found 20 percent with at least one screening GM contamination.

Bhutan still does not produce any GM food/feed; however, GMOs can now be imported upon authorization.

Cuba's long limit on GMO foods helped the country enormously by not interfering with the sun's vitamin D and so aided Cuba's low Covid statistics. Genetically modified foods continue to carry a stigma on the island, not least because former President Fidel Castro harshly criticized transgenic foods and strongly promoted organic crops. Nevertheless, in June 2020, due to the food crisis brought on by cessation of tourism and increased U.S. sanctions, Cuba officially permitted such crops as a "complement to conventional agriculture." This permit will be applied to maize and soybeans and possibly other foods in the future, including sugarcane.

Kyrgyzstan does continue its ban on cultivation and importation, with some allowances for transitional agriculture.

Madagascar continues its ban on both cultivation and importation, although legal regulation is problematic.

Peru's ten-year moratorium on the importation and cultivation of GMOs expires in 2021. Ignoring this ban is a campaign to permit importing soy and corn, although there is fierce opposition because the ban has supported the 2.2 million small-scale farmers who provide 75 percent of the country's produce.

Russia, heavily sanctioned, recently adopted an approval procedure for release of genetically modified organisms into the environment, currently, eighteen GM food lines and fourteen GM feed lines are approved and registered in Russia.

Venezuela, under sanctions, has remained steadfast in denouncing genetically engineered foodstuffs. Yet currently, Venezuela remains reliant on food and feed imports of GM soybean and maize crops from Brazil, Argentina, and the United States.

Zimbabwe quietly lifted a ban on imports of genetically modified corn in January 2020 for the first time in twelve years. A scarcity of corn meal that could lead to famine necessitated genetically modified corn imports from South Africa. Zimbabwe continues to carefully quarantine the grain.

- All is not hopeless, but certainly caveat emptor applies. We can, if we can afford it, buy and eat organic. This is easier in some countries than others. Only sixty-four countries around the world require genetically modified foods to be labeled. These countries include all of the European Union members, Australia, and Japan, among others. GMOs are not currently labeled in the United States or Canada; however, a smattering of products will start to be labeled in the United States thanks to the National Bioengineered Food Disclosure Standard.
- More and more products in the U.S. are self-labeling organic or non-GMO. The cell phone app Buycott helps identify non labeled products and even occasionally those mislabeled as organic.
- Picking and choosing products from countries that ban some or all of their homegrown produce also helps. For example, pasta grown and made in Italy is non-GMO – don't rely on Italian-sounding names though, read the labels.
- Water is quite problematic. Most standard water filters are not designed to remove glyphosates, herbicides, and pesticides. Once glyphosate enters water it does not degrade easily, so while in the United States worst case is water sources contaminated by agricultural runoff, and distance is not a factor here, suburban lawns and gardens also receive more pesticide applications per acre than agriculture and, of thirty commonly used lawn pesticides, seventeen are detected in groundwater, and twenty-three have the potential to leach. A filter specifically designed to remove glyphosate is probably the best bet as much of our bottled water is simply packaged tap water.

All is not completely lost. Cuba has a long history both of abhorring genetically modified crops and pesticides and relying on her people to make the decisions, so there is hope.

And Mexico, under a surprise December 2020 presidential decree by AMLO, as leftwing President Andrés Manuel López Obrador is affectionately known, bans the herbicide glyphosate, plans to replace sixteen tons of GM corn with native varieties and, most importantly, will forbid the importation of genetically modified crops. Not surprisingly since then, Mexico's Supreme Court saw four appeals brought by all of the expected villains, including Bayer-Monsanto, Syngenta, PHI and Dow. What is a surprise given the global state of things is that the Supreme Court unanimously upheld the president's ban, which will go into effect January 2024.

It should be noted that part of Mexico, the Zapatista controlled area around Chiapas, has not permitted genetically modified cultivation since the 1990s.

In conclusion, information on all countries is constantly changing due to the push and pull of environmentalists in the one direction and Bayer and the other giant pesticide and genetic engineering corporations in the other. Also, Bayer et al. have other toxic chemicals waiting in the wings, with some already being tested as the bans on glyphosate are being considered.

Remain informed, let your government representatives know you care and watch what you eat.

11

Consequences of Glyphosate's Effects on Animal Cells, Animals, and Ecosystems

By Robin T. Falk Esser, PhD

Is Glyphosate Only Toxic to Plants, as Claimed by the Pesticide Industry[1] and the EPA[2]?

Glyphosate has been heralded as exerting its herbicidal effects by disrupting metabolic pathways only found in plant cells. Glyphosate was thought to be nontoxic to animal life, because these affected pathways are not found in animal cells (Herman, 1999; Kennedy, 2017).[3] However, since Earth's environment is comprised of myriad interacting complex systems, each with an untold number of variables, it is not surprising that the animal kingdom would *not* be given a free pass. It stands to reason that glyphosate might exert more than the *one* disrupting effect—just maybe, it might end up doing something else. Indeed, glyphosate-based herbicides (GlyBH) have been found to cause endocrine disruption, birth defects, tumors, liver damage, and kidney damage, at concentrations below the allowed acceptable daily levels, in mice and rats (Benedetti, 2004; Prasad, 2009); GlyBH is also toxic, at environmentally relevant concentrations, to a variety of other species, such as frogs, fish, and daphnia, the water flea (Sandrini, 2013; Kennedy, 2017).

Other studies have shown that glyphosate causes significant morphological changes in the tail of tadpoles (Relyea, 2012). Even though the precise mechanisms by which GlyBH exerts all these effects may have not yet been definitively elucidated in some, or even most, cases, the reality of the harmful effects has been demonstrated.

What Are Some Underlying, Molecular Mechanisms by Which Glyphosate Exerts Its Toxicity in Animals' Cells?

Studies have shown that glyphosate induces bone marrow DNA damage and cell death in mice, within twenty-four to seventy-two hours, following one dose of either 25 or 50 mg glyphosate per kilogram of body weight (Prasad, 2009). Furthermore, glyphosate has been shown to be a cholinesterase inhibitor at environmental concentrations in mussels and fish (Sandrini, 2013). Cholinesterase is an enzyme found in animals' cells (including, of course, humans' cells), which is needed during normal metabolism to break down the important neurotransmitter, acetylcholine. If acetylcholine were not rapidly broken down by cholinesterase, it would over-stimulate nerves, muscles, and exocrine glands. Glyphosate also appears to substitute for the amino acid glycine in a number of important metabolic pathways, including those that affect kidney function (S. Seneff, chapter in this book; Seneff, 2018).

Does Glyphosate's Solubility in Water Reduce Its Toxicity, as Claimed by the Pesticide Industry and the EPA?

The chemical structure of glyphosate has also been used to argue against glyphosate bioaccumulating[4] in organisms, helping to justify its widespread worldwide distribution. Substances bioaccumulated or biomagnified[5] up the food chain present a much greater threat to an ecosystem, since those animals occupying the top of the food chain experience devastatingly high pesticide concentrations, as Rachel Carson first pointed out in 1962 (Carson, 1962).

Fat-soluble pesticides (which dissolve better in fats than water), such as chlorinated hydrocarbon insecticides (including DDT and chlordane), carbamates, organophosphates, and pyrethroids are bioaccumulated and biomagnified, as well as persist in the environment, thus wreaking greater destruction than water-soluble pesticides, such as glyphosate,[6] which was an early rationalization for its mass application. However, water-soluble pesticides, since they dissolve well in water, are easily carried by rainwater and runoff, from the area where they were sprayed, into ground water and streams. There, they damage untargeted plants and animals. For example, GlyBH has been found to be toxic to freshwater benthic macroinvertebrates,[7] significantly reducing the biodiversity of the ecosystem (Rzymski, 2013). Other field studies have found environmental concentrations of GlyBH to be toxic to both aquatic invertebrates and vertebrates, as well as phytoplankton, zooplankton, bacteria, and protists (Perez, 2011).

A 2005 University of Pittsburgh study added Roundup at Monsanto's recommended dose to ponds filled with frog and toad tadpoles. Two weeks later, 50 to 100 percent of the tadpole populations of both species of tadpoles had been killed (Reylea, 2005). Does the EPA's requirement that Roundup display a warning label, cautioning that it not be applied near aquatic habitats (No Spray Coalition, 2003), rectify this "problem"? No, because, as described above, GlyBH will be transported by rainwater, runoff, and groundwater to aquatic ecosystems. Further, how probable is it that the user may not read, or may simply disregard, the label? The protection afforded by our environmental laws is only as strong as its weakest link.

Glyphosate-based Herbicides Actually Show Bioaccumulation, Despite Glyphosate's High Solubility in Water!

Surprisingly, even though GlyBH is water soluble and therefore would not be expected to be bioaccumulated by organisms, as described above, it is nevertheless found to be bioaccumulated by animal cells as well as plant cells! This arises from the presence of POEAs (polyoxyethylene amines, such as polyethoxylated tallow amine) in the herbicide mix. POEAs, which act as surfactants, are added to water-soluble glyphosate, to increase the uptake of glyphosate by the plant leaf cuticle, the leaf's waxy surface. The leaf cuticle absorbs fat-soluble substances best. POEAs increase glyphosate uptake, increasing the efficiency of the spraying process *and* increasing Monsanto's profits. POEAs fulfill their intended role admirably—so much so that glyphosate bioaccumulation actually occurs, despite glyphosate being a water-soluble molecule, which would not be expected to show bioaccumulation. Results from the environment indicate that GlyBH is bioaccumulated in terrestrial snails, water hyacinth, carp, and tilapia, as well as in animal animal-cell lab studies (Contardo-Jara, 2009; Hedberg, 2010; Druart, 2011; Wang, 1994).

In What Other Ways Are Added Substances, Mixed with Glyphosate in GlyBH, Toxic to Animal Cells? Are the Additives Inert, as Claimed by the Pesticide Industry?

Any substance in GlyBH, other than glyphosate, is referred to as a *formulant* or *adjuvant*. The GlyBH herbicide mixes are called *formulations* (Defarge, 2018). Formulants, including POEAs, are listed as inert ingredients, but POEAs have been demonstrated to be active, not inert. POEAs, petroleum-based oxidized molecules, present in glyphosate-based herbicides, as described above, have been found to be highly toxic to animals, on their own. They have been reported to be one thousand times more toxic than glyphosate (Schmidt, 2017; Matthews & Associates, 2018). A variety of heavy metals, including arsenic, chromium, nickel, lead, and cobalt are present, as formulants, in GlyBH. These may have come from unintentional contaminations during manufacturing, such as impurities in petroleum, or from industrial by-products or waste. However, formulants could also have been intentionally added for their herbicidal actions, to act as surfactants (as POEAs do), or to serve as nanoparticles,[8] solvents, or antifoam agents (Defarge, 2018; Kookana, 2014; Perez, 2011). The recent finding that arsenic is present in GlyBH[9] (Defarge, 2018), at doses well above the supposedly "safe" limit, hammers a final nail into the coffin already occupied by the "Roundup is safe when used asdirected" argument. Arsenic was found to be present at concentrations that were five to over fifty times greater than the permitted level in both Europe and the United States, in eight out of eleven formulations studied. In six out of eleven formulations, arsenic levels were above the permitted levels even after performing the prescribed dilutions for agricultural or garden use, of 15 percent (Defarge, 2018). Arsenic was once widely used as the pesticide of choice (Li, 2016), and this may be one reason why it is "found" in GlyBH. That is, it may have been purposely added to GlyBH by Monsanto, for its herbicidal actions (Defarge, 2018). It also may be present as an impurity, as mentioned in the last paragraph. If Monsanto added arsenic to its glyphosate herbicide mix, or alternatively, knew it was present and did not report it to the EPA, it would be an intentional breech of the U.S. environmental standards set by the EPA, because any use of arsenic other than in pressure-treated wood was banned in 2009, due to its toxic effects on humans and wildlife (ATSDR, 2009).

Did Monsanto Lie to the Public and the EPA by Omitting the Presence of Arsenic in Roundup? Do EPA Loopholes Allow the Pesticide Industry to Circumvent EPA Regulations?

Because it classified arsenic among the inert formulants, Monsanto was not required to even list it on the ingredient list. Any substance which is not an active ingredient automatically receives the classification of "inert" (Lerner, 2016). Non-active ingredients are not required to be listed, according to the theory they could be "trade secrets" which the manufacturer should not be forced to divulge (Lerner, 2016). Therefore, arsenic has not been listed as an ingredient on the Roundup label (Perez, 2011). This appears to be an EPA-sanctioned loophole. Moreover, is it possible that Monsanto never knew there was any arsenic in its glyphosate herbicide (Roundup)? This would be an interesting area for present or future investigation. We can hope that one day a whistleblower, an insider with a conscience, will leak an old internal memo, its existence as yet only a matter of speculation, implicating Monsanto in an intentional attempt to lie to the EPA and the public. Logically, it seems doubtful that Monsanto would not have done in-house analyses of Roundup. And these in-house analyses would be expected to have "discovered" the presence of arsenic, as well as measuring its concentration as to be above allowed standards.

Another EPA loophole is the fact that safety tests, done either by the industry itself or by EPA scientists, to set safe exposure levels for GlyBH (Roundup) are only done on the declared active ingredient (Williams, 2000). Therefore, the toxic effects of the many formulants, which include POEAs, arsenic, and several other heavy metals, are never investigated in determining the glyphosate herbicide's safety or toxicity. Of course, there are other factors that mediate against the actual safe levels ever seeing the light of day. Requiring the industry to police itself by having it conduct its own safety testing is an obvious weak link. Furthermore, the EPA regulators are often former industry employees, chosen ostensibly because they are the people who are most knowledgeable about the field. This has been referred to as the "revolving door"—the unhealthy intimacy between government agencies and the industries that they have been designated to regulate. One obvious truth that seems to have eluded both Monsanto and the EPA is that for pesticide safety tests to be valid, they must be done on the pesticide mix as a whole, not on only the active ingredients, since this is what the public and wildlife are exposed to. Furthermore, individual safety tests on each individual formulant, at environmentally relevant concentrations, are warranted.

The Case for GlyBH-Triggered Domesticated-Animal Deaths

From 2011 to 2013, in the area of Milan, Italy, glyphosate herbicide mix was the cause of 14 percent of all pesticide (including rodenticides) poisonings of dogs, cats, horses, goats, and sheep. (Caloni, 2015). Other articles by the same research group traced data records back to 2006, finding similar results. Another study documents glyphosate poisoning, including deaths, of cats and dogs in the United Kingdom, from 1999 to 2013 (Bates, 2013). These authors take issue with the pesticide industry's and government's claim that safety studies show a low toxicity for glyphosate herbicides. They point out, as described above, that the safety studies were all done on glyphosate alone, rather than on the actual pesticide mix that the animals are exposed to.

The Massive Decline of Monarch Butterflies Linked to Pesticide Use

The study that most decisively demonstrated a correlation between glyphosate herbicide mix and monarch butterfly populations, important pollinators, was done on data from 1999 to 2003 (Wiley, 2017). Then, in 2015, the monarch population increased, while the application of GlyBH continued to increase (Wiley, 2017; Khan, 2015). GlyBH is designed to wipe out milkweed, which is the sole source of food for the monarch caterpillar larvae. The pesticide industry has clung to this observation as evidence that GlyBH is actually not toxic.

The fallacy in reasoning here is that other factors, such as weather changes, may have triggered an increase in the monarch butterfly population, and the 2015 increase would have been even sharper if GlyBH had not been present (Khan, 2015). The population of monarch butterflies has declined relatively steadily over the past three decades or more, by 90 percent, and has continued to decline following the temporary 2015 spike for the past three years, 2016–2018 (Rice, 2018). Monarch butterfly populations vary from year to year due to weather events. For example, the massive hurricanes of Fall 2017 may have been responsible for part of the population decrease. Climate change does put more energy into the atmosphere and could be an important factor in producing the 2017 hurricane season. Monarch butterflies migrate over very long distances, taking three generations or more to complete a migration. For migrating animals whose populations are in decline, habitat destruction or fragmentation is often a key factor. However, there is no question that pesticide poisoning plays a critical role. GlyBH is, of course, not the only pesticide being applied to our landscapes. Neonicotinoids, such as clothianidin, manufactured by Bayer, are now one of the most widely used pesticides in the world, and like GlyBH, are water soluble (Latham, 2015). This means they become widely distributed via runoff and ground water, poisoning non-target organisms. Neonicotinoids have been implicated in monarch butterfly population decreases (Latham, 2015).

Could Joint and Several Liability Be Applied to Impacted Watersheds or Migratory Flyways, as It Has Been to Superfund Sites?

Pesticides are ubiquitous in our environment, as are other toxins, making it difficult to conclusively prove that species harm is due to only one factor. This indeed is the rationale routinely employed by industry legal teams when trying to extricate their clients from any legal or monetary responsibility for the various deaths they have caused. Whether it is the tobacco, chemical, nuclear, or oil industries—or even small-time polluters—the argument is always the same: "How do you know it's my pollutant that caused your illness, since you are exposed to so many?" Monsanto, Bayer, and Dow Chemical will direct this question toward activists advocating for the wildlife who cannot speak for themselves. The Joint and Several Liability rule of CERCLA §107 makes any one polluter of a declared superfund site entirely responsible, legally and financially, for the entire cleanup, even though there were many polluters (EPA Enforcement at Federal Facilities).[10] Can we argue for Joint and Several Liability to be applied to the pesticide polluters of a particular watershed? Is it plausible that a polluted watershed or even our entire environment, planet Earth, be treated, legally, as one giant Superfund site?

Can the Dangerous Decline in the Honeybee Species Be Attributed to Pesticides?

Some will be surprised to learn that honeybees pollinate one third of the food we eat (Batts, 2016). It is difficult to grasp the immense importance of the honeybee as a pollinator. Ninety percent of the world's food is provided by only one hundred crops, and seventy-one of these 100 crops require honeybees for their pollination.[11] Honeybees produce, via their pollination, $20 billion in crops per year (Batts, 2016). GlyBH, at recommended doses found in the environment, decreases both sensitivity to sucrose and short-term memory, which makes it harder for bees to find not only find food, but also find their hive. These bees, exposed to the environmentally encountered Roundup levels, were involved in much higher frequencies of Colony Collapse Disorder, probably because they couldn't remember how to get back their hive (Herbert, 2014). But Monsanto and the EPA have claimed that GlyBH—Roundup—is not harmful to honeybees at recommended doses (Porterfield, 2015). The EPA has released findings indicating that neonicotinoids[12] weaken, disorient, and kill honeybees (Meyer, 2018; Latham, 2015).

The plight of the honeybees not only impacts our food supply, but also, of course, has dire consequences for the entire biosphere, because of the dependence of flowering plants on pollination.

What Can Be Done to Turn Around the Plight of Wildlife Exposed to GlyBH and Other Pesticides?

A recent example, reported in 2015, of people taking direct action that could have brought about a positive outcome but for a simple (but profound) error, occurred when people living in the monarch migration path planted milkweed in their gardens and elsewhere to counteract the decrease in milkweed caused by pesticides. This grassroots effort had the potential to help restore the monarch butterfly caterpillars' food source. Unfortunately, the type of milkweed that was commonly available to buy was a tropical variety—the native variety was harder to find. When it was planted in the southern United States, such as in Texas and the Gulf States, it didn't die off in the winter. So the monarch butterflies never migrated. One outcome of not migrating is that they became infected more seriously with a parasite. The parasite infection in the non-migrating butterflies became more extensive, because the milkweed didn't die out. So, the parasite infections in the monarch caterpillars became much more serious and widespread. These infected caterpillars were weakened, preventing them from being able to fulfill their section of the migratory route. Instead, they died. (Wade, 2015). Since this occurred, native species of milkweed are now being planted throughout the Monarch butterfly migratory route.

The unexpected environmental effect of planting tropical milkweed is one simple example of Chaos Theory in action, as is the effect of the widespread use of pesticides. In the former case, an unpredicted outcome results from an attempt to rectify a system, which further perturbed the system, perhaps moving it so far from its equilibrium or steady state that it is no longer able to right itself. The widespread application of pesticides for profit, decimating the honeybee populations, that in turn are responsible for Earth's pollination, is an example of this idea, that neatly fits "the butterfly effect"[13]—a popular environmental-science theory, which proposes that because of the many interacting complex systems on planet Earth, a small perturbation in just one variable, hardly noticed, can evoke an unpredictable tidal wave of cataclysmic collapse, spreading throughout the ecosystem or biome, like the ripples from skipping a stone that spreads out in all directions over the surface of a pond.

Works Cited

Agency for Toxic Substances and Disease Registry, C. (2009). *Arsenic Toxicity: What Are the Standards and Regulation for Arsenic Exposure?* (2009). Retrieved May 11, 2018, from CDC-ATSDR-Environmental Health & Medicine Education: https://www.atsdr.cdc.gov/csem/csem. asp?csem=1&po=8www.atsdr.cdc.gov/csem.asp?csem=1&po=8.

Bates, N., and N. Edwards. (2013). "Glyphosate Toxicity in Animals." *Clinical Toxicology* 51, no. 10, (2013): Vol 51; Issue 10; p 1243.

Batts, VikkiK. (2016, June 27). "Study Shows Honeybees Are Starving

Because of Roundup." (June 27, 2016). www.glyphosate.news/2016-06-27-study-shows-honeybees-are-starving-because-of-roundup.html. Retrieved May 11, 2018, from Glyphosate.news: glyphosate.news

Benedetti, A. Le. (2004, November 2). et al. "The Effects of Sub-chronic Exposure of Wistar Rats to the Herbicide Glyphosate-Biocarb." *Science Direct Toxicology Letters;* 153, no. 2 (November 2, 2004): *Volume 153; Issue 2*, pp. 227–232.

Caloni, F. et al. (2015). "Suspected Poisoning of Domestic Animals by Pesticides." *Science of the Total Environment* 539 (January 2016):, 331–336.

Canal, Tami. "EPA Finally Admits What Has Been Killing Bees for Decades." March against Monsanto (January 10, 2016), www.march-against-monsanto.com/epa-finally-admits-what-has-been-killing-bees-for-decades.

Carson, R. (1962). *Silent Spring.* Cambridge, MA Mass: Houghton Mifflin, 1962.

Contardo-Jara, V. E. Klingelmann and C. Wiegande. (2009). "Bioaccumulation of Glyphosate and Its Formulation Roundup Ultra in *Lumbriculus variiegatus* and Its Effects on Biotransformation and Antioxidant Enzymes." *Environmental Pollution*, 157, no. (1 (January 2009): 57–63.

Defarge, N., J. Spiroux de Vendômois, and G. E. Séralini e. (2018). "Toxicity of Formulants and Heavy Metals in Glyphosate-Based Herbicides and Other Pesticides." *Toxicology Reports*, 5 (2018): 156–163.

Druart, C., M. Millet, R. Scheifler, O. Delhomme, and A. de Vaufleury e. (2011). "Glyphosate and Glufosinate-Based Herbicides: Fate in Soil, Transfer to, and Effects on Land Snails. *Journal of Soils and Sediments*, 11, no. 8(2011): 1373–1384.

EPA Enforcement at Federal Facilities. (n.d.). Superfund Liability. Retrieved May 12, 2018, from EPA Enforcement: https://www.epa.gov/enforcement/superfund-liability.

Hedberg, Daniel, and Margareta Wallin. (2010). "Effects of Roundup and Glyphosate on Intracellular Transport, Microtubules and Actin Filaments in *Xenopus laevis* Melanophores." *Toxicology In Vitro*, 24, no. 3 (April 2010): 795–802.

Herbert, L. T., D. E. Vázquez, A. Arenas, and W. M. Farina. e. (2014). "Effects of Field-Realistic Doses of Glyphosate on Honeybee Appetitive

Behaviour." *Journal of Experimental Biology.*, Oct 1: 217, (Pt. 19) (October 2014): 3457–64.

Herman, K., and L. Weaver. (1999). "The Shikimate Pathway." *Annual Review of Plant Biology* 50 (June 1999): 473–503.

Kennedy, D. C. (2017). "Glyphosate Fate and Toxicity to Fish with Special Relevance to Salmon and Steelhead Populations in the . . . Skeena River Watershed." T. Buck Suzuki Environmental Foundation (November 2017)(TBSEF),www.bucksuzuki.org/images/uploads/docs/Glyphosate _report_Final_Nov_21_2017.pdf Burnaby, BC: BioWest Environmental Research Consultants.

Khan, C. (2015). "Monarch Butterfly Population Rejuvenating After Last Year's Record Low." radio program. *All Things Considered* (March 4, 2016): NPR All Things Considered.

Kookana, R. S. et al. (2014). "Nanopesticides: Guiding Principles for Regulatory Evaluation of Environmental Risks." Journal *of Agricultural and Food Chemistry*, 62, no. 19 (April 2014): 4227–4240.

Latham, J. (2015). "New Research Links Neonicotinoid Pesticides to Monarch Butterfly Declines." Independent Science News, (Un)sustainable Farming, Biotechnology, *News* (April 4, 2015), www.independent sciencenews.org/news/new-research-links-neonicotinoid-pesticides-to -monarch-butterfly-declines/.

Lerner, S. (2016, May 17). "New Evidence about the Dangers of Monsanto's Roundup." Retrieved May 11, 2018, from The Intercept (May 17, 2016): www.https://theintercept.com/2016/05/17/new-evidence -about-the-dangers-of-monsantos-roundup.

Li, Y. et al. (2016). "Chronic Arsenic Poisoning Probably Caused by Arsenic-Based Pesticides: Findings from an Investigation Study of a Household." *International Journal of Environmental Research and. Public Health*, 13, no. 1 (January 2016): 133.

Matthews & Associates. (2018). "Roundup More Toxic than Glyphosate." Retrieved May 11, 2018, from Matthews & Associates: Lawyers Working for People (website), accessed May 11, 2018: www.https://dmlawfirm .com/roundup-toxic-glyphosate.

Meyer, N. (2018). EPA finally admits what has been killing bees for decades. Retrieved May 11, 2018, from MARCH AGAINST

MONSANTO: https://www.march-against-monsanto.com/epa-finally -admits-what-has-been-killing-bees-for-decades.

NoSpray.org. (2003). "Glyphosate/Roundup Spraying," *NoSprayNewz*. The Power Hour (website), accessed Retrieved May 11, 2018, from The Power Hour: http://www.thepowerhour.com/news/glyphosate _roundup.htm.

Perez, Gonzalo Luis, María Solange Vera, and Leandro MirandaG. e. (2011). "Effects of Herbicide Glyphosate and Glyphosate-Based For-mulations on Aquatic Ecosystems." In A. Kortekamp, *Herbicides and Environment* London: InTech (peer-reviewed web publisher), 2011, (pp 343–368), www.intechopen.com/books/herbicides-and-environment.

Porterfield, A. (2015, Nov 3). "Glyphosate is no bee killer." Retrieved May 11, 2018, from Genetic Literacy Project (November 3, 2015):, www.http://geneticliteracyproject.org/2015/11/03/glyphosate-is-no-bee-killer/.

Prasad, S. et al. (2009). "Clastogenic Effects of Glyphosate in Bone Marrow Cells of Swiss Albino Mice." *Journal of Toxicology* (2009), Article ID 308985, 6 pages.

Relyea, R. (2012). "New Eeffects of Roundup on Amphibians: Predators Reduce Herbicide Mortality: Hervicides Induce Antipredator Mor-phology." *Ecological Applications* 22, no. 2, Mar 22 (2) p: 634–647.

Reylea, R. e. (2005). "The Lethal Impact of Roundup on Aquatic and Terrestrial Amphibians." *Journal of Ecological Applications*, Vol 15, Issue 4 (2005): 1118–1124.

Rice, D. (2018). "Monarch Butterfly Population Dwindled for Second Straight Year in Mexico." USA Today, March 6, 2018.

Rzymski, P. et al. (2013). "The Effect of Glyphosate-Based Herbicide on Aquatic Organisms—A Case Study." *Limnological Review*, 13, no. 4 (2013): 215–220.

Sandrini, J. et al. (15 April 2013). "Effects of Glyphosate on Cholinesterase Activity of the Mussel Perna and the Fish Danio rerio and Jenynsia mul-tidentata." *Aquatic Toxicology Volumes* 130–131 (April 2013): 171–173.

Schmidt, D. (2017, February 27). "Roundup is More Toxic than Glyphosate Alone, Lawsuit Claims." Retrieved May 11, 2018, from Natural Health 365: https://www.naturalhealth365.com/roundup-monsanto-2151.html.

Seneff, S., and L. Orlando &. (2018). "Glyphosate Substitution for Gly-cine During Protein Synthesis as a Causal Factor in Meso-american

Nephropathy. *Journal of Environmental & Analytical Toxicology* 8, no. 1 (2018): 541, Jan 14.

United States Environmental Protection Agency. "Superfund Liability," https://www.epa.gov/enforcement/superfund-liability.

Wade, L. (2015, Jan 13). "Plan to save monarch butterfly backfires." Retrieved May 11, 2018, from Sciencemag.org:Science Magazine website (January 15, 2013), http://www.sciencemag.org/news/2015/01/plan-save-monarch-butterflies-backfires.

Wang, Y. et al. (1994). "Dissipation of 2,4-d Glyphosate and Paraquat in River Water. *Water Air Soil Pollution*, 72, no. (1–4 (January 1994): 1–7.

Wiley. (2017, May 17). "How Herbicide Use and Climate Affect Monarch Butterflies." Retrieved May 11, 2018, from ScienceDaily (May 17, 2017): www.sciencedaily.com/releases/2017/05/170517090525.htm.

Williams, G., R. Kroes, and I. Munro e. (2000). "Safety Evaluation and Risk Assessment of the Herbicide Roundup and Its Active Ingredient, Glyphosate, for Humans." Regulatory *Toxicology and Pharmacology*, 31, no. 2 (April 2000): 117–165.

12

Unsafe at any Dose? Glyphosate in the Context of Multiple Chemical Safety Failures

By Jonathan Latham, PhD

Glyphosate is an agrochemical for which there is multiple evidence of harm (International Agency for Research on Cancer, IARC). Banning it would therefore seem like a rational decision and obviously in the best interests of public health—obvious, except that glyphosate, like all chemicals released into the environment, is not by itself the problem. Its application all over the world is a symptom of something much larger. Just as scratching an itch does not make a bite go away or prevent more bites, banning glyphosate will only open up space for other agro-chemicals. If glyphosate were banned, chemicals such as Thorazine, 2,4-D, and dicamba would replace it, and campaigning efforts will, at best, have been wasted. At worst, they will have resulted in greater harm. The only way to eradicate glyphosate without making things even worse is to understand its place in the system of corporate food production (as opposed to "cultivation"—Martha Herbert's useful distinction found elsewhere in this book) and act accordingly.

The principle at work is that all complex systems such as global food production contain the possibility of change; but in order for change to occur, it is necessary for an external change-maker to carefully select those points at which pressure can productively be exerted and distinguish them from those that may be more obvious but are ultimately ephemeral. The system will not change of its own accord. Possibly a little pressure at the right place can achieve what historically has not been done because the energy has been misapplied. Perhaps much more can be achieved than the simple banning of glyphosate. But first it will be necessary to understand the processes by which chemicals become widely used, which is best done through their history. We can begin with the lesson of BPA. Piecemeal, and at long last, chemical manufacturers have begun removing the endocrine-disrupting plastic bisphenol-A[1] (BPA) from products they sell. Sunoco no longer sells BPA[2] for products that might be used by children under three. France has a national ban[3] on BPA food packaging. The European Union has banned it from baby bottles.[4] These bans and withdrawals are the result of epic scientific research and some intensive environmental campaigning. But in truth these restrictions are not victories for human health, nor are they even losses for the chemical industry.

For one thing, the chemical industry now profits from selling premium-priced BPA-free products. These are usually made with the chemical substitute BPS, which current research suggests is even more of a health hazard than BPA.[5] But since BPS is far less studied, it will likely take many years to build a sufficient case for a new ban.

But the true scandal of BPA is that such sagas have been repeated many times. Time and again, synthetic chemicals have been banned or withdrawn, only to be replaced by others equally harmful, and sometimes worse. Neonicotinoids, for example, which the International Union for the Conservation of Nature (IUCN) credits with creating a global ecological catastrophe,[6] are modern replacements for long-targeted organophosphate pesticides. Organophosphates had previously supplanted DDT and the other organochlorine pesticides; many bird species are only now recovering from their effects.

So if chemical bans are ineffective (or worse), what should anyone who wants to protect against flame retardants, pesticides, herbicides, endocrine disruptors, plastics, and so on—but who doesn't expect much help from their government or the polluters themselves—do?

What would effective grassroots strategies for the protection of people and ecosystems from toxic exposures look like? Ought their overarching goal be a reduction in total population exposures and/or fewer chemical sales? Or should they aim for sweeping bans, such as of entire chemical classes? Or bans on specific usages (e.g., in all food or in all of agriculture)? Or on chemical use in particular geographic locations (e.g., in and around schools)? Or perhaps a better demand would be the dismantling (with or without replacement) of existing regulatory agencies, such as the culpable EPA. Or should chemical homicide be made a statutory crime? Or all of these together? And last but not least, how can such goals be achieved given the finances and politics of our age?

To make such decisions, the first task is to strip away the mythologies that currently surround the science of toxicology and the practice of chemical risk assessment. When we do this, we find that chemical regulations don't work. The chief reason, which is easy to demonstrate, is that the elementary experiments performed by toxicologists are incapable of generating predictions of safety that can usefully be applied to other species, or even to the same species when it exists in other environments, or to that species if it were to eat a different diet. Numerous scientific experiments have shown this to be so, as will be shown below. It means that the most basic element of supposedly scientific chemical risk assessment is scientifically invalid because extrapolation from specific experiments to the lived reality is not possible. For this reason, and many others too, the protection that those performing chemical risk assessments offer is a pretense. As I will show, risk assessment is not a reality; it is a complex illusion.

This diagnosis may seem depressing and make meaningful chemical control even more unlikely. However, it instead reveals a promising new vista of political opportunities to end pollution and create a sustainable world. Especially in the world of chemical pollution, the truth *can* set you free.

The ensuing discussion makes no significant effort to distinguish human health effects from effects on ecological systems. While these are often treated under separate regulatory jurisdictions, in practice, risks to people and ecosystems are difficult if not impossible to separate.

The story of the toxicological alarms surrounding BPA, which are diverse and unusually well-substantiated, make an excellent starting point for this task.

Ignoring the Full Toxicity of BPA

According to the scientific literature, exposure to BPA in adulthood has numerous effects. It leads to stem-cell and sperm-cell defects (humans), prostate cancer (humans), risk of breast cancer (humans and rats), blood-pressure rises (humans), and liver tumors and obesity (humans and mice) (Grun and Blumberg 2009,[7] Bhan et al. 2014,[8] Prins 2014[9]). However, fetuses exposed to BPA suffer from a significantly different spectrum of harms, ranging from altered organ development (in monkeys) to food intolerance (in humans) (Ayyanan et al. 2011,[10] Menard 2014,[11] vom Saal et al. 2014[12]). Also in humans, early BPA exposures can lead to effects delayed until much later in life, including psychiatric, social, and behavioral abnormalities indicative of permanently altered brain functions (Braun et al. 2011,[13] Perera et al. 2012,[14] Evans et al. 2014[15]).

The above examples are just a representative handful drawn from a much larger body of at least two hundred publications (some have estimated a thousand publications) finding harmful effects of BPA. The sheer quantity of results, diversity of species tested, consequences found, and scientific methodologies used represent a massive accumulation of scientific evidence that BPA is harmful (reviewed in Vandenberg et al. 2012).[16] The evidence against BPA being safe, in short, is as close to unimpeachable as science can manage.

Nevertheless, such a large evidence base indicates that anti-BPA campaigning has been only partially successful. All the bans[17] and the commercial withdrawals still ignore the implications of some of the most alarming scientific findings of all. For example, bans on baby bottles will not prevent fetal exposure, nor will they prevent harms that result even from very low doses of BPA.

Ignoring the Toxicity of BPS

The chemical most frequently used to make BPA-free products is called BPS. As its name implies, BPS is very similar in chemical structure to BPA (see Figure 1). However, BPS appears to be absorbed by the human body significantly more readily than BPA and is already detectable in 81 percent of Americans (Liao et al. 2012).

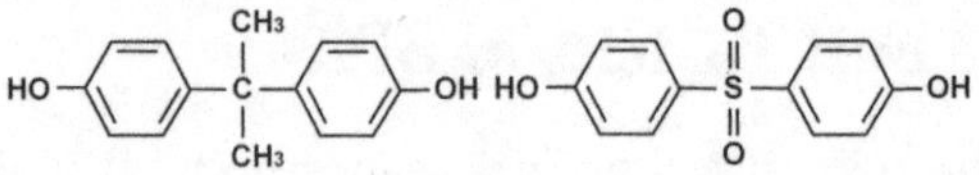

Figure 1

Research into the toxicology of BPS is still at an early stage, but BPS is now looking likely to be even more toxic than BPA (Rochester and Bolden, 2015). Like BPA, BPS has been found to interfere with mammalian hormonal activity. To a greater extent than BPA, BPS alters nerve-cell creation in the zebra fish hypothalamus and causes behavioral hyperactivity in exposed zebra fish larvae (Molina-Molina et al. 2013; Kinch et al. 2015). These latter results were observed at the extremely low chemical concentrations of 0.0068 uM. This is one-thousand-fold lower than the official U.S. levels of acceptable human exposure. The dose was chosen by the researchers since it is the concentration of BPA in the river that passes their laboratory.

Chemical Substitutions Are Business As Usual

The substitution of one synthetic chemical for another, wherein the substitute later turns out to be hazardous, is not a new story. Indeed, a great many of the chemicals that environmental campaigners nowadays oppose (such as Monsanto's best-selling herbicide Roundup) are still considered by many users to be "newer" and "safer" substitutes for chemicals (such as 2,4,5-T) that were once more widely used.

Thus, when the European Union banned the herbicide atrazine, Syngenta replaced it with terbuthylazine.[18] Terbuthylazine is chemically very similar and, according to University of California researcher Tyrone Hayes,[19] it appears to have similar ecological and health effects.

The chemical diacetyl was forced off the market for causing "popcorn lung."[20] However, it has been largely replaced by dimers and trimers of the same chemical. Unfortunately, the safety of these multimers is highly dubious since it is believed that, in use, they break down into diacetyl[21]—the very chemical that had been banned.

The Bt pesticides produced inside GMO crops are considered (by farmers and agribusiness) to be safer substitutes for organochlorine, carbamate, and organophosphate insecticides. These chemicals replaced DDT, which was banned in agriculture following Rachel Carson's *Silent Spring*. DDT was itself the replacement for lead arsenate.[22] Those chemicals are all examples of what are sometimes called "regrettable substitutions."[23] Others have called it "incrementalism." The problem with calling it "incrementalism" is that sometimes the new ones turn out to be even worse.

Chemical bans (or often manufacturer withdrawals) that precede such substitutions are nevertheless normally celebrated as campaigning victories. In a narrow sense they are, but the chemical manufacturers know that substitution is an ordinary part of business. Because weeds and pests become resistant and patents run out, they are usually looking for substitutes irrespective of any environmental campaigning.

Manufacturers also know that, since approvals and permits initially rely primarily on data supplied by the applicant (which is often anyway incomplete),[24] problems with safety typically manifest only later, as independent data and practical experience accumulate. Given this current system, it is almost inevitable that older (or more widely used) chemicals typically have a dubious safety record while newer ones are considered safer.

"Bad Actors": The Rotten Apple Defense in Toxicology

In these cycles of substituting one toxin for another, BPA is likely to become a classic.

Environmental health nonprofits become active participants in this toxic treadmill when they implicitly treat certain chemicals as rotten apples. Some even explicitly refer to particular chemicals as "bad actors."[25] The chemical "bad actor" framing strongly implies that the methods and institutions of chemical regulation are not at fault.

But we can ask, "In what chemical or biological sense can BPA or glyphosate be termed a bad actor? Is there, for example, a specific explanation for how it slipped through the safety net?"

The very short answer to this question relies on the aforementioned results: BPA impairs mammalian hormonal and reproductive systems, disrupts brain function, affects stem-cell development, causes obesity and probably cancer, and causes erectile dysfunction. Many hundreds of research papers attest that BPA's harmful effects are numerous, diverse, prolonged, reproducible, and found in many species. In short, they are easy to detect (e.g., vom Saal et al. 2014).[26]

So while hundreds of scientists outside the regulatory loop have found strong evidence for harm, the formal chemical regulatory system (FDA in the United States; EFSA in Europe) has never flagged BPA, even though, astonishingly and ironically, long before it was thought of as a plastic, BPA first came to the attention of science[27] in specific searches for estrogen-mimicking (i.e., hormone-disrupting) compounds. And despite the overwhelming nature of the published evidence, regulators *still* resist concluding that BPA is a health hazard.[28] And so the clear answer to the "bad actor" question is that there is no special reason why BPA should have slipped through the regulatory process; instead, the case of BPA strongly suggests a different explanation: a dysfunctional regulatory system.

Framing the problem of pollution as being caused by a few "bad actor" chemicals is equally inconsistent with the facts in other cases too. Chemical regulatory systems initially approved chemicals but have sometimes later banned or restricted them (and always under public pressure): atrazine, endosulfan, Roundup (glyphosate), lindane, methyl bromide, methyl iodide, 2,4,5-T, chlorpyrifos, DDT, and others. Many additional chemicals are strongly implicated as harmful by extensive and compelling independent scientific evidence that has so far not been acted on. And of course, chemical regulators have graduated whole classes of "bad actors": the organophosphate pesticides, PCBs, organochlorine pesticides, chlorofluorocarbons, neonicotinoids, phthalates, flame retardants, perfluorinated compounds, and so on.

How many bad actors ought it to take before we instead indict the whole show?

Chemical Regulation in Theory and Practice: The Limits of Toxicology

An alternative approach to judging regulatory systems by their results is to analyze them directly and assess their internal logic and rigor. Thus one can ask what is known about the technical limitations of toxicology and the overall scientific rigor of chemical risk assessment. And, second, one can direct attention to the social and institutional practices of chemical regulation. Are chemical risk assessments, for example, being applied by competent and well-intentioned institutions?

The technical limitations of chemical risk assessment are rarely discussed in detail (but see Buonsante et al. 2014).[29] A full discussion would be lengthy, but some of the most important limitations are outlined in the paragraphs below.

The standard assays of toxicology, called LD-50 or ED-50, involve the administration (usually oral feeding) of chemicals in short-term tests of up to ninety days to defined strains of organisms (most often rats or mice). These test organisms are of a specified age and are fed standardized diets.

The results are then extrapolated to other doses, other age groups, and other environments. Such experiments are used to create *estimates of harm*. Together with *estimates of exposure* they form the essence of chemical risk assessment. When specific chemicals are flagged as warranting further scrutiny, other techniques may be brought to bear, which may include epidemiology, cell-culture experiments, and biological modeling. But the basis of risk assessment is always the estimation of exposure and the estimation of harm. To say that both estimates are prone to error is an understatement.

Part I: Limits to Estimating Chemical Exposures

Fifty years ago, no one knew that many synthetic chemicals would evaporate at the equator and condense at the poles, from where they would enter polar ecosystems. Neither did scientists appreciate that all synthetic fat-soluble compounds sufficiently long-lived would bioaccumulate as they rose up the food chain and thus reach concentrations inside organisms sometimes many millions of times above background levels. And until recently was it not understood that sea creatures such as fish and corals would become major consumers of the plastic particles flushed into rivers.[30] These misunderstandings are all examples of historic errors in estimating real-world exposures to toxic substances.

A general and broad limitation of these estimates is that real-world exposures are very complex. For instance, commercial chemicals are often impure or not well defined. Thus PVC plastics are a complex mixture of polymers and may be further mixed with cadmium or lead (in varied concentrations).[31] One implication of this is that it is impossible for experiments contributing to risk assessment to be realistic: actual exposures are always unique to individual organisms and vary enormously in their magnitude, duration, variability, and speed of onset, all of which influence the harm they cause.

Additionally, many regulatory decisions do not recognize that exposures to individual chemicals typically come from multiple sources. This failing is often revealed following major accidents or contamination events. Regulatory agencies will assert that actual accident-related doses do not exceed safe limits. However, such statements usually ignore the reality of multiple sources; because regulations function in effect as permits to pollute, many affected people may have already been receiving significant exposures to that chemical prior to the accident.

Returning to the specific case of BPA, no one appreciated until 2013 that the main route of exposure to BPA in mammals is absorption through the mouth (under the tongue) and not the gut (across the villi of the small intestine). The mouth is an exposure route whose veinous blood supply *bypasses* the liver, and this allows BPA to circulate unmetabolized in the bloodstream (Gayrard et al. 2013).[32] Oral exposure accounts for most of the BPA absorbed into the human body and the unexpectedly high BPA concentrations. Before this was known, many toxicologists explicitly denied the plausibility of measurements showing high BPA concentrations in human blood. They had assumed that BPA was absorbed via the gut and rapidly degraded in the liver.

Part II: Limits to Estimating Harms

Similarly significant obstacles are faced in estimating harm. Many of these obstacles originate from the obvious fact that organisms and ecosystems are enormously biologically diverse.

The solution adopted by chemical risk assessment is to extrapolate. It is also assumed that extrapolation allows the results of one or more experiments to cover other species and other environmental conditions.

Many of the assumptions required for such extrapolations, however, have never been scientifically validated. Lack of validation is most obvious for species not yet discovered or those that are endangered. But in other cases, they are actively known to be invalid[33] (e.g. Seok et al. 2013).[34]

For example, in their responses to specific chemicals, rats often do not extrapolate to humans.[35] Indeed, they often do not extrapolate even to other rats. Thus individual strains of rats respond differently, but also young and old rats give different responses. So do male and female rats (vom Saal et al. 2014).[36] So too do rats fed nonstandard diets (Mainigi and Campbell 1981).[37]

Even more extreme extrapolations are employed in ecological toxicology. For example, data on adult honeybees is typically extrapolated to every stage of the bee life cycle, to all other bee species, and sometimes to all pollinators, without the experimenters citing any supporting evidence. Such extrapolations may seem absurd, but they are the primary basis of the claim that chemical risk assessment is comprehensive.

There are many other limits to estimating harm. Until it was too late, scientists were not aware that a human with an eighty-year lifespan could have a window of vulnerability to a specific chemical as short as four days.[38] Neither was it known that the effects of chemicals could be strongly influenced by the time of day they are ingested.[39]

Another crucially important limitation is that, for budgetary and practical reasons, toxicologists necessarily focus on a limited number of specific *endpoints*. An endpoint is whatever characteristic the experimenter chooses to measure. Typical endpoints are death (mortality), cancers, organism weight, and organ weights, but endpoints can even be more subtle measures like neurotoxicity. Whole politics is associated with the choice of endpoints, which reflects their importance in toxicology, including allegations that endpoints are sometimes chosen for their insensitivity rather than their sensitivity. But the inescapable point is that no matter what endpoints are chosen, there is a much vaster universe of unmeasured endpoints, such as learning defects, immune dysfunction, reproductive dysfunction, and multigenerational effects. Ultimately, most potential harms don't get measured by toxicologists and so are missing from risk assessments.

Another example of the difficulty of estimating real-life harms is that organisms are exposed to mixtures of toxins (Goodson et al. 2015).[40] The issue of toxin mixtures is extremely important (Kortenkamp 2014).[41] All real-life chemical exposures occur in combinations, either because of previous exposure to pollutants or because of the presence of natural toxins. Many commercial products, moreover, such as pesticides, are only available as formulations (i.e., mixtures) whose principal chemical purpose is to enhance the potency of the product, such as the glyphosate coformulant POE-tallowamine, now widely banned. Risk assessments, however, just test the "active ingredient" alone (Richard et al. 2005).[42]

Consider too that all estimates of harm depend fundamentally on the assumption of a sigmoid dose-response relationship that, for each chemical, can be linearized for the purpose of analysis. This assumption is necessary to estimate harms of doses that are higher, lower, or even in-between tested doses. This assumption is rarely tested, yet for numerous toxins (notably endocrine-disrupting chemicals) a linear dose-response relationship has been disproven. Thus, a key question for any risk assessment is whether the assumption is reliable for the novel compound under review (reviewed in Vandenberg et al. 2012).[43]

Replacing Doubts with False Certainty

To summarize, the process of chemical risk assessment relies on estimating real-world exposures and their potential to cause harm by extrapolating from one or a few simple laboratory experiments. The resulting estimates come with enormous uncertainty. In many cases, the results have been extensively critiqued and shown to be either dubious or actively improbable (Chandrasekera and Pippin 2013).[44] Yet extrapolation continues—even though we know that the various errors must multiply—because the alternative is to actually measure these different species using different mixtures and under different circumstances. Given the challenges this would entail, the continued reliance on simplistic assumptions is understandable.

Nevertheless, one might have thought that such important limitations and assumptions would be frequently noted as caveats to risk assessments. They should be, but they are not. Following the United Kingdom's traumatic and disastrous outbreak of BSE (mad cow disease) in the 1980s, during which most of the UK population was exposed to infectious prions following highly questionable scientific advice, this exact recommendation was made in the Phillips report.[45] Lord Phillips proposed that such caveats should be *specifically* explained to non-scientific recipients of scientific advice. In practice, however, Phillips changed nothing. Only lip service, at best, is applied to this recommendation. Collecting and explaining the necessary caveats would be extraordinarily time-consuming and would undermine—in fact destroy—the credibility of the risk assessment process.

When an unusual scientific document does discuss the limitations of chemical risk assessment (such as this description of the failure of interactions between pesticides to extrapolate between closely related species),[46] it rapidly becomes obvious just how much the knowledge and understanding available to us are dwarfed by actual biological and overall system complexities. As any biologist ought to expect, ordinary life situations will multiply errors and overwhelm risk assessment standard assumptions.

For good reason, many scientific experts are therefore concerned about the number and quantity of man-made chemicals in our bodies. Recently, the International Federation of Gynecology and Obstetrics linked chemical exposure to the emergence of new diseases and disorders. They specifically mentioned obesity, diabetes, hypospadias, and reproductive dysfunction and noted: "The global health and economic burden related to toxic environmental chemicals is in excess of millions of deaths" (Di Renzo et al. 2015).[47] The Federation acknowledged this to be an underestimate. And the estimate does not count disabilities. Given that we all have man-made toxins in our bodies, that many of these chemicals have no safe dose, and that the obvious consequences are frequently found in otherwise normal members of the population, there is little choice but to conclude that we all are poisoned, our lives lessened, and many of us will die prematurely from the effects of those chemicals.

Conflicts of Interest in Chemical Risk Assessment

In addition to the technical difficulties, there is also the problem that the scientists who produce scientific knowledge often have financial (and other) conflicts of interest. Conflicts, we know, lead to biases that affect science well before it is incorporated into risk assessment (e.g., Lesser et al. 2007).[48]

A fascinating example of apparent unconscious bias comes from a recent survey of scientific publications on the non-target effects of pesticidal GMO (Bt) crops in outdoor experiments. It was commissioned by the Dutch government (COGEM 2014).[49] The report observed that researchers who found negative consequences of GMO Bt crops were disregarding their own findings, even when these were statistically significant. Even more interesting to the Dutch authors was that the rationales offered for doing so were oftentimes illogical. Typically, researchers were using experimental methods specialized for detecting ecotoxicological effects that were transient or local, but when such effects were found, the researchers dismissed the significance of their own results for being either transient or local. The COGEM report represented prima facie evidence that researchers within a whole academic discipline were avoiding conclusions that would throw doubt on the wisdom of using GMO Bt crops. Apparently, the Bt researchers had a prior ideological commitment to finding no harm.

Corporate Capture and Institutional Dysfunction

Chemical regulation occurs primarily within a relatively small number of governmental or "independent" regulatory institutions.

Of these, the United States Environmental Protection Agency (EPA) is the most prominent and widely imitated example. The EPA has a variety of institutional and procedural defects that prevent it being an effective regulator. Perhaps the best known of these is to allow self-interested chemical corporations to conduct the experiments and provide the data for risk assessment. This lets them summarize (or even lie about)[50] the results. As was once pointed out by Melvin Reuber, former EPA consultant, it is extraordinarily easy for an independent commercial testing operation to bias or fix the result of a typical toxicology study for the benefit of a client.[51]

How the EPA first allowed corporations to generate and submit their own regulatory data is a story well worth knowing.

In the 1980s, Industrial Bio-Test Laboratories (IBT)[52] was the largest independent commercial testing laboratory in the United States. FDA scientist Adrian Gross discovered that IBT (and other testing companies) were deliberately, consistently, and illegally misleading the EPA and the FDA[53] about their results. Aided by practices such as the hiring of a chemist from Monsanto—the manufacturer of PCBs—to carry out the PCB toxicity studies, IBT created an illusion of chemical safety for numerous pesticides and other chemicals. Many are still in use. They include Roundup, atrazine, and 2,4-D, all commonly used in U.S. agriculture. Between them, Canadian regulators drew up a list of 106 questionable chemical registrations, and the FDA identified 618 separate animal studies as being invalid due to "numerous discrepancies between the study conduct and data."[54] Both regulators suppressed their findings.

Senior IBT managers were jailed, but what the scandal had revealed was that whenever results showed evidence of harm—which was often—misleading regulators was standard practice.

More remarkable even than the scandal was the EPA's response. Instead of bringing testing in-house, which would seem the logical response to a system-wide failure of independent commercial testing, EPA instead created a Byzantine system of external reporting and corporate summarizing for the results of chemical toxicity studies submitted to it. The result is a secretive and bureaucratic maze that ensures no EPA employee ever sets eyes on the original laboratory experiments, or even the primary data. Only a handful of EPA staff can access even the summarized results of "independent" chemical testing. This system has the consequence of excluding any formal possibility that whistleblowing on the part of Federal employees or FOIA requests (from outsiders) might reveal fraudulent or otherwise problematic toxicology tests. After the IBT scandal, the EPA thus calculatedly turned a blind eye to any potential future wrongdoing in the world of chemical testing with the full knowledge that the chemical regulatory system it oversaw was systemically corrupt.

Probably more familiar to readers is what is called "regulatory capture."[55] This takes many forms, from the offering to public servants of favors and future jobs, to the encouragement of top-down political interference with regulatory agencies. The culminating effect is to ensure that political will within agencies to protect the public is diluted or lost.

Regulatory capture can become a permanent feature of an institution. For example, OECD member countries have an agreement called the Mutual Acceptance of Data (MAD).[56] MAD is appropriately named. It has the effect of explicitly excluding from regulatory consideration most of the peer-reviewed scientific literature (Myers et al. 2009a).[57] The purported goal of MAD was to elevate experimental practices by requiring certification via Good Laboratory Practice (GLP),[58] which was a procedure introduced after the IBT scandal (Wagner and Michaels 2004).[59] GLP is a mix of management and reliability protocols standard in industrial laboratories but rare in universities and elsewhere. However, the consequence of accepting MAD has been to specifically exclude from regulatory consideration evidence and data not produced by industry.

The MAD agreement explains much of the regulatory inaction over BPA. Because of MAD, the FDA (and also its European equivalent, the European Food Safety Authority) have ignored the hundreds of peer-reviewed BPA studies—since they are not GLP—in favor of just two by industry. These two industry studies, whose credibility and conclusions have been publicly challenged by independent scientists, showed no ill effects of BPA (Myers et al. 2009b).[60]

Whistleblowing at the EPA

Various EPA whistleblowers have described in detail the specifics of their former organization's capture by branches of the chemical industry.

Whistleblower William Sanjour[61] has described how regulatory failure was ensured by the organizational structure imposed on the EPA at its Nixon-era inception. The structure of EPA is inherently conflicted since it has the dual functions of both writing *and* enforcing regulations. Unwillingness to enforce high standards led his superiors to order Sanjour to write deliberate loopholes into those regulations.[62] More recently, the EU's EFSA was similarly caught proposing loopholes for new regulations on endocrine disrupting chemicals. Inserting loopholes is standard practice in the writing of chemical safety regulations.[63]

In the same article,[64] Sanjour proposed that since corporate capture renders them useless, the public would be better off with no regulatory agencies. In a similar vein, former EPA pesticide scientist Evaggelos Vallianatos[65] called his former employer, at book length, the "polluter's protection agency."[66] Another EPA whistleblower, David Lewis,[67] this time at EPA's Office of Water, has shown in court-obtained documents that EPA scientists buried evidence and even covered up deaths so as to formulate regulations that would permit land application of sewage sludge. This sludge was routinely contaminated with pathogens, heavy metals, industrial chemicals, pharmaceuticals, flame retardants, and other known hazardous substances. The corruption around sewage sludge regulations extended well beyond the EPA. It encompassed other federal agencies, several universities, the National Academy of Science,[68] and municipalities. David Lewis eventually obtained a legal judgment that the City of Augusta, Georgia, had "fudged"[69] the toxicity testing of its own sewage sludge in order to meet EPA guidelines. The city had done so at the request of EPA.

In another recent case, DeSmogBlog obtained,[70] through a Freedom of Information Act request (FOIA), internal documents showing how EPA offered access to its fracking study plans: "[Y]ou guys are part of the team here," one EPA representative wrote to Chesapeake Energy as they together edited study planning documents in October 2013, "please write things in as you see fit."[71]

Even more recently, EPA whistleblower and chemist Dr. Cate Jenkins and the nonprofit Public Employees for Environmental Responsibility

(PEER)[72] successfully sued the EPA[73] for suppressing information about toxic effects on 9/11 first responders. The case ended with a judgment showing that the EPA had, among numerous egregious acts, created fake email accounts (including for EPA head Lisa Jackson) to evade accountability. According to Judge Chambers, the EPA: "Failed, and failed miserably, over an extended course of time in complying with its discovery obligations and . . . Court discovery orders."

Judge Chambers also found that the EPA worked a "fraud on the Court" through numerous "false claims" and inaccurate claims of privilege, which upon examination applied to "none of the documents provided." The judge also found that the EPA deliberately and illegally destroyed an unknown number of documents which should have been under a litigation hold.

The ultimate effect of these institutional defects is that chemical risk assessments in the United States and the European Union have a safety bar for approval that is so low that regulators virtually never decline to approve a chemical. In contrast, the exact same institutions use standards for taking any chemical *off* the market that are so high that such an event nearly never happens. Yet if both standards were based purely on science, as they claim to be, both bars would be at the same height.

This double standard represents the overwhelming bias in the system. At every stage of chemical risk assessment—from the funding of research[74] to the ultimate decision to approve a chemical—the process is dominated by commercial concerns and not by science (as was recently shown yet again).[75]

Chemical Risk Assessment: Can the Show Be Salvaged?

It therefore seems clear that to frame individual chemicals as "bad actors" is incorrect. Chemical risk assessments themselves are the problem. Thus we can perfectly explain why approved chemicals accumulate red flags when exposed to the scientific process, and also why those that replace them are no less harmful. Specific chemicals like BPA are thus the messengers and shooting them one by one is not only pointless, it is counterproductive. It distracts and detracts from the infinitely more important truth, that the institutions, the methods, and thus the entire oversight of chemical regulation is failing in what it claims to do, which is to protect us from harm.

Importantly, chemical regulatory systems are not just broken, they are unfixable. Even with the best intentions, such as the full cooperation of all the institutions mentioned here and of the entire academic research community, remedying the technical problems would be a task that is beyond Herculean.

Consider just one of these, the testing of a chemical in combination with others. The testing of mixtures is an improvement, often suggested by NGOs. Thousands of scientific studies show that this is an important consideration. The pesticide chlordecone, for example, increases the toxicity of an "otherwise inconsequential" dose of the common contaminant carbon tetrachloride by sixty-seven-fold in rats (Curtis et al. 1979).[76]

To test mixtures properly, however, would be astonishingly expensive and also enormously costly toward experimental animals. According to the U.S. National Toxicology Program, standard thirteen-week studies of the interactions between just twenty-five chemicals would require thirty-three million experiments costing $3 trillion. This is because each chemical needs to be tested against all possible combinations of the others. To study mixtures of all eleven thousand chlorinated chemicals in commerce would require 10^{3311} experiments. This is more experiments than atoms in the universe. Our entire planet would have to devote itself to animal experimentation, and the work would still not be done by the end of time (Yang 1994). Even then, we would only know the toxicity of organochlorines toward a single test species. Would the results be extrapolative to any other species? Well, we could buy another planet and test it!

Imagine also that an adequate test for synthetic chemicals were devised and it were run by competent institutions. Would any chemical pass? The multiple harms of the single chemical BPA, plus the frequency with which chemical substitutes turn out later to be harmful, and plenty of other data, suggests it is likely that few chemicals would pass. This conclusion, of course, contradicts the crucial presumption of innocence that underlies all chemical regulation. What is so improbable, after all, about proposing that all man-made chemicals cause dysfunction at low doses in a significant subset of all the biological organisms on earth?

Strategizing for Success

The implications of this are many, but the one of specific importance to environmental health campaigners is that organizing for a ban on a specific hazardous chemical, even of glyphosate, is a strategic error. If chemical risk assessment is ineffective, then demanding a ban is pointless because it would result only in the substitution of a chemical that is no better. Even worse, if chemical risk assessment is ineffective, such campaigns undermine the wider cause of inciting caution about chemical use because they imply, falsely, that chemical regulations protect the public or limit pollution.

If the public heard that chemical regulations were effective from the chemical industry alone they probably would not believe it. However, they hear it from the entire environmental movement. Why, they perhaps reason, would the environmental movement pretend chemical testing was effective if it wasn't? And indeed, mainstream environmental nonprofits, for the most part, traditionally reinforce this message still further whenever they call for *more* testing.

In the light of this understanding, if they accept the accumulated scientific evidence, environmental and public health advocates who campaign for bans or restrictions on single chemicals have an opportunity to substantively rethink their strategies and reframe their activities. This doesn't necessarily mean abandoning discussion of individual chemicals. What it does mean is explicitly framing those specific chemicals not as "bad actors" but as symptoms of a much bigger problem of incompetent and dysfunctional regulation, with all that implies.

It is a challenge. But it is also a tremendous opportunity. Having facts that are starker and analysis that is more scientific and more rigorous creates a superior and more powerful basis upon which to organize and strategize. It brings more ambitious environmental health goals within reach. Advocates can choose from a broader range of possible approaches and more easily engage a broadened segment of the population. They can place clear and obvious intellectual distance between their own realistic strategies for protecting the public and the planet and the plainly inadequate views of the chemical industry. For example, it is surely easier to explain to a layperson the generic absurdities of chemical risk assessment than it is to explain the toxicological intricacies of Roundup or 2,4-D, especially one chemical (of 80,000) at a time. They say the truth can set you free, but in the world of toxic campaigning it is a strategy that has hardly been tried yet.

In the late 1990s, Greenpeace USA adopted the novel campaigning position that all chlorinated hydrocarbons should be banned, in part on the grounds that every one so far investigated had proven toxicologically problematic. In threatening thousands of products of the chemical industry with a strategic goal that had a realistic chance of significantly enhancing the quality of our environment, Greenpeace took chemical campaigning to a new level.

Greenpeace was hit by an unprecedented campaign of corporate espionage.[77] Their offices were bugged, and their computers were hacked; they were infiltrated by phony volunteers, and more. The chemical industry was spooked. Greenpeace eventually backed off, but by raising the stakes and making their case with science, they had shown a way.

The book *Pandora's Poison*[78] elaborates on some of the ambitious ideas for eradicating pollution that Greenpeace tried but never in the end adequately road tested. It is time to learn those lessons and move chemical hazard campaigning from being mere gesture politics to a place more in keeping with the high human and ecological stakes. That is, far outside the comfort zone of the chemical industry.

* * *

Jonathan Latham, PhD, is the executive director of the Bioscience Resource Project and the editor of *Independent Science News*.

References

Ayyanan, A., O. Laribi, S. Schuepbach-Mallepell, C. Schrick, M. Gutierrez, T. Tanos, G. Lefebvre, J. Rougemont, O. Yalcin-Ozuysal, and C. Brisken. "Perinatal Exposure to Bisphenol A Increases Adult Mammary Gland Progesterone Response and Cell Number." *Molecular Endocrinology* (2011); DOI: 10.1210/me.2011–1129.

Bae, S., and Y. C. Hong. "Exposure to Bisphenol A From Drinking Canned Beverage Increases Blood Pressure: Randomized Crossover Trial." *Hypertension* 62, no. 2 (February 2015): 313–19.

Bhan, A., I. Hussain, K. I. Ansari, S. Bobzean, L. I. Perrotti, and S. S. Mandal "Bisphenol-A and Diethylstilbestrol Exposure Induces the Expression of Breast Cancer Associated Long Noncoding RNA HOTAIR In Vitro and In Vivo. *Journal of Steroid Biochemistry and Molecular Biology* 141 (May 2014): 160–70.

Braun, J.M., A. E. Kalkbrenner, A. M. Calafat, K. Yolton, X. Ye, K. N. Dietrich, and B. P. Lanphear "Impact of Early Life Bisphenol A Exposure on Behavior and Executive Function in Children." *Pediatrics*, 128 (2011): 873–82.

Buonsante, Vito A., H. Muilerman, T. Santos, C. Robinson, and A. C. Tweedale (2014) "Risk Assessment's Insensitive Toxicity Testing May Cause It to Fail." *Environmental Research* 135 (2014): 139–47.

Chandrasekera, P. C., and J. J. Pippin. "Of Rodents and Men: Species-Specific Glucose Regulation and Type 2 Diabetes Research. *ALTEX* 31, no. 2 (2014): 157–76.

Curtis, L. R., W. Lane Williams, and Harihara M. Mehendale (1979) "Potentiation of the Hepatotoxicity of Carbon Tetrachloride Following Preexposure to Chlordecone (Kepone) in the Male Rat." *Toxicology and Applied Pharmacology* 51, no. 2 (November 1979): Pages 283–93.

Evans, S. F., R. W. Kobrosly, E. S. Barrett, S. W. Thurston, A. M. Calafat, B. Weisse, R. Stahlhut, K. Yolton, and S. H. Swan. (2014) "Prenatal Bisphenol A Exposure and Maternally Reported Behavior in Boys and Girls." *NeuroToxicology* 45 (2014): 91–99.

Gayrard, V., M. Z. Lacroix, S. H. Collet, C. Viguié, A. Bousquet-Melou, P. L. Toutain, and N. Picard-Hagen "High Bioavailability of Bisphenol A from Sublingual Exposure." *Environmental Health Perspectives* 121, no. 8 (August 2013): 951–56.

Goodson, W. H., et al. (2015) "Assessing the Carcinogenic Potential of Low-Dose Exposures to Chemical Mixtures in the Environment: The Challenge Ahead. *Carcinogenesis* 36, Supplement 1 (June 2015): S254–96.

Kinch, C., K. Ibhazehiebo, J. H. Jeong, H. R. Habibi, and D. M. Kurrasch "Low-Dose Exposure to Bisphenol A and Replacement Bisphenol S Induces Precocious Hypothalamic Neurogenesis in Embryonic Zebrafish." *Proceedings of the National Academy of Sciences* 112, no. 5 (February 2015): 1475–80, doi: 10.1073/pnas.1417731112.

Kortenkamp, A. (2014) "Low Dose Mixture Effects of Endocrine Disrupters and Their Implications for Regulatory Thresholds in Chemical Risk Assessment." *Current Opinion in Pharmacology* 19 (December 2014): 105–11.

Lesser, L. I., C. B. Ebbeling, M. Goozner, D. Wypij, and D. S. Ludwig.

(2007) "Relationship between Funding Source and Conclusion among Nutrition-Related Scientific Articles." *PLOS Medicine* 4, no. 1 (January 2007): e5, DOI: 10.1371/journal.pmed.0040005.

Liao, C., F. Liu, H. Alomirah, V. Duc Loi, M. Ali Mohd, H. B. Moon, H. Nakata, and K. Kannan "Bisphenol S in Urine from the United States and Seven Asian Countries: Occurrence and Human Exposures." *Environmental Science and Technology* 46 (2012): 6860–6866.

Mainigi, K. D., and T. C. Campbell (1981) "Effects of Low Dietary Protein and Dietary Aflatoxin on Hepatic Glutathione Levels in F-344 Rats." *Toxicology and Applied Pharmacology* 59 (1981): 196–203.

Melzer, D., Nicholas J. Osborne, William E. Henley, Ricardo Cipelli, Anita Young, Cathryn Money, Paul McCormack, Robert Luben, Kay-Tee Khaw, Nicholas J. Wareham, and Tamara S. Galloway "Urinary Bisphenol: A Concentration and Risk of Future Coronary Artery Disease in Apparently Healthy Men and Women." *Circulation* 125 (2012): 1482–90.

Menard, S., L. Guzylack-Piriou, M. Leveque, V. Braniste, C. Lencina, M. Naturel, L. Moussa, S. Sekkal, C. Harkat, E. Gaultier, V. Theodorou, and E. Houdeau. "Food Intolerance at Adulthood after Perinatal Exposure to the Endocrine Disruptor Bisphenol A." *FASEB Journal* 28 (2014): 4893–900.

Molina-Molina, J. M., Esperanza Amaya, Marina Grimaldi, José-María Sáenz, Macarena Real, Mariana F. Fernández, Patrick Balaguer, and Nicolás Olea. "In Vitro Study on the Agonistic and Antagonistic Activities of Bisphenol-S and Other Bisphenol-A Congeners and Derivatives via Nuclear Receptors." *Toxicology and Applied Pharmacology* 272 (2013): 127–136.

Myers, J. P., et al. "Why Public Health Agencies Cannot Depend on Good Laboratory Practices as a Criterion for Selecting Data: The Case of Bisphenol A." *Environmental Health Perspectives* 117 (2009a): 309–15.

Myers, J. P, T. H. Zoeller, and F. vom Saal. "A Clash of Old and New Scientific Concepts in Toxicity, with Important Implications for Public Health." *Environmental Health Perspectives* 117 (2009b): 1652–55.

Perera, F., Julia Vishnevetsky, Julie B. Herbstman, Antonia M. Calafat, Wei Xiong, Virginia Rauh, and Shuang Wang. "Prenatal Bisphenol A

Exposure and Child Behavior in an Inner-City Cohort." *Environmental Health Perspectives* 120 (2012): 1190–94.

Prins, G. S., Wen-Yang Hu, Guang-Bin Shi, Dan-Ping Hu, Shyama Majumdar, Guannan Li, Ke Huang, Jason Nelles, Shuk-Mei Ho, Cheryl Lyn Walker, Andre Kajdacsy-Balla, and Richard B. van Breemen. "Bisphenol A Promotes Human Prostate Stem-Progenitor Cell Self-Renewal and Increases In Vivo Carcinogenesis in Human Prostate Epithelium." *Endocrinology* 155, no. 3 (March 2014): 805–17.

Richard, S., S. Moslemi, H. Sipahutar, N. Benachour, and G. E. Seralini. "Differential Effects of Glyphosate and Roundup on Human Placental Cells and Aromatase." *Environmental Health Perspectives* 113, no. 6 (June 2005): 716–20.

Rochester, J. R., and A.L. Bolden (2015) "Bisphenol S and F: A Systematic Review and Comparison of the Hormonal Activity of Bisphenol A Substitutes." *Environmental Health Perspectives* 123, no. 7 (July 2015): 643–50, DOI:10.1289/ehp.1408989.

Rubin, B. S., M. K. Murray, D. A. Damassa, J. C. King, and A. M. Soto. "Perinatal Exposure to Low Doses of Bisphenol A Affects Body Weight, Patterns of Estrous Cyclicity, and Plasma LH Levels." *Environmental Health Perspectives* 109 (2001): 675–680.

Seok, J., et al. "Genomic Responses in Mouse Models Poorly Mimic Human Inflammatory Diseases. *Proceedings of the National Academy of Sciences* 110, no. 9 (February 2013): 3507–12.

Tarapore, P., Jun Ying, Bin Ouyang, Barbara Burke, Bruce Bracken, and Shuk-Mei Ho. "Exposure to Bisphenol A Correlates with Early-Onset Prostate Cancer and Promotes Centrosome Amplification and Anchorage-Independent Growth In Vitro." *PLoS One* 9, no. 3 (March 2014): e90332, DOI: 10.1371/journal.pone.0090332.

Vandenberg, Laura N., Theo Colborn, Tyrone B. Hayes, Jerrold J. Heindel, David R. Jacobs Jr., Duk-Hee Lee, Toshi Shioda, Ana M. Soto, Frederick S. vom Saal, Wade V. Welshons, R. Thomas Zoeller, and John Peterson Myers (2012) "Hormones and Endocrine-Disrupting Chemicals: Low-Dose Effects and Nonmonotonic Dose Responses." *Endocrine Reviews* (2012), DOI: http://dx.doi.org/10.1210/er.2011–1050.

vom Saal, F., Catherine A. VandeVoort, Julia A. Taylor, Wade V. Welshons, Pierre-Louis Toutain, and Patricia A Hunt. "Bisphenol A (BPA)

Pharmacokinetics with Daily Oral Bolus or Continuous Exposure via Silastic Capsules in Pregnant Rhesus Monkeys: Relevance for Human Exposures." *Reproductive Toxicology* 45 (2014): 105–16.

Wagner, Wendy, and David Michaels. "Equal Treatment for Regulatory Science: Extending the Controls Governing the Quality of Public Research to Private Reseach." *American Journal of Law & Medicine.* 30 (2004): 119.

Yang, R. S. H. (1994) "Toxicology of Chemical Mixtures Derived from Hazardous Waste Sites or Application of Pesticides and Fertilizers." In *Toxicology of Chemical Mixtures*, ed. R. S. H. Yang. 99–117. Cambridge, MA: Academic Press, 1994.

Yuan, Z., S. Courtenay, R. C. Chambers, and I. Wirgin. "Evidence of Spatially Extensive Resistance to PCBs in an Anadromous Fish of the Hudson River." *Environmental Health Perspectives* 114 (2006): 77–84.

13

Glyphosate on Trial: The Search for Toxicological Truth

By Sheldon Krimsky, PhD

In 2015, the International Agency for Research on Cancer (IARC), an independent research group under the auspices of the World Health Organization, issued its toxicological evaluation of the herbicide glyphosate. IARC concluded: "Glyphosate is probably carcinogenic to humans."[1] The release of its report created a firestorm of activity throughout the world. Glyphosate is the most widely used herbicide on the planet. Since it first came into use in 1974, it has been evaluated many times by a number of governmental bodies including the U.S. Environmental Protection Agency (EPA), the European Union, and the European Food Safety Authority (EFSA), where it was found safe for human use with no strong evidence that it was a probable human carcinogen.[2]

Monsanto, manufacturer of one formulation of glyphosate called Roundup™, sold both for farm and domestic use, took aggressive action against the findings of the new study by funding counter studies, reviews, data reanalysis, lawsuits, and media campaigns to protect its product from IARC's negative report. On March 28, 2017, California's Office of Environmental Health Hazard Assessment (OEHHA) announced it was adding Roundup™ as well as other glyphosate-based weed killers to the state's Proposition 65 list of cancer-causing chemicals. In a letter to Monsanto, OEHHA wrote, "Proposition 65 required the listing of certain chemicals and substances" when found to be cancer causing agents. "Under the statute, case law and regulations, chemicals identified by IARC as carcinogens with sufficient evidence of carcinogenicity in humans or animals must be listed under Proposition 65."[3] Under the law, California was not required to, nor did it, do additional testing or risk analysis. Monsanto challenged OEHHA's listing; its challenge was defeated at the trial court and was appealed. If the final judicial approval is given for the glyphosate listing, companies will have to label the glyphosate-based weed killer as a "probable human carcinogen." On April 19, 2018, a California Appellate Court ruled that Monsanto's glyphosate herbicide can be labeled as a probable human carcinogen under Proposition 65.

Meanwhile, the toxicology of glyphosate is being debated within the scientific community. The goal of this chapter is to examine the different interpretations of the scientific literature on the health and safety of glyphosate and its formulations among herbicides. Why does one World Health Organization agency reach a decision that glyphosate is a probable human carcinogen while others, including the EPA, conclude, with equal confidence, that it is not a human carcinogen? How should consumers respond?

Before I answer these questions, I begin with the backstory—the story not found in the scientific literature but that can have a profound effect on it. The backstory is as important as the science because the evaluation of glyphosate is embedded in an intensely corporatized climate of agricultural chemicals. We have learned that some scientists are paid by companies to reach conclusions that support their profit margins. Unfortunately, there are always a minority of scientists who are willing to act as shills for a corporate benefactor.

Like the hidden story behind tobacco[4] and lead science,[5] the purpose of the corporate funding of glyphosate research was not to find the truth but to protect the product from regulation commensurate with its risk. We have learned much about rogue tobacco and lead science when the corporations involved in their manufacture and promotion, complying with court orders, released discovery documents during litigation. Similarly, Monsanto is being sued by hundreds of plaintiffs with claims that Roundup™ has harmed them, in some cases claiming it afflicted them with non-Hodgkin's lymphoma, a form of blood cancer.

The discovery documents released by Monsanto reveal a backstory of corporate malfeasance.[6] The documents provide evidence that Monsanto was engaged in ghostwriting (that is, writing articles for established scientists to submit as sole authors without revealing the company's role),[7] hiring contract research companies to undertake invalid toxicology studies, exerting undue influence on regulatory agencies,[8] threatening lawsuits against publishers and journals, creating so-called expert panels comprised of individuals who sign on to a study vetted and funded by Monsanto without acknowledging the company's role, hiding the fact that its own formulation for Roundup™ was never tested for its carcinogenic effects, and campaigning against the credibility of published scientific studies and their authors that reached conclusions it disliked.

These are just a portion of the corporate malfeasance cases revealed in the discovery documents. Similar attacks have been reported in Europe of scientists who have reported adverse effects of glyphosate or Roundup™.[9]

It should be understood that the controversy over the toxicity of glyphosate or Roundup™ is not a debate over neutral sectors of the scientific community at odds over the intricacies of toxicological methods. The issues are replete with political overtones. Yet, we can discuss the front story of glyphosate toxicology as it appears in the scientific literature. There is much to be learned about the way honest science should proceed.

First, I shall discuss the background of IARC and its standing in the world health community. Second, I shall give a short history of glyphosate and its use as a weed killer. Third, I shall discuss the basis of IARC's decision summarizing the evidence it used. Fourth, I shall summarize the criticisms of IARC's decision, including its evidence and its methods of analysis, and the role of corporate science in the support of glyphosate. Finally, I shall conclude by examining why different agencies reach different conclusions and where that leads the cautious consumer and environmentalist. My conclusions about which science is trustworthy is based on the transparency of those doing the research, the ethical standards of the agencies interpreting the weight of evidence, whether the research is done without funding from stakeholders, and the precautionary principle—when sufficient, albeit not definitive, evidence dictates regulatory protection.

IARC's Origins

IARC was established in May 1965 by a resolution of the World Health Assembly, under Article 18(k) of the World Health Organization (WHO) constitution, as a specialized cancer agency of WHO. Its fifty-year anniversary volume noted, "The International Agency for Research on Cancer (IARC) is the outcome of an initiative by a group of leading French public figures, who succeeded in persuading President [Charles] de Gaulle to accept a project to lighten humanity's ever growing burden of cancer."[10]

An open letter signed by thirteen leaders in the fields of cancer research, physics, journalism, engineering, architectures, and religion was delivered to Elysee Palace on November 7, 1963; it jump-started the initiative for a world cancer agency. *Le Monde* described the event in a headline: "Pour développer la lutte contre le cancer des personnalités françaises lancent un appel en faveur d'une institution internationale de recherche pour la vie"[11] (In developing the struggle against cancer some French personalities have issued a call for an international organization to look for ways to save lives.)

The idea behind the letter was that nations should re-direct military budgets to fund the battle against cancer and to promote international collaboration in cancer research. It called for an allocation of 0.5 percent of the military budgets of participating nations. The initial participating countries were Germany, France, Italy, the United Kingdom, and the United States, soon followed by Australia and the Soviet Union. By 1972, ten nations signed on to IARC.

IARC launched its monographs program in 1971. Its aim was "to develop an instrument capable of evaluating the best evidence available at a given time on carcinogenic agents, in order to provide a sound scientific basis for cancer prevention."[12]

Science and ethics were at the core of IARC's approach to understanding cancer. On the science side, the agency has supported the best expert knowledge in publishing statistical methods guiding its analysis of case-control studies and cohort studies, two of the seminal methods for understanding cancer etiology.[13] IARC emphasized two innovative features of its work: the systematic approach to examining and evaluating each agent by the same procedures, and the proposition that the soundest way to reach the "truth" about the carcinogenicity of an agent is through open discussion and reciprocal cross-checking by leading experts. "Given the imperfect nature of all human knowledge, the truth is always approximate, but it can be explicitly stated and qualified by the degree of confidence attached to the statement."[14] Trust was critical for gaining confidence in the science.

On the ethics side, IARC noted, "In practice, scientific judgment can be distorted by secondary interests and goals extraneous to, and interfering with, the primary goal of pursuing scientific, reasonable truth, such as financial incentives or advocacy standpoints. Hence, the experts chosen to participate in evaluations had to be as free as possible of such conflicting interests." This becomes a critical element in developing trust in its risk assessment of financially lucrative chemicals that have been on the market for decades and for which there is a powerful constituency for supporting its continuous use. That was certainly the case with DDT, PCBs, asbestos, and benzene.

IARC's monographs program was founded by Lorenzo Tomatis and introduced in 1971 to develop the best scientific evidence to evaluate whether a chemical was carcinogenic. Tomatis was IARC's director from 1982 to 1993. He articulated IARC's precautionary approach in evaluating a chemical's toxicology. He wrote, "In the absence of absolute certainty, rarely if ever reached in biology, it is essential to adopt an attitude of responsible caution in line with the principles of primary prevention, the only one that may prevent unlimited experimentation on the entire human species."[15] Each monograph is produced by a Working Group of international experts who meet in Lyon for seven to ten days. With the help of IARC's professional staff, the Working Group reviews the scientific literature on the carcinogenicity of human exposures to the chemical under study.

Evidence of carcinogenicity is classified into one of five categories: carcinogenic, probably carcinogenic, possibly carcinogenic, not classifiable, and probably not carcinogenic to humans. "Intensive discussions and repeated revisions of the monograph text take place during what is nowadays an eight-day long meeting."[16]

History of Glyphosate

Glyphosate, or chemically named N-(phosphonomethyl) glycine, is a derivative of the amino acid glycine. It is a white odorless crystalline solid. While working at the pharmaceutical company Cilag, Swiss chemist Henri Martin discovered the compound glyphosate in 1950.[17] It never was advanced as a drug. Johnson and Johnson acquired Cilag and sold its research samples, including that of glyphosate, to Aldrich Chemical.[18] Two decades later, Monsanto scientists were investigating compounds as potential water-softening agents. They synthesized over one hundred related analogs. Some of these compounds, closely related to glyphosate, had herbicidal properties. Monsanto scientist John Franz worked on the compounds to determine what gave them their herbicidal properties, and glyphosate was synthesized in 1970.

Monsanto eventually patented glyphosate under the trade name Roundup. The first U.S. approval for glyphosate came in 1974. In 1985, the EPA classified glyphosate as a Group C chemical, which means under its designation that it is a possible carcinogen to humans based on rodent studies. Glyphosate was re-registered in the United States in 1993 as a Class E chemical on the finding that it "does not pose unreasonable risks or adverse effects to humans or the environment."[19]

Since that time, it has become the most widely used herbicide in agriculture, and the second most widely used herbicide in home gardens next to 2,4-D. When Monsanto developed herbicide resistance in plants, it linked the genetically engineered crops to their formulation of glyphosate, under the trade name Roundup.

The EPA's mandate under the Federal Insecticide, Fungicide, and Rodenticide Act (FIFRA) is to evaluate pesticides for registration. Unlike IARC, there are no legal requirements that the reports and studies companies submit to EPA are published or peer reviewed. Also, the EPA only evaluates the active ingredient, glyphosate, and not the whole formulation with adjuvant chemicals, or Roundup™. Around 1996 when Roundup Ready seeds entered the commercial markets, glyphosate use increased dramatically.[20]

IARC reviews Glyphosate Toxicity

IARC was engaged in a study of the toxicology of glyphosate in 2015. It published the results of its assessment of glyphosate along with other chemicals in a 2017 document titled *Some Organophosphate Insecticides and Herbicides*,[21] reflecting the views and expert opinions of an IARC Working Group on the "Evaluation of Carcinogenic Risks to Humans," which met in Lyon in March 2015. The glyphosate assessment appeared in volume 112 of its monograph series on cancer.

IARC's finding that glyphosate is a "probable human carcinogen" was immediately criticized in the media. Scientific articles were soon published citing shortcomings or errors in its assessment of the ubiquitous herbicide. To fully understand the different viewpoints, it is important to understand the procedures taken by IARC in its review. Because toxicology involves the selection and interpretation of scientific studies, which may differ among toxicologists, and because biases can enter the process, gaining trust in the institution that undertakes the analysis contributes to one's confidence in the outcome.

IARC begins preparing for an evaluation of a chemical in its Monograph Programme a year before the meeting is scheduled. In the case of glyphosate, IARC established a Working Group consisting of eighteen scientists from France, Chile, Italy, Australia, Canada, Finland, the Netherlands, New Zealand and the United States. Two were listed as unable to attend the meeting where the consensus statement was developed; however, later reports indicated that seventeen members were present. Eight members of the working group were from the United States and worked at government agencies, universities, and a consulting group. Only one of these was listed as not attending the meeting. Also invited was one retired scientist from the Centers for Disease Control and Prevention in the specialist category, four representatives from national health agencies, and six observers, including one academic scientist who acted as an observer for the Monsanto Company. There are five categories of participants to the Monograph meeting: the Working Group, invited specialist(s), representatives of national and international health agencies, observers with relevant scientific credentials, and the IARC staff secretariat.

The Working Group with the help of the IARC staff is solely responsible for issuing the final risk assessment of the chemical. IARC is autonomous to a significant degree, while a part of the World Health Organization and the United Nations. There are strict conflict of interest rules for the members of the Working Group and others involved in the discussions and assessment. Each participant, including the IARC secretariat, is required to disclose pertinent research, employment, and financial interests related to the subject matter of the meeting, covering the past four years or anticipated in the future. IARC officials evaluate the declarations of interest to determine if a conflict of interest warrants modification of participation. All financial interests are disclosed in the published Monograph.

Once the Working Group is established, some members are asked to prepare working papers in their areas of expertise. Four subcommittees were formed to revise and summarize the findings of the working papers and strive to reach a consensus.

The Working Group will accept certain types of studies as part of its assessment of the human carcinogenicity of chemicals: cohort studies, case-control studies, correlation or ecological studies, intervention studies, and case reports. A cohort study is a longitudinal study of a group of people who share a common experience or environmental exposure compared to a similar group whose members do not share that experience or exposure. Investigators follow the people in these groups to determine the risk factor from the exposure, in this case glyphosate. A case-control study is always retrospective. It starts with an outcome, i.e., a disease like non-Hodgkin's lymphoma, and then it traces the outcome back to investigate exposures of a group that has the disease and also a similar group that does not have the disease, to determine the possible cause. Ecological studies investigate outcomes based on populations defined either geographically or tempo-rally. Thus, farmers using glyphosate in one region of the country or during one time period may be compared to farmers who do not use glyphosate in another region or another time period.

Any risk factors are averaged for both of these populations and com-pared, using statistical methods. Intervention studies involve two groups, one of which receives some intervention and another that does not. In the case of glyphosate, one group of farmers may receive special protective cloth-ing and masks while another group of farmers use standard protections. Or one group has stopped its exposure to glyphosate while the other continues to be exposed. The disease outcomes of these groups are compared. The Working Group reviews medical reports of individual cases of disease and examines the historical background and exposures of the patients.

In a legal deposition, the chair of the glyphosate Working Group, Aaron Blair, was asked about the importance to IARC of a paper by DeRoos et al. (2003), which studied people exposed to glyphosate in Nebraska, Iowa, Minnesota, and Kansas between 1979 and 1986. The paper found a dou-bling of the risk for non-Hodgkin's lymphoma.[22] The exchange follows:

> Q.: Is this one of the pieces of evidence upon which your commit-tee based their opinion [where] there was a positive association between exposure to glyphosate and non-Hodgkin's lymphoma

from exposure to Roundup or glyphosate outside the realm of chance?

A.: [from Aaron Blair] Yes.

The exchange goes to the heart of the matter. The chairman of the IARC Working Group affirms that his committee was swayed by statistically significant studies linking glyphosate or its formulations to cancer. There is not the slightest hesitancy in his response.

IARC imposes some strict criteria on the types of data it will accept in its Monograph assessment. "IARC evaluations rely only on data that are in the public domain and available to independent scientific review. Data from government agency reports and doctoral theses that are publicly available can also be considered. The evaluation of glyphosate by the Working Group included only industry studies that met these criteria. However, they did not include data from summary tables in online supplements to published articles, which did not provide enough detail for independent assessment."[23] This is a critical piece of information because Monsanto argued that some industry data were not included in IARC's glyphosate Monograph that may have been included in other assessments. The EPA, on the other hand, does accept registrant generated studies that are unpublished and not peer reviewed.[24]

Primary Evidence

IARC began its analysis by indicating that the study of primary significance for the Working Group review of glyphosate was a prospective cohort study conducted in Iowa and North Carolina called the Agricultural Health Study. It was the only cohort study to have published findings on the exposure of people to glyphosate and the associated risk of cancer at many different sites. Among the goals of the study was to identify and quantify cancer risks among men, women, and minorities associated with their direct exposure to pesticides and other agricultural agents. When the total cohort was assembled in 1997, there were about seventy-five thousand adult study subjects. Prior studies of farmers throughout the world indicated that they had higher rates of non-Hodgkin's lymphoma than non-farmers.

A questionnaire developed by the National Institutes of Health was administered to pesticide users. About fifty pesticides were in the survey, which included questions about crops grown, livestock raised, pesticide application methods, and personal protection equipment used.[25]

The IARC Working Group identified seven reports from the Agricultural Health Study (AHS) and several reports from case-control studies related to glyphosate. These studies were considered important because of the relative size of the study populations. The studies that grew out of the AHS surveys and state cancer registries allowed investigators to develop relative risks and calculate correlations between glyphosate and disease endpoints. In reviewing the studies, the Working Group reported no association between exposure to glyphosate and prostate, breast, colorectal, skin, and pancreatic cancers.[26] There was one cancer endpoint which IARC found correlated with glyphosate exposure: non-Hodgkin's lymphoma (NHL). The report stated: "Two large case-control studies of NHL from Canada and the U.S. A., and two case-control studies from Sweden, reported statistically significant increased risks of NHL in association with exposure to glyphosate."[27] The risks the Monograph cited persisted even in studies that adjusted for exposure to other pesticides.

The animal carcinogenesis studies were mixed—some showed a significant increase in cancer, while others found no significant increase in tumors at any site. After referencing 269 scientific studies in its ninety-four-page glyphosate Monograph, IARC reported that there is limited evidence in humans for the carcinogenicity of glyphosate from the positive associations it found between the herbicide and NHL. By "positive association," IARC meant that there were studies that showed an excess risk for people exposed. It also stated that "there is sufficient evidence in experimental animals for the carcinogenicity of glyphosate."[28] Based upon the weight of human and animal data IARC concluded "Glyphosate is probably carcinogenic to humans."[29]

Divergent Scientific Results

Why did other government health agencies reach a different conclusion on the carcinogenicity of glyphosate? There are three main reasons to expect divergence in the findings of the toxicology for the herbicide:

- The other agencies used different sets of evidence because there were different selection criteria for choosing the studies or IARC did not have the latest data.
- They used different methods for aggregating or weighing the evidence.
- They interpreted the same evidence differently.

A group of scientists from two agencies that undertook an assessment of glyphosate and reached a much lower risk value than that of IARC published a paper that tried to explain the different outcomes. They were from the European Food Safety Authority (EFSA) and the German Federal Institute for Risk Assessment. According to the authors, "Uses of different data sets, particularly on long-term toxicology/carcinogenicity in rodents, could partially explain divergent views. . . . The EU evaluation, which considered studies not available to IARC, also updates the toxicological profile of glyphosate proposing new toxicological reference values."[30]

After reviewing the evidence of both IARC and the evaluations of the other European agencies, the authors concluded that the same epidemiological studies were used in all the assessments. "The same weak evidence in humans for the carcinogenicity of glyphosate was interpreted differently by IARC and EFSA. IARC considered the association between exposure to glyphosate and non-Hodgkin lymphoma as *limited* evidence in humans, while in the EU assessment, most experts considered the evidence as very limited and insufficient for triggering the classification."[31] IARC used the information about glyphosate's carcinogenicity in animals, and for two mechanisms of action, namely genotoxicity and oxidative stress, to buttress the plausibility of the limited evidence in humans.

According to the authors, the source of the divergence in assessing glyphosate is in putting the pieces of all the studies together and in reaching the weight of the evidence. The EU evaluation from EFSA and FAO was based on the likelihood of getting NHL from low dose exposures. IARC was less interested in the probability of contracting NHL, but on the possibility. "Definitions for limited and sufficient evidence in humans and animals are identical for IARC and the UN-GHS [United Nation's Globally Harmonized System of Classification and Labeling of Chemicals]; however, differences in criteria and methodological considerations for weighing and assessing the evidence can lead to divergent interpretations between the

IARC assessment and regulatory evaluations following UN-GHS criteria, even when based on the same evidence."[32]

Critics of IARC claimed that data available to the Working Group showed no correlation between glyphosate exposure and NHL; some of this was from a pooled study (data are aggregated for more statistical power), which had more exposed subjects in them. At least one of the unpublished studies found no evidence of a link between glyphosate and cancer. Even though the chair of the Working Group knew of the data, it was not shared with the other members or considered as part of IARC's analysis. This information was revealed when the media got hold of court documents in a trial with the plaintiff suing Monsanto, attributing glyphosate as the cause of its clients' NHL illnesses.[33] The deposition of Aaron Blair was one of the documents. Dr. Blair was questioned by Monsanto's attorneys for over three hours on the issue of why he and the Working Group did not use the new data that was available before they met. Monsanto's lawyers argued that farmers were acquiring NHL before glyphosate was on the market dating back to the 1960s, and that there were many factors, other than glyphosate, that could have explained the excess of NHL among farmers. Under oath, Blair testified that he was involved in a study with data that showed a two-fold increase in the risk of NHL of farmers handling glyphosate for greater than two days a year. He could not report this study to IARC because it had not been published by March 2015 when the Working Group had met. That speaks to the even-handedness of the process.

The plaintiff's attorney posed the following question to Dr. Blair after he was cross-examined by the defendant's attorney as an independent expert witness, not as a plaintiff's expert:

Q.: Has anything you have been shown by Monsanto's lawyers in the three hours and forty minutes that he questioned you changed the opinion that you had at the IARC meeting about glyphosate and non-Hodgkin lymphoma?
A.: No.[34]

IARC provided an official response to many of the allegations appearing in the media. It emphasized that "the IARC Monographs evaluations are based on the systematic assembly and review of all publicly available and pertinent studies by independent experts, free from vested interests."[35] In response to claims that it omitted certain critical information, IARC stated that it had reviewed about 1,000 studies and cited 269 references. In response to the question about why IARC's outcome for glyphosate was at odds with the assessments of major regulatory agencies, it noted that "many regulatory agencies rely primarily on industry data from toxicological studies that are not available in the public domain. In contrast, IARC systematically assembles and evaluates all relevant evidence available in the public domain for independent scientific review." One of the strongest criticisms of the IARC raised in litigation by the defense was that it neglected to incorporate all the data of the Agricultural Health Study (AHS), a U.S. two-state health study of farmers' and their families' exposure to agricultural chemicals. The IARC reported that the AHS study was large and well-conducted but that it was incorrectly described as the "most powerful" study. "The weakness of the study is that people were followed up for a short period of time, which means fewer cases of cancer would have had time to appear."[36]

On December 12, 2017, the Environmental Protection Agency issued its glyphosate assessment.[37] This was a follow-up to a series of risk assessments for the herbicide. The agency classified glyphosate as a possible human carcinogen in 1985. Then in 1986 it asked the FIFRA Advisory Panel (SAP) to evaluate the carcinogenic potential of glyphosate. Based on the available evidence, SAP recommended that the herbicide be classified as a Group D classification (not classifiable as to its human carcinogenicity) but asked that EPA obtain new studies. With the addition of rodent studies, in 1991 the EPA's Carcinogenicity Peer Review Committee classified glyphosate as a Group E chemical (evidence of non-carcinogenicity for humans), downsizing its risk. In 2015, the EPA's Cancer Assessment Review Committee concluded that glyphosate is not likely to be carcinogenic to humans.

In its most recent report, the EPA stated, "Due to study limitations and contradictory results across studies of at least equal quality, a conclusion regarding the association between glyphosate exposure and the risk of NHL [non-Hodgkin's lymphoma] cannot be determined based on available data."[38] The EPA's conclusion was based on six human studies, which were also a part of the IARC's review. Unlike the EPA, the IARC used animal studies as a central part of its decision about the probability glyphosate causes non-Hodgkin's lymphoma in humans.

Conclusion

Several historical points should be stated before we put a coda to this story of glyphosate risk assessment. First, regulatory agencies are notoriously lobbied by chemical manufacturers to prevent or lower the regulation of the chemical products. The political climate can determine the effectiveness of the lobbying efforts. Second, when regulatory agencies in the United States undertake a chemical assessment, the panels established have been replete with conflicts of interest usually unnoticed by the public.[39] Third, if a regulatory body issues an order limiting the exposure or use of a substance, litigation by the manufacturer almost always follows, derailing or slowing up the process. It has taken between twenty-five and fifty years to remove highly toxic products such as asbestos, PCBs, and lead from the marketplace in the United States. In the forty-year period since chemical regulations for industrial, non-agricultural chemicals were fully in place in the United States, of the estimated eighty-five thousand chemicals in commerce a mere five have been fully regulated or banned for use.[40] Although the law covering agricultural chemicals is stronger than the law covering industrial non-agricultural chemicals, many of the same biases and influences apply equally. Fourth, no single test can determine conclusively that a chemical is unsafe or safe enough. Each of the tests has its own limitations. Epidemiological studies address human exposures, but they do not produce causality, as one would get in a controlled experiment. Animal studies have all sorts of confounding factors, even as they can offer causal explanations for effects. And there are always questions about whether humans and animals react similarly to chemicals.

I began this paper with a description of the IARC's structure and principles. No regulatory body in the United States can meet this agency's independence, diversity, and integrity. It does not mean that it cannot miscalculate a risk of a chemical. Given that an entire private industry-funded sector is devoted to producing uncertainty in risk assessments, critiquing methods of experiments that demonstrate toxicity, and influencing regulatory bodies to establish a risk standard that would permit all chemicals regardless of what public health scientists say about their hazards, confidence that an agency is using the best, peer-reviewed, unbiased science not funded by the product manufacturers would help societies navigate across the dueling claims as reflected in the search for toxicological truth of glyphosate and Roundup™. Critics of the IARC would like to have another twenty-five years of research to narrow every area of uncertainty before the product could be adequately regulated. There is another way to think about it. Glyphosate, as well as many other chemicals and their formulations, are put on the market when there are large gaps of uncertainty about their safety as indicated by the questions being posed a quarter century after glyphosate was approved. Why not spend twenty-five years closing all the gaps of uncertainty *before* a chemical like glyphosate is allowed to be used in the first place?

My review of the case suggests that the attack on the IARC's assessment of glyphosate is driven by financial rather than scientific interests. The IARC acknowledges that there remain uncertainties, but with the scientific evidence at hand, the precautionary principle impels one to accept its finding and for regulatory bodies to act on it. The EPA and its advisory panels keep requesting more and better human data on the relationship between glyphosate and non-Hodgkin's lymphoma because glyphosate is assumed safe unless definitively proved otherwise by evidence that could take decades to obtain. The IARC's approach operates from the precautionary principle where strong suggestive evidence is enough to stop the usage of a chemical until a definitive answer is found.

* * *

Sheldon Krimsky is a Lenore Stern professor of humanities and social sciences at Tufts University.

14

Reuters vs. U.N. Cancer Agency: Are Corporate Ties Influencing Science Coverage?

By Stacy Malkan

Ever since it classified the world's most widely used herbicide as "probably carcinogenic to humans,"[1] a team of international scientists at the World Health Organization's cancer research group have been under withering attack by the agrochemical industry and its surrogates.[2]

In a front-page series titled "The Monsanto Papers," the French newspaper *Le Monde* (6/1/17) described the attacks as "the pesticide giant's war on science." The newspaper reported, "To save glyphosate, the firm [Monsanto] undertook to harm the United Nations agency against cancer by all means."[3]

One key weapon in industry's arsenal has been the reporting of Kate Kelland, a veteran Reuters reporter based in London. With two industry-fed scoops and a special report, reinforced by her regular beat reporting, Kelland has aimed a torrent of critical reporting at the WHO's International Agency for Research on Cancer (IARC), portraying the group and its scientists as out of touch and unethical, and leveling accusations about conflicts of interest and suppressed information in its decision-making.

The IARC working group of scientists did not conduct new research but reviewed years of published and peer-reviewed research before concluding that there was limited evidence of cancer in humans from real-world exposures to glyphosate and "sufficient" evidence of cancer in studies on animals. The IARC also concluded there was strong evidence of genotoxicity for glyphosate alone, as well as glyphosate used in formulations such as Monsanto's Roundup brand of herbicide, whose use has increased dramatically as Monsanto has marketed crop strains genetically modified to be Roundup Ready.[4]

But in writing about the IARC decision, Kelland has ignored much of the published research backing the classification, and focused on industry talking points and criticisms of the scientists in seeking to diminish their analysis. Her reporting has relied heavily on pro-industry sources, while failing to disclose their industry connections; contained errors that *Reuters* has refused to correct; and presented cherry-picked information out of context from documents she did not provide to her readers.

Raising further questions about her objectivity as a science reporter are Kelland's ties to the Science Media Centre (SMC), a controversial nonprofit public-relations agency in the United Kingdom that connects scientists with reporters and gets its largest block of funding from industry groups and companies, including chemical industry interests.

The SMC, which has been called "science's PR agency,"[5] launched in 2002 partly as an effort to tamp down news stories driven by groups like Greenpeace and Friends of the Earth, according to its founding report.[6] The SMC has been accused of playing down the environmental and human health risks of some controversial products and technologies, according to multiple researchers[7] who have studied the group.

Kelland's bias in favor of the group is evident, as she appears in the SMC promotional video and the SMC promotional report, regularly attends SMC briefings,[8] speaks at SMC workshops,[9] and attended meetings in India[10] to discuss setting up an SMC office there.

Neither Kelland nor her editors at *Reuters* would respond to questions about her relationship with SMC, or to specific criticisms about her reporting.

Fiona Fox, director of the SMC, said her group did not work with Kelland on her IARC stories or provide sources beyond those included in SMC's press releases. It is clear, however, that Kelland's reporting on glyphosate and IARC mirrors the views put forth by SMC experts and industry groups on those topics.

Reuters Takes on Cancer Scientist

On June 14, 2017, *Reuters* published a special report[11] by Kelland accusing Aaron Blair, an epidemiologist from the U.S. National Cancer Institute and chair of the IARC panel on glyphosate, of withholding important data from its cancer assessment.

Kelland's story went so far as to suggest that the information supposedly withheld could have changed IARC's conclusion that glyphosate is probably carcinogenic. Yet the data in question was but a small subset of epidemiology data gathered through a long-term project known as the Agricultural Health Study[12] (AHS). An analysis of several years of data about glyphosate from the AHS had already been published and was considered by IARC, but a newer analysis of unfinished, unpublished data was not considered, because IARC rules call for relying only on published data.

Kelland's thesis that Blair withheld crucial data was at odds with the source documents on which she based her story, but she did not provide readers with links to any of those documents, so readers could not check the veracity of the claims for themselves. Her bombshell allegations were then widely circulated, repeated by reporters at other news outlets (including *Mother Jones*)[13] and immediately deployed as a lobbying tool[14] by the agrochemical industry.

After obtaining the actual source documents, Carey Gillam, a former *Reuters* reporter and now research director of U.S. Right to Know (the nonprofit group where I also work), laid out multiple errors and omissions in Kelland's piece.[15] The analysis provides examples of key claims in Kelland's article, including a statement supposedly made by Blair, that are not supported by the 300-page deposition of Blair[16] conducted by Monsanto's attorneys, or by other source documents.

Kelland's selective presentation of the Blair deposition also ignored what contradicted her thesis—for example, Blair's many affirmations of research showing glyphosate's connections to cancer, as Gillam wrote in a Huffington Post article (June 18, 2017).[17] Kelland inaccurately described Blair's deposition and related materials as "court documents," implying they were publicly available; in fact, they were not filed in court, and presumably were obtained from Monsanto's attorneys or surrogates. (The documents were available only to attorneys involved in the case, and plaintiff's attorneys have said they did not provide them to Kelland.)

Reuters has refused to correct the errors in the piece, including the false claim about the origin of the source documents and an inaccurate description of a key source, statistician Bob Tarone, as "independent of Monsanto." In fact, Tarone had received a consultancy payment[18] from Monsanto for his efforts to discredit IARC.

In response to a usRTK request to correct or retract the Kelland article, *Reuters* global enterprises editor Mike Williams wrote in a June 23 email, "We have reviewed the article and the reporting on which it was based. That reporting included the deposition to which you refer, but was not confined to it. The reporter, Kate Kelland, was also in contact with all the people mentioned in the story and many others, and studied other documents. In the light of that review, we do not consider the article to be inaccurate or to warrant retraction."

Williams declined to address the false citing of "court documents" or the inaccurate description of Tarone as an independent source.

Since then, the lobbying tool *Reuters* handed to Monsanto has grown legs and run wild. A June 24 editorial[19] by the *St. Louis Post Dispatch* added errors[20] on top of the already misleading reporting. By mid-July, right-wing blogs were using the *Reuters* story to accuse the IARC of defrauding U.S. taxpayers,[21] pro-industry news sites were predicting the story would be "the final nail in the coffin"[22] of cancer claims about glyphosate, and a fake science news group[23] was promoting Kelland's story on Facebook with a phony headline claiming that the IARC scientists had confessed to a cover-up.[24]

Bacon Attack

This was not the first time Kelland had relied on Bob Tarone as a key source, and failed to disclose his industry connections, in an article attacking the IARC.

An April 2016 special investigation[25] by Kelland, "Who Says Bacon Is Bad?," portrayed IARC as a confusing agency that is bad for science. The piece was built largely on quotes from Tarone, two other pro-industry sources whose industry connections were also not disclosed, and one anonymous observer.

The IARC's methods are "poorly understood," "do not serve the public well," sometimes lack scientific rigor, are "not good for science," "not good for regulatory agencies," and do the public "a disservice," the critics said.

The agency, Tarone said, is "naïve, if not unscientific"—an accusation emphasized with capital letters in a sub-headline.

Tarone works for the pro-industry International Epidemiology Institute and was once involved with a controversial cell phone study,[26] funded in part by the cell phone industry, that found no cancer connection to cell phones, contrary to independently funded studies[27] of the same issue.

The other critics in Kelland's bacon story were Paulo Boffetta, a controversial ex-IARC scientist who wrote a paper defending asbestos while also receiving money to defend the asbestos industry in court, and Geoffrey Kabat, who once partnered with a tobacco industry-funded scientist to write a paper defending secondhand smoke.

Kabat also serves on the advisory board of the American Council on Science and Health (ACSH), a corporate front group.[28] The day the *Reuters* story hit, ACSH posted a blog item bragging that Kelland had used its advisor Kabat as a source to discredit IARC.[29]

The industry connections of her sources, and their history of taking positions at odds with mainstream science, seems relevant, especially since the IARC bacon exposé was paired with a Kelland article about glyphosate[30] that accused IARC advisor Chris Portier of bias because of his affiliation with an environmental group.

The conflict-of-interest framing served to discredit a letter, organized by Portier and signed by 94 scientists,[31] that described "serious flaws" in a European Union risk assessment that exonerated glyphosate of cancer risk.

The Portier attack, and the good science/bad science theme, echoed through[32] chemical industry PR channels[33] on the same day the Kelland articles appeared.

IARC Pushes Back

In October 2016, in another exclusive scoop,[34] Kelland portrayed IARC as a secretive organization that had asked its scientists to withhold documents pertaining to the glyphosate review. The article was based on correspondence provided to Kelland by a pro-industry law group.[35]

In response, IARC took the unusual step of posting Kelland's questions and the answers they had sent her,[36] which provided context left out of the *Reuters* story.

IARC explained that Monsanto's lawyers were asking scientists to turn over draft and deliberative documents, and in light of the ongoing lawsuits against Monsanto, "the scientists felt uncomfortable releasing these materials, and some felt that they were being intimidated." The agency said they had faced similar pressure in the past to release draft documents to support legal actions involving asbestos and tobacco, and that there was an attempt to draw deliberative IARC documents into PCB litigation.

The story didn't mention those examples, or the concerns about draft scientific documents ending up in lawsuits, but the piece was heavy on critiques of the IARC, describing it as a group "at odds with scientists around the world," which "has caused controversy" with cancer assessments that "can cause unnecessary health scares." The IARC has "secret agendas" and its actions were "ridiculous," according to a Monsanto executive quoted in the story.

The IARC wrote in response[37] (emphasis in original): "The article by *Reuters* follows a pattern of consistent but misleading reports about the IARC Monographs Programme in some sections of the media beginning after glyphosate was classified as *probably carcinogenic to humans*."

The IARC also pushed back on[38] Kelland's reporting about Blair, noting the conflict of interest with her source Tarone and explaining that the IARC's cancer evaluation program does not consider unpublished data and "does not base its evaluations on opinions presented in media reports," but on the "systematic assembly and review of all publicly available and pertinent scientific studies, by independent experts, free from vested interests."

PR Agency Narrative

The Science Media Centre—which Kelland has said[39] has influenced her reporting—does have vested interests and has also been criticized for pushing pro-industry science views. Current and past funders[40] include Monsanto, Bayer, DuPont, Coca-Cola, and food and chemical industry trade groups, as well as government agencies, foundations, and universities.

By all accounts, SMC is influential in shaping how the media covers certain science stories, often getting its expert reaction[41] quotes in media stories and driving coverage with its press briefings.[42]

As Kelland explained in the SMC promotional video,[43] "By the end of a briefing, you understand what the story is and why it's important." That is the point of the SMC effort: to signal to reporters whether stories or studies merit attention, and how they should be framed.

Sometimes, SMC experts downplay risk and offer assurances to the public about controversial products or technologies; for example, researchers have criticized SMC's media efforts on fracking, cell phone safety, chronic fatigue syndrome, and genetically engineered foods.[44]

SMC campaigns sometimes feed into lobbying efforts. A 2013 Nature article[45] (July 10, 2013) explained how SMC turned the tide on media coverage of animal/human hybrid embryos away from ethical concerns and toward their importance as a research tool—and thus stopped government regulations.

The media researcher hired by SMC to analyze the effectiveness of that campaign, Andy Williams of Cardiff University, came to see the SMC model as problematic, worrying that it stifled debate.[46] Williams described SMC briefings[47] as tightly managed events pushing persuasive narratives.

On the topic of glyphosate cancer risk, SMC offers a clear narrative in its press releases. The IARC cancer classification, according to SMC experts[48], "failed to include critical data," was based on "a rather selective review" and on evidence that "appears a bit thin" and "overall does not support such a high-level classification." Monsanto[49] and other industry groups[50] promoted the quotes.

SMC experts had a much more favorable view of risk assessments conducted by the European Food Safety Authority (EFSA[51]) and the European Chemicals Agency (ECHA[52]), which cleared glyphosate of human cancer concerns.

The EFSA's conclusion[53] was "more scientific, pragmatic and balanced" than IARC's, and the ECHA report[54] was objective, independent, comprehensive, and "scientifically justified."

Kelland's reporting in *Reuters* echoes those pro-industry themes and sometimes used the same experts, such as a November 2015 story[55] about why European-based agencies gave contradictory advice about the cancer risk of glyphosate. Her story quoted two experts directly from an SMC release,[56] then summarized their views: "In other words, IARC is tasked with highlighting anything that might in certain conditions, however rare, be able to cause cancer in people. EFSA, on the other hand, is concerned with real life risks and whether, in the case of glyphosate, there is evidence to show that when used in normal conditions, the pesticide poses an unacceptable risk to human health or the environment."

Kelland included two brief reactions from environmentalists: Greenpeace called the EFSA review "whitewash," and Jennifer Sass from the Natural Resources Defense Council said IARC's review was "a much more robust, scientifically defensible and public process involving an international committee of non-industry experts." (An NRDC statement[57] on glyphosate put it this way: "IARC Got It Right, EFSA Got It from Monsanto.")

Kelland's story followed up the environmental group comments with "critics of IARC . . . say its hazard identification approach is becoming meaningless for consumers, who struggle to apply its advice to real life," and ends with quotes from a scientist who "declares an interest as having acted as a consultant for Monsanto."

When asked about the criticisms of pro-industry bias of the SMC, Fox responded:

> We listen carefully to any criticism from the scientific community or news journalists working for UK media, but we do not receive criticism of pro-industry bias from these stakeholders. We reject the charge of pro-industry bias, and our work reflects the evidence and views of the 3,000 eminent scientific researchers on our database. As an independent press office focusing on some of the most controversial science stories, we fully expect criticism from groups outside mainstream science.

Expert Conflicts

Scientific experts do not always disclose their conflicts of interest in news releases issued by SMC, nor in their high-profile roles as decision-makers about the cancer risk of chemicals like glyphosate.

Frequent SMC expert Alan Boobis, professor of biochemical pharmacology at Imperial College London, offers views in SMC releases on aspartame[58] ("not a concern"), glyphosate in urine[59] (no concern), insecticides and birth defects[60] ("premature to draw conclusions"), alcohol,[61] GMO corn,[62] trace metals,[63] lab rodent diets[64] and more. The ECHA decision[65] that glyphosate is not a carcinogen "is to be congratulated," according to Boobis, and the IARC decision[66] that it is probably carcinogenic "is not a cause for undue alarm," because it did not take into account how pesticides are used in the real world.

Boobis declared no conflicts of interest in the IARC release or any of the earlier SMC releases that carry his quotes. But he then sparked a conflict-of-interest scandal[67] when news broke that he held leadership positions with the International Life Sciences Institute (ILSI), a pro-industry group,[68] at the same time he co-chaired a U.N. panel that found glyphosate unlikely to pose a cancer risk[69] through diet. (Boobis is currently chair of the ILSI Board of Trustees and vice president ad interim[70] of ILSI/Europe.)

ILSI has received six-figure donations[71] from Monsanto and CropLife International, the pesticide trade association. Professor Angelo Moretto, who co-chaired the U.N. panel on glyphosate along with Boobis, also held a leadership role in ILSI.[72] Yet the panel declared[73] no conflicts of interest.

Kelland did not report on those conflicts, though she did write about[74] the findings of the "UN experts" who exonerated glyphosate of cancer risk, and she once recycled a Boobis quote from an SMC press release[75] for an article about tainted Irish pork.[76] (The risk to consumers was low.)

When asked about the SMC conflict of interest disclosure policy, and why Boobis' ISLI connection was not disclosed in SMC releases, Fox responded: "We ask all researchers we use to provide their COIs and proactively make those available to journalists. In line with several other COI policies, we are unable to investigate every COI, though we welcome journalists doing so."

Boobis could not be reached for comment but told the *Guardian*, "My role in ILSI (and two of its branches) is as a public sector member and chair of their boards of trustees, positions which are not remunerated."

But the conflict "sparked furious condemnation from green MEPs and NGOs," the *Guardian* reported, "intensified by the [UN panel] report's release two days before an EU relicensing vote on glyphosate, which will be worth billions of dollars to industry."[77]

And so goes it with the tangled web of influence involving corporations, science experts, media coverage, and the high-stakes debate about glyphosate, now playing out on the world stage as Monsanto faces lawsuits[78] over the chemical due to cancer claims and seeks to complete a $66 billion deal with Bayer.[79]

Meanwhile, in the United States, as *Bloomberg* reported on July 13, 2017, "Does the World's Top Weed Killer Cause Cancer? Trump's EPA Will Decide."[80]

Update: Reuters' Kate Kelland Again Promotes False Narrative about IARC and Glyphosate Cancer Concerns

Continuing her record of industry-biased reporting about the International Agency for Research on Cancer (IARC), Reuters reporter Kate Kelland again attacked the science panel with an October 19, 2017, story that the panel edited a draft scientific document before issuing the final version of its assessment on glyphosate that found glyphosate a probable human carcinogen.[81]

The American Chemistry Council, the chemical industry trade group, immediately issued a press release[82] praising Kelland's story, claiming her story "undermines IARC's conclusions about glyphosate" and urging policy makers to "take action against IARC over deliberate manipulation of data."

Kelland's story quoted a Monsanto executive claiming that "IARC members manipulated and distorted scientific data" but failed to mention the significant amount of evidence that has emerged from Monsanto's own documents[83] through court-ordered discovery that demonstrate the many ways the company has worked to manipulate and distort data on glyphosate over decades.

The story also failed to mention that most of the research IARC discounted was Monsanto-financed work that did not have sufficient raw data to meet IARC's standards. And though Kelland cites a 1983 mouse study and a rat study in which IARC failed to agree with the original investigators, she failed to disclose that these were studies financed by Monsanto and the investigators IARC failed to agree with were paid by Monsanto. She also failed to mention the critical information that in the 1983 mouse study, even the EPA toxicology branch did not agree with Monsanto's investigators[84] because the evidence of carcinogenicity was so strong, according to EPA documents. It said in numerous memos that Monsanto's argument was unacceptable and suspect, and it determined glyphosate to be a possible carcinogen.

By leaving out these crucial facts, and by twisting others almost inside out, Kelland has authored another article that serves Monsanto quite well but victimized innocent members of the public and policy makers who rely on trusted news outlets for accurate information. The only encouraging point to be taken from Kelland's story is that this time she admitted Monsanto provided her with the information.

As we have previously reported, Kelland's earlier reporting on IARC has been deeply problematic; her stories have contained errors that *Reuters* refused to correct, made blatantly misleading claims about documents that were not provided to the public, and relied on industry-connected sources who were presented as independent sources.

* * *

This chapter was originally published in Fairness & Accuracy in Reporting.[85]

* * *

See also these related stories and documents:

- This Monsanto document[86] describes the company's public relations plan to "orchestrate outcry with IARC decision"[87] in anticipation of the agency's cancer rating on glyphosate. The PR plan names Sense About Science, a London based lobbying group, as a "Tier 2" ally and suggests the group could lead media outreach efforts against IARC. Sense About Science is the sister group of Science Media Center,[88] a corporate-funded PR firm in London with close ties to Kelland (as documented in the story below).
- "Reuters' Kate Kelland IARC Story Promotes False Narrative," by Carey Gillam, June 28, 2017.[89]
- "Monsanto Spin Doctors Target Cancer Scientist in Flawed Reuters Story," by Carey Gillam, June 19, 2017.[90]

15

Genetic Engineering, Pesticides, and Resistance to the New Colonialism

Argentina is a leading producer of soy for export and cattle feed, and almost all of it is "Roundup Ready" soy, resistant to the herbicide glyphosate, and manufactured and patented by Monsanto.

Javiera Rulli, a biologist who works on issues of agriculture and food sovereignty in Argentina, lives and works in a Kolla community in the Yungas, the tropical montane forest region in the northwest of Argentina. She says that after the United States and Brazil, Argentina is the world's largest producer of genetically modified crops. Although it was once known as the world's grain barn, Argentina has become a soy dictatorship of the International Monetary Fund and the World Bank, with a growing external debt.

Said Rulli, "The implementation of large-scale intensive agriculture has brought about a loss of agricultural biodiversity and the destruction of local economies," Rulli wrote. "The industrial agriculture has resulted in the concentration of land in the hands of big landowners and giant corporations, resulting in the expulsion of rural workers and small- and medium-sized producers. As a result, today more than half of the population survives under the poverty level."[1]

Rulli, who spoke in New York in 2005 at a forum co-sponsored by the No Spray Coalition, explained how the massive glyphosate spraying goes hand-in-hand with the militarization of neighboring countries such as Paraguay and is directly related to the genetically engineered soy expansion: "Growers of GM soy from Brazil crossed the border and attacked a peasant community, Tekojoja, in Caaguazu, Paraguay, in order to drive them off their lands and to claim them for themselves for planting genetically engineered soy. They evicted 270 people, burned down 54 of the houses and all of the non-GMO crops. Two local farmers were killed—Angel Cristaldo and Lus Torres—many people were injured and 130 were arrested, among them many women and children."[2]

Genetically engineered foods, and the herbicides and pesticides required to maintain them, are weapons not only in Javiera Rulli's Argentina but also in the U.S. government's arsenal. U.S. capital imposes the technology on other countries through its aid packages as well as the International Monetary Fund's and World Bank's structural adjustment programs (SAPs).

Former Secretary of State Henry Kissinger approves of the IMF's structural adjustment and U.S. Aid for International Development (USAID) programs. He once portrayed America aid this way: "To give food aid to a country just because they are starving is a pretty weak reason." For Kissinger, food is to be used as a weapon in the achievement of U.S. foreign policy objectives. And so, the United States systematically dumps cheap genetically engineered produce saturated with pesticides on foreign markets, undermining local producers and forcing them to purchase the patented seeds from the company manufacturing them, along with the pesticides needed to kill off the plants' competitors.[3] Driven from their lands, local producers become dependent on the United States and its corporations, and a number of them try to flee across the border to the United States.

In his 2001 book, *A Cook's Tour,* chef Anthony Bourdain had a very different take on Kissinger, one worth savoring:

> Once you've been to Cambodia, you'll never stop wanting to beat Henry Kissinger to death with your bare hands. You will never again be able to open a newspaper and read about that treacherous, prevaricating, murderous scumbag sitting down for a nice chat with Charlie Rose or attending some black-tie affair for a new glossy magazine without choking. Witness what Henry did in Cambodia—the fruits of his genius for statesmanship—and you will never understand why he's not sitting in the dock at The Hague next to Milosevic.

Prior to the advent and initial proliferation of genetically modified organisms in the 1980s—a technology intricately tied to the manufacture of pesticides and in particular Monsanto's Roundup—the tentacles of globalization expanded outward. U.S. foreign policy gathered the disparate needs of its corporations and unified them into a set of objectives that would expand the system, objectives that were enforced by U.S. military power. U.S. Marine

Corps Commander Major General Smedley Darlington Butler explained his own role and that of the U.S. military in that period:

> I spent thirty-three years and four months in active military service as a member of this country's most agile military force, the Marine Corps. I served in all commissioned ranks from Second Lieutenant to Major-General. And during that period, I spent most of my time being a high class muscle-man for Big Business, for Wall Street and for the Bankers. In short, I was a racketeer, a gangster for capitalism.
>
> I suspected I was just part of a racket at the time. Now I am sure of it. Like all the members of the military profession, I never had a thought of my own until I left the service. My mental faculties remained in suspended animation while I obeyed the orders of higher-ups. This is typical with everyone in the military service.
>
> I helped make Mexico, especially Tampico, safe for American oil interests in 1914. I helped make Haiti and Cuba a decent place for the National City Bank boys to collect revenues in. I helped in the raping of half a dozen Central American republics for the benefits of Wall Street. The record of racketeering is long. I helped purify Nicaragua for the international banking house of Brown Brothers in 1909–1912. I brought light to the Dominican Republic for American sugar interests in 1916. In China I helped to see to it that Standard Oil went its way unmolested.
>
> During those years, I had, as the boys in the back room would say, a swell racket. Looking back on it, I feel that I could have given Al Capone a few hints. The best he could do was to operate his racket in three districts. I operated on three continents.

In 1934, Butler testified before the U.S. Congress that he had been offered millions of dollars to lead an insurrection and stage a fascist coup against President Franklin Delano Roosevelt. And he named names. The coup was to be funded by corporate behemoths headed by DuPont and J.P. Morgan.[4]

Fifty years later, DuPont—one of the key manufacturers of pesticides in the United States—patented the first genetically engineered mouse, following upon a 1980 U.S. Supreme Court decision that ruled that a private entity could legally patent a genetically engineered organism[5] and use it for private profit. Thus, DuPont inaugurated an era of neocolonialism and enclosure of the living cell and of life *within.*

Biopiracy was given a literal "lease on life," a new form of enclosure as revolutionary in its consequences as were the British Enclosure Acts of the seventeenth and eighteenth centuries in relation to the old forms of emerging capitalism. There, lands that for centuries had been owned by no one and used in common were expropriated and privatized, launching a new era dominated by corporate capital that in 1801 was consolidated into British law. In what was later to become Germany, women accused of being witches were burned at the stake; those who made the accusations were, in many cases, allowed to confiscate the property of those they had accused, providing material incentives for lying and theft. New laws today allow private interests to "own" the genetic material of another's biological cell and use it for their own private profit.

Genetic engineering—the artificial recombination, private patenting, and production of altered or foreign DNA sequences of living cells—vanquishes Alexander the Great's complaint of having "no new worlds to conquer." It provides global capital with the basis from which to offset, for the time being, capitalism's accumulation crisis—a term that no longer refers only to the "primitive accumulation through conquest" of nature "out there" that Karl Marx described so vividly but also to the colonization of life itself, the nature *within.* The development of pesticides, and in particular Monsanto's Roundup, has played a crucial role in enabling this era's enclosures of the living cell.

Hundreds of years ago, the idea of "nature" became joined in the United States with that of wilderness. The poet of the American Revolution, Philip Freneau, was a tree-hugger in the late 1700s, who became "outraged when he learned that the trees of New York City were soon to become an endangered species. The city passed an ordinance that stated that after June 10, 1791, 'no tree was to stand within the city limits,'"[6] and Freneau went to war in verse against the government of New York City. One line in Freneau's most widely discussed poem stands out: "Trees now to grow is held a crime." Freneau, who had been publishing popular poems about trees he loved, such as "The Dying Elm," rallied the people of New

York City in support of the city's trees. The mass poetic uprising forced the city to repeal the Act.

Across the Atlantic in Europe fifty years after Freneau defeated New York City's attempt to clear all trees from its area, the twenty-four-year-old editor of Cologne's *Rheinische Zeitung* wrote forceful editorials in defense of trees against privatization, and in favor of the rights of peasants to collect dead wood from the forest floor. Just as biotech corporations today patent genetic sequences, humans in power have long tried to privatize nature for profit. In Europe, this included lands that had been unrestricted by law and used in common for millennia. The editor, Karl Marx, railed against the state's jackbooted stormtroopers' expropriation of the commons. Ecological justice was central to Marx's outlook from his earliest writings, and Marx—outraged by the cutting down of forests for private profit and the enclosure of lands used in common—denounced the state's criminalization of peasants who took dead wood for heating and cooking. Marx named that expropriation "primitive accumulation." He pointed out that by 1842, 85 percent of all prosecutions in the Rhineland dealt with a new crime: the "theft" of dead wood lying on the ground, which the state enforced only on peasants while allowing wealthy businessmen and corporations to strip whole forests with impunity. Marx, especially, explained how such "enclosures" came to receive social acceptance and sanction in law.

How did it happen that people allowed trees on public lands to be privatized and reshaped to serve the needs of capital? Why didn't people revolt? (Well they *did*, according to Silvia Federici, in *Caliban and the Witch: Women, the Body and Primitive Accumulation*.) We can ask the same questions today: How did our once-public universities, hospitals, beaches, libraries, drinking water, parks, individual DNA sequences, and even prisons in the United States suddenly become privatized? Private mercenary armies now make up a large percentage of soldiers fighting in Iraq and Afghanistan; the water tables are so polluted that drinking water is now sold in plastic bottles, their sources owned by the world's largest corporations and stolen from local communities left without clean drinking water,[7] while the plastic wastes have accumulated in the Pacific Ocean forming a floating plastic island three times the size of Texas.

In the last fifty years, fully 80 percent of the world's forests have been chopped down. Forests prevent floods; they maintain healthy soil; they defuse hurricanes and detoxify drinking water. They oxygenate the air and serve as habitats for thousands of species. In the United States, less than 5 percent of the old-growth forests remain. In Argentina and Brazil, huge swaths of primeval rainforest are being cut down to enable genetically engineered soy to be monocropped for animal feed and biofuels exported to the United States, Japan, China, and Europe. In Indonesia, millions of acres of forest have been burned for palm oil production, mining, and cattle grazing. In Mexico the Lacandona forest—the home of the ancient Mayan people and the Zapatista rebellion—is under siege by international paper companies as much as by federal troops. Under the Clinton/Gore administration, more trees were clear-cut in the United States than by any of their predecessors in recent history—or even by all of them combined. Meanwhile, the *New York Times* cuts down 60,000 trees per week to publish its Sunday edition.

Anne Petermann, executive director of the Global Justice Ecology Project, writes that "plantations of genetically engineered eucalyptus trees have caused massive deforestation in Brazil, Chile and Portugal and have replaced biodiverse native grasslands in South Africa. They are the top tree being genetically engineered for traits including herbicide resistance, faster growth, reduced lignin and freeze tolerance. The USDA hopes to approve GE freeze tolerant eucalyptus trees for vast plantations across the coastal plain and gulf coast of the Southern us," which will, ironically, lead to expanded deforestation in the Southeast.

"Pharming," Genetic Engineering, and the U.S. Military

Monsanto, Dow, Bayer, DuPont, and other agribusiness and pharmaceutical corporations work closely with the Pentagon and the Veterans' Administration in developing herbicides like 2,4,5-T, a component of Agent Orange, sprayed extensively by the U.S. military in Vietnam[8] and responsible for contamination of water supplies, rice paddies, and forests. They've caused horrible ailments in soldiers, chromosomal damage, and deformities in children, with effects lasting to this day. Monsanto and Dow intentionally falsified key data on the effects of Agent Orange on human health.[9] Admiral Elmo Zumwalt, commander of U.S. naval forces in Vietnam and member of the Joint Chiefs of Staff, charged that the government's exoneration of

Agent Orange in Vietnam was "politically motivated . . . to cover up the true effects of dioxin, and manipulate public perception."[10]

Earlier this year, Vietnam—not the United States!—banned the spraying of herbicides paraquat and 2,4-D. "This is particularly a significant victory as many rural farmers, women and children are poisoned by those herbicides," said Nguyen Thi Hoa, deputy director of Centre for Sustainable Rural Development.[11]

In the early 1990s, Novartis (the gigantic corporation invented by combining pharmaceutical companies Ciba-Geigy and Sandoz) patented a way of encoding each corn plant to produce *Bacillus thuringiensis,* an otherwise naturally occurring insecticide, resistance. Now, every cell of genetically engineered "BT corn" is turned into a mini-pesticides factory. Not only does BT corn kill insects beneficial to crops, it speeds up the development of resistance in so-called pests; and because so much agriculture in the U.S. is monocropped, diseases are able to spread quickly across entire fields. Resistance to Roundup and to Bt crops is now widespread.[12] (See "Monsanto: Origins of an Agribusiness Behemoth" by Brian Tokar earlier in this book for more on the Bt toxin.)

Monsanto covered it up, just as it used a similar playbook to hide the results of tests that showed that aspartame, the chief ingredient in the sugar substitutes NutraSweet and Equal, caused brain lesions in laboratory rats.[13] (The rats, of course, are unwilling participants in these tests, as are all animals encaged for experimentation and testing of chemical poisons.) The same happened with its manufacture of PCBs, until Congress finally banned PCBs in 1979.[14] Monsanto's factory in Sauget, Illinois, discharged an estimated 34 million pounds of toxins into the Mississippi River. The facility was a major producer of chloronitrobenzenes, bioaccumulative teratogens, which have been detected at levels as high as one thousand parts per billion in fish over one hundred miles downstream almost forty years later and are ubiquitous in the global ecosystem.[15]

Monsanto also manufactures the herbicide butachlor (trade names: Machete, Lambast), which poses both acute and chronic health risks, especially via contamination of water supplies. Although manufactured in Iowa, butachlor was never registered for use in the United States, nor did it gain a food residue-tolerance permit. In 1984, the EPA rejected Monsanto's registration applications due to "environmental, residue, fish and wildlife, and toxicological concerns."[16] Typically, Monsanto refused to submit additional data requested by the EPA. Despite being banned for use in the United States, Monsanto is allowed to manufacture Machete here and sell it abroad. Dozens of countries in Latin America, Asia, and Africa use the product primarily on paddy rice, which is then imported back into the United States and consumed here.[17]

A few of the large not-for-profit environmental groups didn't want to jeopardize their funding (nor their salaries) by joining in the No Spray Coalition's efforts to ban *all* chemical pesticides *and* the genetic engineering of plants. They did not want to examine how the United States uses genetically engineered crops (and now trees), and the pesticides they require, to disrupt the economies of other countries, forcing them into dependency, nor to make an issue of the fact that almost all U.S. food aid to the victims of the Tsunami in the South Pacific and to earthquake victims in Pakistan is genetically engineered and saturated with pesticides. One result of the U.S. "police action" in Somalia in 1992 was the planting of thousands of acres of genetically modified cassava.[18] Following the U.S. bombing of Iraq in 2004, U.S.-appointed administrator of the Coalition Provisional Authority, L. Paul Bremer, issued Order 81, officially titled Amendments to Patent, Industrial Design, Undisclosed Information, Integrated Circuits and Plant Variety Law. The edict prohibited farmers from saving seeds from genetically engineered crops and made it illegal for them to replant those seeds. Bremer's edict was part and parcel of the IMF "structural adjustment program." This new law protected the privatization of Monsanto's genetically engineered seeds and opened up Iraqi agriculture to the cultivating of GMOs, where it had been unavailable before the bombing and edict.[19]

US foreign policy has long fingers and finds ever new ways to squeeze profits, power, and control from ancient dust—and it does so in ways that most of us have never thought of. For example, the manufacture of synthetic vanilla by the United States has been wreaking havoc on the economies of Madagascar, the Comoros Islands, and Reunion Island, which depend on natural vanilla exports as their primary source of income. Here's another example: lauric acid, a substance used by industry in soaps, cosmetics, chocolate, and other foods, is historically derived from palm kernel oil and coconut oil in the Philippines and a handful of small countries where it is the main means of economic stability. Today, it is increasingly produced in the

United States and Canada from canola (rapeseed) that has been genetically spliced with DNA from the California bay plant.[20]

These are business decisions behind U.S. government policy, with ramifications for political and social matters. More significant than coconuts and canola, foremost among the food products through which the International Monetary Fund and World Bank control other countries are corn and rice. The sale under NAFTA and other trade agreements of cheap genetically engineered corn to Mexico threatens to undermine stable indigenous communities centered around the dozens of varieties of local corn.[21] David Quist, a graduate student at the University of California at Berkeley, and Ignacio Chapela, professor of Environmental Science, Policy, and Management at the same university, discovered that indigenous corn varieties from Oaxaca contained DNA from GMOs.[22]

S'ra DeSantis, at the Institute for Social Ecology in Vermont, explained that since so many producers are dependent on their small plots of land to feed their communities, "the DNA from GMOs could interfere with the expression of unique physical characteristics and genetic predispositions, making the indigenous corn less suitable for its particular environment. As indigenous corn varieties lose their ability to produce in southern Mexico, yields will decrease and the campesinos' livelihoods will be undermined."[23] The export of GMOs to Mexico sets in motion a chain of events that enables large corporate farmers to displace Mexico's indigenous farmers, supplanting them with genetically engineered and pesticide-saturated corn from abroad, despite that country's attempts to prevent it. Foreign corporations are then able to consolidate their hold on the food production apparatus as well as the market.

The Zapatistas in the Lacondona forest in Chiapas, Mexico, recognized this threat from the start and, focusing on opposing the importation of corn and the destruction it would cause to their local, self-sufficient economies, launched their famous rebellion in southern Mexico on January 1, 1994—the day the North Atlantic Free Trade Agreement (NAFTA) was to take effect.

The Zapatistas had it right, and not only about the economic consequences of hybrid and genetically engineered corn imports from the United States. They understood that by genetically engineering crops, the U.S. government and corporations would increase by a factor of ten the power of using food as a weapon.

In 1998, protesters in Europe prevented a barge filled with genetically engineered corn from docking in Switzerland. The Swiss government agreed with the protesters and refused it entry. Like the fabled "garbage barges" of decades past that circled the globe finding no takers for their trash, the genetically engineered corn sailed up and down the Rhine in search of a depot.[24] (We might note that the number of salmon caught each year in the Rhine 120 years ago at the time Marx was railing against increasing capitalist control of the forests was around 150 thousand; by the time John F. Kennedy made his famous proclamation in Germany, "*Ich bin ein Berliner,*" there were exactly *zero* salmon in the Rhine, all having been wiped out over the decades by the poisons dumped or that had run-off into that river.)[25] Similarly, protesters stepped up efforts to keep genetically engineered food out of Germany.[26] As a result, Germany is today one of the countries most adamant in opposing the planting of genetically engineered seeds and the importing of genetically engineered food into Europe (although the German government has also succumbed to corporate pressures to allow some planting of genetically modified seeds, not for food, but for "industrial purposes" such as extracting ethanol for automobile fuel).

It was Fidel Castro—accidental environmentalist that he was during the last two decades of his life—who sharply criticized the misuse of the world's available land to genetically engineer corn and sugar for biofuels (to produce ethanol for cars), which Fidel noted was not only immoral but would result in the elimination of the world's forests and, in Fidel's words, lead to "the internationalization of genocide."

The GMO industry, however, pressed for biofuels, and some mainstream environmental groups joined them in their advocacy. They focused on what they felt was a "socially responsible" end-use of the product in automobiles, taking the place of gasoline, rather than on the consequences of bribing peasants to plant and maintain GMO crops for export to be used as fuel. Though healthy food is a true necessity, the industry portrayed genetically engineered and highly herbicided biofuels as a savvy ecological answer to humanity's continued dependence on fossil fuels. As a result, Monsanto, Syngenta (merger of part of Novartis with AstraZeneca), Eli Lilly, Bayer, Aventis, and four or five other giant pharmaceutical and biotech companies, along with the IMF and World Bank, gained allies among some environmentalists. With the help of a few "progressive" groups whose interest in the issue was in making incremental moves away from fossil fuels, Monsanto et al. was able to in effect dictate agricultural and economic policies throughout the world.

Africa has become a major GMO seeding ground. Backed by the might of the U.S. military, the planting of bioengineered crops and the consequent patenting of seeds are intended to drive indigenous people from the lands they'd traditionally shared. The IMF has been quite explicit about this. In 1981, U.S. officials ordered that all the pigs in Haiti be destroyed to prevent the spread of swine flu, they said, even though Haiti's 1.3 million pigs seemed to be immune to the disease.[27] At the same time, the IMF offered its new "structural adjustment program" for Haiti, openly stating its goal of confiscating one-third of all rural lands from peasants, even though their families had grown food on them for many generations, and turning them over to agribusiness conglomerates to produce export crops: sugar, coffee, and cotton. All the pigs were killed, and formerly self-sufficient peasants were driven from their lands. They traveled to cities, often to other countries, in search of work. Some were sold as slaves to the sugar plantations in the Dominican Republic. Others ended up in sweatshops in assembly zones, where even the pretense of environmental regulation, unions, minimum wages, and health and safety conditions is anathema. Export zones (in Mexico, they are known as *maquiladoras*) are the inner circle of Dante's inferno, neoliberal regions of vast and depraved exploitation where toxic wastes—the effluents of affluence—pour into the streets, drinking water is contaminated, and incinerators burn endlessly, filling the air with lead, dioxin, cadmium, asbestos, mercury, and PCBs. Glyphosate is just one of their problems.

Genetic engineering is an essential component of the new globalization of capital. It provides private corporations—and the governments they control—with the means to conquer those parts of life that have thus far stood outside of their domain: the inner workings of the living cell.

Welcome to Dr. Frankenstein's universe,[28] or, as some have termed its application to agriculture, "FrankenFoods." Giant corporations are now legally allowed to patent genetically modified organisms (GMOs) and release them into the wild, while claiming ownership of the seeds those plants generate. In some cases, the seeds are designed to germinate only when sprayed with certain pesticides.[29] Farmers are forbidden by contract to replant seeds from crops engineered by Monsanto unless they pay the company royalties; Monsanto has set up its own police force to go onto farmers' lands and confiscate crops to test for patented DNA sequences—even if caused by genetically engineered pollen that drifted onto pristine fields. Where farmers throughout time immemorial had saved seeds from existing crops for replanting in future seasons, farmers today are increasingly forced to lease their seeds, plants, and animals from biotech conglomerates.

Unlike defective products of other technologies—say, faulty automobiles, for instance—genetically altered organisms cannot be "recalled" once they've been released. Genes spliced from different and previously self-contained species become part of the natural world and, through reproduction of the organisms they're part of, spread on their own. Once released, they are irretrievable and self-replicating. Herbicide-resistant qualities can spread to other plants.[30] Rapid-growth capacities have already become the hallmark of genetically engineered salmon, which then compete with other fish and alter what remains of the ecological balance in the area. Antibiotic resistance can spread to bacteria such as staphylococcus, diphtheria, salmonella, bubonic plague, cholera, typhoid, and a whole range of dangerous diseases. And genes for new and virulent toxins can, accidentally or purposefully, spread to wild plants. Researchers at Riso National Laboratory in Denmark have found that plants—whose natural immunities develop over many years through the interaction of many varieties, species, and microbes as part of an interdependent (if fragile) ecology—spontaneously cross-fertilize. Genetically engineered canola (rapeseed), for instance, passes its genes for herbicide resistance to compatible weeds; the same is true of other plants. The offspring resulting from the crossbreeding of genetically engineered plants and "weeds" not only exhibit herbicide-resistant traits themselves but also are capable of passing on herbicide and antibiotic resistance to subsequent generations. Engineering on the genetic level introduces dangers of a qualitatively different sort than whole organisms that are spliced or hybridized. At what point do the dangers become irreversible? Pesticides are designed for monocropped fields of genetically engineered crops that would otherwise be unable to compete with other plants, incurring diseases that a healthy and natural crop would ward off.

Meanwhile, new "trait-controlled" technology—dubbed by Pat Mooney of Rural Advancement Foundation International (RAFI, now ETC group) "Traitor" technology—engineers seeds that grow into crops that will die unless treated with a particular chemical, thus turning chemical pesticides into "vitamins" necessary for plant survival and enabling the manufacturers to increase their control over the world's food supply.

Throughout the world, as people begin to understand the implications of genetic modification of agriculture, the new genetic technologies—and their dependence on monocropped, pesticide-saturated, and water-guzzling factory farms—triggered widespread opposition. Ecological activists in England pulled up experimental fields of genetically engineered crops. In India, farmers burned their own fields, and tens of thousands committed suicide in despair over being required to pay for patented seeds. Some areas are fighting for an outright ban and, in the meantime, for labels on all foods containing genetically engineered products, under the debatable assumption that if people knew what was actually in their food they'd not buy it. There are already significant limits on GE growing and imports in dozens of countries. There are GMO and pesticide-free zones and moratorium resolutions throughout the United States, as well as in hundreds of European municipalities.[31]

The tens of thousands of activists blockading delegates to the World Trade Organization meeting in Seattle in November 1999 listed opposition to genetic engineering and the global domination of local farming by foreign corporations as among their main concerns. That opposition spread throughout the world.

Once perceived as "flakes" or "environuts," resisters to genetically engineered agriculture are turning out to be heroes, far-sighted activists willing to challenge the self-serving rationalizations offered by the pharmaceutical industry and its bought politicians. The European Union has attempted to enforce a moratorium on purchases and planting of genetically modified crops, as have many small nations inspired by Zambia's 1999 rejection of GMO "emergency" food from the U.S. Aid for International Development (USAID) and the International Monetary Fund, following that country's blocking of the World Trade Organization's formation in 1999 over that very issue. The protests have now reached into the inner sanctums of the GE corporations themselves. A few years back, the Gerber Company announced that it would no longer use genetically engineered crops in its baby foods. The irony is that Gerber is owned by Nestlé, which doesn't shun the use of GMOs in its other products. (And there have been questions raised as to whether Gerber is using engineered microbes to produce the vitamins it uses in its baby foods.)[32] Similarly, the company serving lunch in Monsanto's cafeteria in England announced it will no longer serve genetically engineered produce.

But even as resistance increases in some areas of the world, and even as independent findings confirm bit by bit what ecological activists have long feared, and even as we learn that Roundup sprayed on genetically modified corn kills milkweed, the food source of Monarch caterpillars, and that the engineered corn kills millions of bees generating a cascading chain reaction through many species, the industry's propagandists, including such shills for Monsanto as former U.S. President Jimmy Carter,[33] are busy trying to bolster the market for GE products in Africa, Latin America, and China.

In the face of Monsanto's plans to open its first plant in Russia,[34] Russian President Vladimir Putin signed Federal Law No 358 on July 3, 2016 prohibiting cultivation of genetically engineered plants and breeding of genetically engineered animals on the territory of the Russian Federation.[35] President Putin had earlier said that "measures should be taken to protect the Russian market and consumers from GMO products, as their use could have unforeseen consequences," news agency RIA Novosti reported.[36]

Leticia Goncalves, who heads Monsanto's operations in Europe and the Middle East, was not perturbed. "We still believe that Ukraine and Russia both are long-term opportunities for our business and we want to make sure we are in a position to accelerate our business growth despite the short-term geopolitical and macroeconomic challenges."[37]

But, contra Monsanto, President Putin envisioned a future in which Russia would become "the world's largest supplier of ecologically clean and high-quality organic food." He's called on the country to become completely self-sufficient in food production by 2020: "We are not only able to feed ourselves taking into account our lands, water resources—Russia is able to become the largest world supplier of healthy, ecologically clean and high-quality food which the Western producers have long lost, especially given the fact that demand for such products in the world market is steadily growing."[38]

Before he died in November 2016, Fidel Castro entered the increasingly global debate by strongly opposing genetic engineering. At a conference in Havana cosponsored by the Mexico-based Center for Global Justice,[39] I introduced a public appeal against the genetic engineering of agriculture, the end of my talk, and a number of participants signed on to it, including several Cuban professors.[40]

Annual "Millions Against Monsanto" marches and protests take place throughout the world;[41] they draw on the effective resistance over two decades to Monsanto's manufacture and dissemination of genetically engineered Bovine Growth Hormone.[42] Recombinant Bovine Growth Hormone (rBGH) is a genetically engineered drug injected into cows to increase the amount of milk produced, and which also increases the levels of cancer-causing chemicals in milk. Milk derived from rBGH contains dramatically higher levels of IGF-1 (Insulin Growth Factor), a risk factor for breast and colon cancer. IGF-1 is not destroyed by pasteurization. High levels of IGF-1 are also linked to hypertension, premature growth stimulation in infants, gynecomastia in young children, glucose intolerance, and juvenile diabetes.[43]

Dr. Samuel Epstein, professor of occupational and environmental medicine at the University of Illinois School of Public Health and chair of the Cancer Prevention Coalition until his death in early 2018, reported that IGF-1 causes cells to divide, induces malignant transformation of normal breast epithelial cells, and is a growth factor for human breast cancer and colon cancer. Yet, as has been the case with virtually all genetically engineered foods and hormones, rBGH was never adequately tested before the FDA allowed Monsanto to market it. A standard test of new biochemical products and animal drugs is based on twenty-four months of testing with several hundred rats. But rBGH was tested for only ninety days on thirty rats. This short-term rat study was submitted to the FDA but never published. The FDA—filled with agribusiness appointees including several from Monsanto itself—had refused to allow anyone outside that agency to review the raw data from this truncated study, saying it would "irreparably harm" Monsanto.

In 1998, Canadian scientists managed to obtain the full studies for the first time. They were shocked to learn that the FDA never even looked at Monsanto's original data on which the agency's approval of rBGH had been based. In reviewing the data, the Canadian scientists discovered that

Monsanto's secret studies showed that rBGH was linked to prostate and thyroid cancer in laboratory rats.[44]

Monsanto had actually cut the study short and omitted any mention of the cancers in its report to the FDA—or so the agency now says. And so, a few companies, which had invested hundreds of millions of dollars developing a product having absolutely no consumer benefit and that poses severe health risks, were able to foist a dangerous product on an unprotected populace with the help of the government. Its manufacturer, the Monsanto Company, has treated disclosure of these effects with the same contempt for public health it exhibited in stonewalling critics of some of its other products. The company has been contracting with farmers to inject cows with rBGH since 1994. Still, many farmers continue to resist.

Farmers have also protested by dumping milk from cows injected with genetically engineered hormones. Consumer activists have exposed the executives of biotech corporations and their supporters in government, holding them up for public ridicule and condemnation. Meanwhile, Monsanto flouted the law at every opportunity. For instance, the company claimed that every truckload of milk in Florida is tested for excessive antibiotics, which are given to rBGH cows to treat mastitis—a swelling of the udders and pus-filled bloody lesions, which occur seven times as often in cows injected with rBGH than in those that are not.[45] But Florida dairy officials and scientists say that Monsanto's claims to have tested every truckload of milk in Florida are simply not true. Another law required Monsanto to notify the FDA about every complaint the company received from dairy farmers such as Charles Knight, who had complained to Monsanto that the rBGH injections were killing his cows. But the Food and Drug Administration said that the company hadn't submitted Knight's complaint, even though required to do so. When confronted, Monsanto officials later admitted they had not reported Knight's complaint because, they said, the company brass didn't realize that Knight was complaining about rBGH.[46] Similarly, Monsanto says that Canada's ban on rBGH had nothing to do with human health concerns. But Canadian government officials say just the opposite and, in fact, they claim that Monsanto had tried to bribe them with offers of $1 to $2 million to gain approval for rBGH. (Monsanto officials don't deny the "offers," but say those funds were offered for "research.")

As a consequence, outraged consumers in state after state demanded legislation requiring labeling, at the very least, of dairy products derived from rBGH-treated cows. Legislation to that effect was, and continues to be, torpedoed as often by Democrats as Republicans at the behest of Monsanto. Instead, legislation is annually introduced to Congress *that limits the liability* of corporations. Al Gore, G. W. Bush, and Hillary Clinton are just a few of those who have received large contributions from the pharmaceutical companies and who have gone to bat for genetically engineered agriculture.

Genetic engineering—tethered at the hip to the manufacture of herbicides like glyphosate, dicamba, Dow's Enlist Duo (a mix of glyphosate and 2,4-D), and more—reduces everything in nature to objects for commercial manipulation and commodification.[47] (The compound Enlist Duo is even more dangerous to humans because it can transform into a gas after being sprayed and carry the glyphosate it contains for long distances.)[48]

All of this was known by the late 1990s, but Roundup is still on the market; Monsanto and its executives enjoy impunity and have not yet faced criminal charges. Under the Democratic Party administrations of Bill Clinton and Al Gore, and Barack Obama and Joe Biden—to say nothing of the Republican Party administration of George W. Bush and Dick Cheney—Monsanto and other corporate giants were given free rein to devise government policy; the company scoffed at calls to implement the precautionary principle, which would put the responsibility onto the manufacturer for proving the chemicals to be safe before being allowed to market them. And, at least thus far in the United States, it has succeeded in blocking grassroots environmental movements that try to ban pesticides and the genetic engineering of crops.

Revolving Door

The FDA official who made the decision to approve rBGH without performing the required long-term health studies, Michael R. Taylor, had been a law partner at the law firm of King & Spaulding and privately lawyered on behalf of Monsanto during the FDA approval process of rBGH. President Bill Clinton appointed Taylor to the Food and Drug Administration, where he fast-tracked rBGH's approval. Taylor soon became Deputy FDA Commissioner and appointed others from Monsanto to positions at the FDA, with President Clinton's blessing.

In March 1994, the Pure Food Campaign and the Foundation for Economic Trends, headed by Jeremy Rifkin, petitioned the FDA and the Department of Health and Human Services to investigate Michael Taylor's conflict of interest. Three members of Congress asked the General Accounting Office to investigate. Within days of the FET complaint, Taylor was mysteriously transferred out of the FDA.

But in 2009, President Barack Obama re-appointed Taylor to the Food and Drug Administration, where he collaborated with the Rockefeller and Bill and Melinda Gates Foundations as "the go-between man for Monsanto and the U.S. government, this time with the goal to open up African markets for genetically-modified (GM) seed and agrochemicals."[49]

In the Clinton era, where the revolving door between government policy-making positions and private for-profit corporations was spinning as fast as Henry David Thoreau was whirling in his grave, dozens of government officials did Monsanto's bidding on the public weal. In 1999, U.S. Vice President Al Gore actually ordered the Environmental Protection Agency to *slow down* its implementation of stricter standards for agricultural pesticides.[50]

Should any of the numerous legal cases make their way to the Supreme Court they will come before, among others, Justice Clarence Thomas. Thomas began his career as a lawyer for . . . Monsanto. Similarly, 2016 presidential candidate Hillary Clinton is also no stranger to promoting Monsanto—for a price. A front group for the corporation paid her $335,000 for giving the keynote address at the BIO convention[51] in San Diego in 2014, where she coached biotech lobbyists on how to overcome consumer fears over GMOs. "Rather than lecture the audience on the need for transparency and improved safety assessments, Clinton coached the audience of biotech devotees to develop "a better vocabulary" to change negative public perception about GMO agriculture: "'Genetically modified' sounds Frankenstein-ish. 'Drought-resistant' sounds like something you'd want," said Clinton. "Be more careful so you don't raise that red flag immediately."[52]

Food Democracy Now also expressed alarm at Hillary Clinton's misuse of her position as head of the State Department to promote GMOs around the world. The organization says that Clinton "abused her authority by having State Department officials threaten leaders of other nations for not wanting to approve GMO crops for sale in their country." On May 14, 2013, the *New York Daily News* reported that State Department officials under

Hillary Clinton were actively using taxpayer money to promote Monsanto's controversial GMO seeds around the world.[53]

Even worse, writes Food Democracy Now, a batch of diplomatic cables released by Wikileaks revealed that officials in Clinton's State Department were actually intervening at Monsanto's request "to undermine legislation that might restrict sales of genetically engineered seeds." Under Hillary Clinton, the U.S. State Department was so gung-ho to promote GMOs that *Mother Jones* writer Tom Philpott called it "the de facto global-marketing arm of the ag-biotech industry, complete with figures as high-ranking as former Secretary of State Hillary Clinton mouthing industry talking points as if they were gospel."[54]

The Monsanto Company is also listed among the entities donating between $1 million and $5 million to the Clinton Foundation, the not-for-profit corporation established by former President Bill Clinton to "strengthen the capacity of people throughout the world to meet the challenges of global interdependence"[55] but which is under investigation by the Justice Department and legitimate attack by activists for pilfering millions of dollars that were supposed to be going to earthquake relief in Haiti.

16

Big Science and the Curious Notion of "Progress"

The radical social movements of the 1960s blossomed into an emerging ecology movement. In the 1980s, the women's liberation movement began issuing profound critiques of science. Previously, Marxists had endorsed Vladimir Lenin's industrial-centered view that socialism equals workers' councils (soviets) plus installation of electrical wiring—a view of socialism, and of science, that was not very different than liberal policy-makers in industrial capitalist countries. But in the 1970s, a number of scientists began to challenge not only the Marxian question of who owns and controls Science (with a capital "S," what I call "Big Science"), but explored, questioned, and challenged the cultural and political assumptions embedded in the scientific method itself and its "search for objectivity," which had been the goal of scientific inquiry since the Enlightenment and the industrial revolution. Radical scientists in the social movements produced magazines such as *Science for the People*, with the aim of demystifying science and re-examining science's core notions. Richard Levins, Martha Herbert, Ivan Illich, and many others published sustained critiques of science, and created an El Niño of sorts that wound its way into the late 1980s. All during this time, anarchists, too, were issuing critique after critique of corporate environmentalism, which led to the school of thought known as social ecology.[1] These incisive intellectual investigations, all occurring at the same time and feeding off of each other and the social and antiwar movements they were part of, revealed the capitalist ideological imperatives concentrated in the very essence of science. The currents cohered into mass-movements against nuclear power, genetic engineering, global warming, the globalization of capital, the robotization of work, the massive application of pesticides, and in favor of animal rights and alternatives to the pharmaceutical-industrial model of health care. And, dialectically, these movements inspired renewed interest in radical critiques of science and industrialization.

The movements to task the idea that science and technology are some-how "neutral" and "objective," and challenged that framework as itself part of an ideological construct and a figment of capitalist mythology. So too with what they saw as capitalism's similarly reified invention of a universal, greedy, and unchanging human nature, which "objective" science first posits and then finds wanting. Challenge *everything!* Once taken for granted by leftists as devoid of politics, the factory form of production became, under this new radical understanding, *dripping* with ideology. Capitalism makes the factory form of production seem necessary and also inevitable; it is a means for achieving a certain kind of rationalized efficiency, and of control-ling nature, including human nature. We can't think of any other ways to do things.

For the purposes of this book, we would do well to re-examine the things we take for granted pertaining to science, pesticides and politics, and especially those ways of thinking of which we are not aware, to bet-ter understand and to change our relationship to our natural environment. It's tempting to not have to do any of that, to say, "The evidence speaks for itself." But evidence rarely speaks for itself. Seemingly objective facts require interpretation. And the interpretations given by science, for the most part, are based on assumptions hidden even to honest scientists, despite their sometimes good intentions and brilliance—to say nothing of those scientists bought by Monsanto and other corporations.

Several generations prior to the critical re-assessments by the women's liberation and ecology movements, physicist Werner Heisenberg (he of the Uncertainty Principle) observed: "Natural science does not simply describe or explain nature. It is part of the interplay between nature and ourselves; it describes nature *as exposed to our method of questioning.*"[2]

Where does "our method of questioning" come from? What does it con-sist of?

As the social movements of the 1960s thru mid-1980s receded, their philosophical contributions were de-fanged, assimilated, and turned into lucrative commodities by an expanding and triumphal capitalism, as was their music, graffiti, and other forms of once revolutionary critique. Heisenberg's observations, consistent with his quantum observations, were barely understandable to most scientists (let alone everyone else). Science regrouped and regained its momentum under the prevailing Western notion that "objective" scientific facts exist somewhere "out there" waiting to be discovered, independent of our methods of questioning.

One who continued to challenge the dominant orthodoxy in observing the ways in which social conditions influenced and interacted with scientific thought was Stephen Jay Gould, a towering figure in late twentieth century science, who observed, "Science is no inexorable march to truth, mediated by the collection of objective information and the destruction of ancient superstition. Scientists, as ordinary human beings, unconsciously reflect in their theories the social and political constraints of their times. As privileged members of society, more often than not they end up defending existing social arrangements as biologically foreordained."[3]

Consequently, scientists generally endorsed the development of expanded technological projects in capitalist as well as in self-described socialist countries. (Lenin was a big fan of Taylor's "time and motion" studies, in which human movements were stripped of their holistic meaning and broken down into their parts, the better to get the most production out of an assembly-line worker.) They argued that the social good such projects would bring about outweighed whatever future negative ramifications they might have, and which, separated from social movements, they rarely bothered to consider.[4] Far too many leftists, for example, failed to be critical of nuclear power plants in the 1950s and 60s, and the misnamed Green Revolution in agriculture, the massive misdiagnosis and drugging of "hyperactive" children, the return of electro-shock "therapy," the mass-spraying of pesticides, and the genetic engineering of organisms and the development of biotechnology. All were rationalized by perceived social benefit, but their effects turned out to be environmentally and socially devastating.

Unasking the Question

Since we are not taught to unearth and to question basic assumptions of Western society, they generally go unnoticed until something—often a social and political movement—forces us to examine them. Biologists Richard Levins and Richard Lewontin explain how this plays out on levels of consciousness that we almost never consider, such as when scientists construct their research from the "coming together of individual atomistic bits, each with its own intrinsic properties," and expect it to cohere into or determine the behavior of the system as a whole. The methodology reflects the way they (and everyone) are taught to think, which is reinforced every day by the things we encounter in industrial capitalism.

For example, a chemical pesticide may appear to be needed to kill weeds in a field of genetically engineered soy. Levins and Lewontin criticize that way of seeing as "reductionist," because it doesn't consider the larger view of why the farm is monocropped and set up in such a way to begin with, which enables diseases to wash right through entire fields, and which thus require pesticides to keep the crops alive and intact. This method is very common in Western sciences today, in which "lines of causality run from part to whole, from atom to molecule, from molecule to organism, from organism to collectivity."[5] It is a way of finding out about the world that entails cutting it up into bits and pieces (conceptually, as well as in actuality) and attempts to reconstruct the properties of the system from the "parts of the parts" so produced, as they futilely try to put Humpty Dumpty together again by piecing together the individual fragments. Levins and Lewontin explain that "those problems that yield to [this kind of] attack are pursued most vigorously, precisely because the method works there. Other problems and other phenomena are left behind, walled off from understanding. . . . The harder problems are not tackled, if for no other reason than that brilliant scientific careers are not built on persistent failure."[6]

Geneticist, cell biologist, and Nobel Prize recipient Barbara McClintock opined, similarly, that the scientific method cannot by itself provide real understanding. "It gives us relationships which are useful, valid, and technically marvelous; however, they are not the truth," she says. Indeed, the *doing* of scientific work—even giving the benefit of the doubt to the chemical industry's scientists and their enablers in government—has become more and more atomized, fragmented, broken down into specialized disciplines and subdisciplines that in many ways are sealed off from each other: Not just biology, chemistry, physics, ecology, for example, but molecular biology, evolutionary genetics, and developmental embryology. The narrowing scope of research allows scientists to focus more intently on particular areas of interest, but it has also inundated us with information concerning individual pieces studied in isolation such that, paradoxically, the more information we gather the less we understand. Examining smaller and smaller isolated parts more often than not hammers into place *a way* of examining the world that precludes the ability to see or understand the whole, and to construct a morality and sense of justice based on it.

One purpose of this book is to restore a more holistic vision that provides a framework for understanding the patterns beneath all the related "facts." Once one accepts the negative categories "pests" and "weeds," it's but a small step to require pesticides to get rid of them. In rejecting the poisoning of living organisms, we've begun to shift the way we think of ourselves *in* the world around us. Such a revolution in consciousness is a prerequisite for completing a revolution in the social system dependent on the poisoning of other organisms—and ourselves.

The Atomization of Work

"You're a very good worker," said the efficiency expert schooled in the time-and-motion studies of Frederick Taylor, as he watched a carpenter plane a piece of wood. "Now if we can just stick a buffer on your elbow you could plane and buff the wood with the same motion."

"Yeah," the carpenter responded, "and if you'd stick a broomstick up your ass you could take your notes and sweep the floor at the same time."

In the movie *Modern Times*, Charlie Chaplin plays an assembly-line worker whose job is to wrench bolts all day as they come flooding down the conveyor belt, faster, ever faster. Charlie has no idea why. He just gets paid for it, and it warps his mind as well as his body.

The film is a blistering indictment of industrial production under capitalism. Like other assembly-line workers, Charlie is a victim of the "science" of mass production. In the early 1900s, Frederick Taylor introduced time-and-motion studies into industry, examining the fragmentary repetitive motions of the industrial labor process with the aim of increasing output and efficiency by subdividing each task and reducing each worker's movements as much as possible to mimic the mechanical motions of a machine.

Harry Braverman, in *Labor and Monopoly Capital*, explains the significance of this qualitative change in the way things were being produced on society in general, and what makes it unique to industrial capitalism:

> The division of labor in society is characteristic of all known societies; the division of labor in the workshop is the special product

of capitalist society. The social division of labor divides society among occupations, each adequate to a branch of production; the detailed division of labor destroys occupations considered in this sense, and renders the worker inadequate to carry through any complete production process. In capitalism, the social division is enforced chaotically and anarchically by the market, while the workshop division of labor is imposed by planning and control. Again in capitalism, the products of the social division of labor are exchanged as commodities, while the results of the operation of the detail worker are not exchanged within the factory as within a marketplace, but are all owned by the same capital. While the social division of labor subdivides *society,* the detailed division of labor subdivides *humans,* and while the subdivision of society may enhance the individual and the species, the subdivision of the individual, when carried on without regard to human capabilities and needs, is a crime against the person and against humanity.[7]

While all societies have historically featured various divisions of labor—some people farming, others hunting, etc.—the atomization of work into repetitive mechanical motions *within* those occupational divisions (what I'm calling, the "parts") was something new,[8] ushering in an entirely new period described by Karl Marx as the transition from the "formal" domination of capital to "the real."

Often lost in studying the specific mechanical function of a "part" is its relationship to other "parts" within the whole. It is not that the whole is more than the sum of its parts, but that by being parts of a particular whole, the *parts* acquire new properties, which they do not have in isolation or as parts of a different whole. And as the parts acquire new properties by virtue of their proximity to each other in the context of the whole, they impart new properties to the whole, which are reflected in changes to the parts, and so on.[9] There are often surprising and unpredictable qualities of any whole (an organism, species, political era, set of numbers, musical notes, or industrial production). *The "whole" shapes and defines the parts and their interactions as much as the parts shape and define the whole,* which in turn affects the parts. For example, in the case of cellular differentiation, the position of each new cell with respect to the surrounding cells, and not its genetic component alone, defines what each cell becomes. This relation is always in motion. I use the term "dialectical" to encapsulate all of this continuous interaction between different levels of complexity.[10]

Levins and Lewontin point out that "part" and "whole" have a special relationship to each other, in that one cannot exist without the other, any more than "up" can exist without "down." What constitutes the parts is defined by the whole that is being considered. Is something a "weed" or a plant? A "pest" or an insect? In what context? As already mentioned, what might be considered a "weed" in one context and marked for extinction could provide the medicines needed to cure the cancers caused by the very same chemicals deployed to exterminate them; what might be considered a "pest" in one context might, in another, serve as food for birds and frogs, pollinate plants, filter water, and remediate toxins in the soil.

On the human level, McClintock explained to her biographer, Evelyn Fox Keller, "one must have the time to look, the patience to 'hear what the material has to say to you,' the openness to 'let it come to you.' Above all, one must have 'a *feeling for the organism.*'" A revolution in consciousness requires more than rationality that just happens to be economically profitable; it requires, writes Keller, "a longing to embrace the world in its very being, through reason and beyond." For McClintock, Keller continues, "reason—at least in the conventional sense of the word—is not by itself adequate to describe the vast complexity—even mystery—of living forms. Organisms have a life and order of their own that scientists can only partially fathom."

Over the years, Fox Keller writes, "a special kind of sympathetic understanding grew in McClintock, heightening her powers of discernment, until finally, the objects of her study have become subjects in their own right; they claim from her a kind of attention that most of us experience only in relation to other persons. 'Every component of the organism is as much of an organism as every other part.'" McClintock adds: "Every time I walk on grass I feel sorry because I know the grass is screaming at me."[11]

Biologist McClintock's feelings for the world she studied led her to embrace a Buddhist perspective, with its affinities to the dialectical one I've outlined. *There can be no independent observer standing outside and apart from what she is observing.* There can be no "true" consciousness of *any* situation that doesn't, at the same time, enter, become part of, and transform it, and thereby one's consciousness *of* it. Consciousness is not a passive reflection of a static totality but an active engagement with that totality of which it, itself, is dynamically a part.

Evelyn Fox Keller and Barbara McClintock are an antidote to those numerous scientists who objectify organisms as "pests" and "weeds," to be exterminated by the wonders of modern science, with its poisonous sprays fueling the economic profits of such corporations as Monsanto, Dow, and DuPont. Keller gives us McClintock's view that "you need to have a feeling for every individual plant. . . . It is the overall organization, or orchestration, that enables the organism to meet its needs, whatever they might be, in ways that never cease to surprise us.[12]

That capacity for surprise gave McClintock, who died in 1992, "immense pleasure. She recalls, for example, the early post–World War II studies of the effect of radiation on *Drosophila* [fruit flies]: 'It turned out that the flies that had been under constant radiation were more vigorous than those that were standard. Well, it was hilarious; it was absolutely against everything that had been thought about earlier. I thought it was terribly funny; I was utterly delighted. Our experience with DDT has been similar. It was thought that insects could be readily killed off with the spraying of DDT. But the insects began to thumb their noses at anything you tried to do to them.'"[13] Could this same toxic assault lead to the rise of future generations of humans resistant to specific poisons (after it kills off most of us)? Perhaps, if one thinks in terms of the species-as-a-whole and ignores the suffering of individual people who are sickened or killed before they're able to reproduce, preventing the process of natural selection from "weeding" our human communities and evolving a new form of humanity resistant to the toxins they're spraying.

With the completion of the human genome project, more surprises were about to overturn the reductionist argument and restore a bit of humility. Stephen Jay Gould explains: "The fruit fly *Drosophila*, the staple of laboratory genetics, possesses between 13,000 and 14,000 genes. The roundworm *C. elegans*, the staple of laboratory studies in development, contains only 959 cells, looks like a tiny formless squib with virtually no complex anatomy beyond its genitalia, and possesses just over 19,000 genes."

It turns out that *Homo sapiens* possess around nineteen thousand genes too,[14] around the same quantity as the "lowly" roundworm! As Gould points out, under the old view of life human complexity could not be generated by nineteen thousand genes, with "one item of code (a gene) ultimately making one item of substance (a protein), and the congeries of proteins making a body," and where "fixing" an aberrant gene, for example, would thus cure a specific human ailment. The old view was wrong, and Gould welcomes our liberation from "the simplistic and harmful idea, false for many other reasons as well, that each aspect of our being, either physical or behavioral, may be ascribed to the action of a particular gene 'for' the trait in question." Gould continues:

> But the deepest ramifications will be scientific or philosophical in the largest sense. From its late 17th century inception in modern form, science has strongly privileged the reductionist mode of thought that breaks overt complexity into constituent parts and then tries to explain the totality by the properties of these parts and simple interactions fully predictable from the parts. . . . But once again—and when will we ever learn?—we fell victim to hubris, as we imagined that, in discovering how to unlock some systems, we had found the key for the conquest of all natural phenomena. . . . Organisms must be explained as organisms, and not as a summation of genes. . . . Moreover, these noncoding regions, disrespectfully called "junk DNA," also build a pool of potential for future use that, more than any other factor, may establish any lineage's capacity for further evolutionary increase in complexity.[15]

Evelyn Fox Keller continues along a similar anti-reductionist path. She quotes Barbara McClintock:

> Our surprise is a measure of our tendency to underestimate the flexibility of living organisms. The adaptability of plants tends to be especially unappreciated. Animals can walk around, but plants have to stay still to do the same things, with ingenious mechanisms. . . . Plants are extraordinary. For instance . . . if you pinch a leaf of a plant you set off electric pulses. You can't touch a plant without setting off an electric pulse. . . . There is

> no question that plants have [all] kinds of sensitivities. They do a
> lot of responding to their environment. They can do almost any-
> thing you can think of. But just because they sit there, anybody
> walking down the road considers them just a plastic area to look
> at, [as if] they're not really alive.[16]

Take the human or plant biological cell. Each cell (the part) contains the same genetic code as every other cell in the individual's body (the whole). How is it that the genes "know" which sequence of chemical reactions to turn on and which to turn off so that the cell becomes a particular kind? The reductionists critiqued by other authors in this book attribute cell differentiation to special genes, called "regulator genes," that tell the other genes what to do and when to do it. Well, one might wonder, what tells *them*?

One gets bogged down when trying to build up a picture of how a complex organism or ecosystem works by adding up and re-assembling the parts, as though they can be separated from each other, from the whole, from their development over time, and from environmental variables and function autonomously.

Take, for example, the Mississippi alligator, a reptile severely affected by the massive use of pesticides. Alligator eggs developing in the temperature range 26–30°C hatch females. Change nothing but the temperature, to 34–36°C, and *the same eggs* will hatch only males. Eggs that hatch in the range 31–33°C produce alligators of either sex, with the probabilities changing from female to male as the temperature rises. What causes the egg's temperature to change? The macro temperature is important—global climate change may play a role here and cause more male alligators to be born. On the other hand, there are counteracting factors, such as cooling rains—also subject to global climate change—and the time of year in which the eggs are laid (which may be changing too). Temperatures vary in the microenvironment immediately surrounding the egg. It turns out that, under normal circumstances, the most important factor in whether the alligator will be male or female is the egg's location within the nest. Eggs surrounded by other eggs tend to be slightly warmer and, thus, tend to hatch males. Eggs around the circumference tend to be slightly cooler and tend to hatch females.

Clearly, *genes* by themselves are not strict determiners as claimed by Richard Dawkins in his popular book *The Selfish Gene*. They depend upon and interact with the surrounding microenvironment—in this case, the temperature of the air in the immediate vicinity—which, in turn, influences environments at other levels, such as the chemistry of the cell, which is the genes' immediate environment. The problem of where to draw the boundary of the immediate environment (its "community")—in this case the gene's—plays a critical role in determining what is "objectively" happening.

In addition, the three-dimensional double-helix configuration of DNA is guided by nontranscribed segments of the genome that geneticists until recently called "junk DNA." How do these interact with the microenvironment in shaping the sequences of which they themselves are a part?

Reframing Everything We Take for Granted

The holistic basis for reframing the way we see pesticides should by now be coming into focus. It's not just about outlining the dangers of pesticides but also *how to think about them and their effects on the complex interactions of living organisms*. Understanding an organism's relationship to the ecosystem in which it lives (as well as the ecosystem *within*) requires ways of seeing that carry beyond the cause and effect linearity to which we are accustomed. The sex of individual alligators, as well as the sexual dispersal over the population, is not determined by one isolated gene but, at the very least, by environmental temperatures working in a sort of feedback loop with the full genetic complement; it is influenced by the interaction of variables from different levels of complexity: temperature, genes, location of the egg in the nest, environment within the eggs, and of course the gross destruction of the alligator's natural habitat.

Stuart Newman, professor of cell biology at New York Medical College, points out that the position and relationship of each new cell with respect to the surrounding cells "bring out" specific qualities that define what each cell does. Will it be a muscle cell? A blood cell? A bone cell? A skin cell? Each kind of cell performs specific functions in the body that differ greatly from other kinds. And yet, within a given organism—indeed, within a given group of similar organisms, called a *species*—each cell is made up of the same number and sequence of chromosomes as every other cell in that species.

The kind of cell each becomes is as strongly influenced by its context and location—its relationship to its surrounding environment—as by the type of parent cells it had.[17]

Interactions among organisms create complex environments that then feed back and reshape the very organisms said to have caused them, transforming the entire relationship. But such holistic thinking does not characterize Monsanto's business model. And the company's approach is reflected in the ways government officials and media think about and externalize the environment. Government officials are influenced by the chemical industry's promises, and the industry's monetary contributions to those politicians' campaign chests grease the wheels in helping politicians accept the industry's products with scant testing, even when the politicians on occasion express concern over the industry's excessively effusive claims.

In the United States, it is common for we, the people, to think in terms of cause and effect, every effect being determined by one or a few causes, every trait being determined by and an expression of one or a few genes. Such was the case for early models of how DNA determines genes, genes determine chromosomes, which determine cells, which determine tissues, which determine organs, which determine organisms, and on out into the multi-layered cosmos. According to the original genetic models of the 1950s and 1960s—which still dominate most collegiate texts—the genetic information of a segment of DNA—a gene—is transcribed into messenger RNA that in turn is translated into a protein, one-to-one-to-one. But then a donkey upset the applecart. "Researchers made the surprising discovery that, in the cells of higher organisms, messenger RNA is altered by enzymes *before its information is translated into protein.*[18] In the language of genetics, pieces of RNA are excised from the molecule and the remaining pieces are fused to make the functional RNA that then serves as the template for protein synthesis. There is no one-to-one correspondence between DNA sequence and proteins."[19] In fact, current research suggests that the subtle spatial relationships among parts of the genome may be as important as the actual sequence. And that previously misnamed "junk" DNA plays a major role in sustaining those relationships and geometries, which appear significantly to guide gene expression.

Rather than fitting together the pieces to describe the Whole as in Western philosophy, a holistic approach attempts to look at entire ecosystems as totalities, with their underlying unity as the starting point, inviting us to examine how the "whole" informs interactions of the "parts." We need to do that with *every* issue. One important effect of that type of approach is the minimization of unintended consequences (which are rampant, as Edward Tenner informs us in his fascinating book *Why Things Bite Back: Technology and the Revenge of Unintended Consequences*). But that's not the only reason to look at things holistically.

Stuart Newman took the implicit critique of strict genetic determinism a step further and explicitly laid out multi-tiered mechanisms of development, cell morphogenesis, and pattern formation that relied on such non-reductionist factors as the position of a cell with respect to other cells: how position affects its internal chemistry, which in turn affects salt levels and other nutrients, which in turn affects the development of the body's organs.[20] According to Newman, genes are more repositories of development that has already happened than active determinants of what is *going to* happen, removing biology from its reductionist framework and bringing to it a powerful interactive or dialectical approach. Elsewhere, Newman writes:

> Both cells and ecosystems can thus be analyzed as highly complex networks of large numbers of components undergoing mutually dependent changes in their relative abundances. But while this way of thinking is common among ecologists, it is not well suited to making precise predictions, and has failed to take hold to any significant extent in cell biology. Instead, the most common intellectual framework of cell and molecular biologists is a reductionist approach. The preferred objects of study are detailed interrelations among small numbers of relatively isolated components. In this paradigm, an understanding of the qualitative properties of the system as a whole, such as the conditions for stable, periodic, and chaotic behaviors, is sacrificed in favor of exact knowledge of a more limited set of phenomena.
>
> Undoubtedly, many scientists, working in this reductionist tradition, were surprised to learn from recent studies of so-called oncogenes, or cancer-associated DNA, that the introduction into cells of the capability of making a normal cellular protein in

slightly greater amounts, or in a slightly altered form than usual, could render that cell cancerous, with all the multifarious behavioral changes implied by that term. In spite of this, many molecular biologists, when asked to consider the impact of introducing new components into complex *ecological* systems, have remained within their reductionist framework and have dismissed the potential for ecological harm from the release of what they consider to be well-characterized entities.[21]

Recap

Reductionist science claims that our "sameness" over time is the result of genes, which predetermine and program each cell. It tries to explain each level of complexity by searching for ever-smaller determining factors. Reductionism is assumed without question in science and it is every bit as ideological *at its core* as religion.[22] Few recognize that the very positing of the existence of scientific progress as value-free is itself value-laden; it is bound to ways of thinking that came to the fore in the West centuries ago during the Enlightenment, and which are reinforced in part by the instrumentalist worldview (which is what I mean when I use the word *ideology*) that came about with the development of capitalism. "The phrase 'time is money' dates from this period, as does the invention of the pocket watch, in which time, like money, could be held in the hand or pocket," Morris Berman tells us in *The Reenchantment of the World*, his stunning critique of the relationship of how we think to the historical development of capitalism. "The mentality that seeks to grasp and control time was the same mentality that produced the world view of modern science. . . . Clearly, then, one can speak of a general 'congruence' between science and capitalism in early modern Europe. The rise of linear time and the mechanical thinking, the equating of time with money and the clock with the world order, were parts of the same transformation, and each part helped to reinforce the others."[23]

The general ways we categorize the world around us (and our own places in it) are part and parcel of the particular social conditions and history of our society. The questions we think to ask—or don't ask—and the ways in which we try to solve them do not stand outside of politics and society. Just the opposite! Together, as "science," they form the ideological *bundle* through which capitalism's hidden philosophical assumptions reflect the production of commodities and utilitarian ways of seeing earth's minerals and human labor. Those assumptions validate themselves and extend into ever-new reaches of our lives. Who owns the genetic sequences of one's biological cell, one's self? With the judicial allowance of corporations to privatize those sequences, where does the self begin and end? If democracy is based on the self-determination of each individual, where lies the boundaries of the self doing the determining?

Thinking about the Process of Thinking, Wherein Subject and Object Switch Places

Indeed, the observer's ability to recognize that interconnection between observer and observed is itself an attribute of the totality—not of the isolated individual—at a certain point in its self-development. The ways we categorize the world around us, our own place in it, and our underlying assumptions that often go unseen and taken to be "natural" are part of the particular social conditions under investigation. Karl Marx addressed the intricacies of that difficult dialectic when he wrote, "Mankind thus inevitably sets itself only such tasks as it is able to solve, since closer examination will always show that the problem itself arises only when the material conditions for its solution are already present or at least in the course of formation."[24]

The fight to protect life from herbicides and adulticides contains a similar triple edge. One set of people say, "We spray pesticides to protect the majority of the community from diseases and to vanquish weeds and mosquitoes." Others say that spraying represents a backward way of thinking about both people and plants, recalling that one person's weed is another's dinner or medicinal source. And always, underlying both, is a history of oppression and exploitation that often goes unrecognized and is taken for granted.

Which way of thinking will prevail at any given time? This is the fight that Rachel Carson felt compelled to engage in and still very much with us today. Government and corporate lies run much deeper today, the issues are more profound, and their propagation is more effective. What we try to show in this book is how to reframe controversies in science, such as the mass use of pesticides, so they reveal a heretofore hidden set of politics and philosophies. Doing so can guide us in cohering an international movement against the use of pesticides, which are poisoning the planet.

17

When Rights Collide: Genetic Engineering & Preserving Biocultural Integrity

By Martha R. Herbert, PhD, MD

All people have the right to a food supply that
has not been genetically engineered.
—Article 3 of the Genetic Bill of Rights

What can we do when two conflicting assertions of rights are in whole or in part mutually exclusive? Material from genetically engineered (GE) crops drifts in the wind and contaminate fields, as well as contaminating evolutionary lines of descent. Wouldn't asserting the right to have access to GMO-free-food be equivalent, if not more fundamental, than, Monsanto's right to genetically modify foods?

In 2005 the Council for Responsible Genetics published the volume, "Rights and Liberties in the Biotech Age: Why We Need a Genetic Bill of Rights"[1] in which it asserted: "All people have a right to a food supply that has not been genetically engineered."[2] The insistence on access to food free of genetic engineering is at the same time a call to restrict, or curtail entirely, GE food.[3]

Meanwhile, Genetically Modified Food (GMF) comes with fellow travelers—noxious pesticides and herbicides, notably the very widely used *Roundup*, which is the Trojan Horse of GMO food, winning welcome and trust through its promises to "feed the world." GM crops are genetically engineered to be more resistant to glyphosate, the main (but not the only) ingredient in Roundup. The same corporation, Monsanto, that manufactures GMO crops, also makes the pesticides that specifically work with them. For people who buy into the promises of GMO foods being the only way to feed the planet, the fellow traveler chemicals are accepted as necessary evils. Holding this trusting view puts a filter on information coming in, with flawed supportive studies read uncritically, and studies documenting harm dismissed. But to those whose view is not thus filtered, it is impossible for them to dismiss how large the dangers of both GMOs and their fellow traveler chemicals loom.

Once we advocate protecting the option of GE-free food we cannot rest on insisting that such food be merely preserved as an option, because that option will quickly be overrun by GMOs, and non-engineered foods will be impossible to grow. Thus, agribusiness should not have the right to implement the genetic engineering of food at all, as it precludes the rights of those who desire non-GMO food.

Fundamentally, the technology has not delivered on its promises,[4] and moreover cannot do so; its purpose is to sell corporate products and expropriate profit from biological processes. It is not about genuinely addressing human concerns.

Cultivation vs. Production

To insist on access to non-GE food is a good start, but it is not sufficient for dealing with the broad ramifications of a genetically modified food supply. The issues here go far beyond health effects and testing/labeling of new food products, and choice in the supermarket aisles. They reach into fundamental questions about how we evaluate technologies. The instrumentalist demonstration that a technology appears to "work" is short sighted if the longer-term consequences and ripple effects of the technology are ignored. The issues also reach into questions about how we organize agriculture and how we keep our fellow living beings alive. Industrial farming, of which GE food is only the most recent example, has forced a transition from food *cultivation* to food *production*.[5] The emphasis on *production* dismisses an enormous range of metabolic, ecological and cultural considerations related to food.[6]

Proponents of GE food promise that genetic engineering will increase food productivity. But they ignore a host of other relevant domains which need their integrity to be maintained—including the metabolism of our bodies, farming communities and cultures with their complex local knowledge systems, and the cultural resonance of cuisine.

Even in narrow productivist terms, GE has inherent profound shortcomings, making it likely to yield not productivity, but its opposite—crop failures, diseases, or blights from unforeseen vulnerability of genetically manipulated strains, cultivated as widespread monocultures. Moreover, although promoters of GE food say increased food yield will solve world hunger, for social and economic reasons (e.g. maldistribution of ample food stocks) productivity is not the true core issue.[7]

Insistence that food be GE free is critically important because metabolic, ecological and cultural sustainability are at stake. Contrary to the widely promulgated belief system that genetic technologies are the only ways for us to solve critical food and health problems, there are other ways to improve the present agricultural system. Science is now able to develop "gentle, thought-intensive technologies"[8] to advance beyond industrial and engineered monocultures. What we need for both physical survival and for a future worth living in is a scientifically sophisticated, context-sensitive and culturally rich recovery of *cultivation*. GMOs, and the vested interests that obstruct balanced debate about them, are obstacles to this deeply needed advance.

Debate Needs to be Open, Broad, Full, and Transparent

The necessity of keeping the food supply GE-free can be defended on many levels, ranging across molecular genetics, cell biology, plant and animal physiology, ecology, economics, health, culture, even aesthetics. GE food proponents often try to narrow the field of concerns by calling up results of "sound science," which consist largely of hastily conducted, short-term industry-funded studies. As we are seeing with the Monsanto Papers, key studies have been concealed from regulators or contain fraudulently reported results. Proponents sometimes attempt to restrict debate to health issues, and then foreclose discussion on the grounds that industry-contracted studies to date show few health risks. Meanwhile GE food was labeled as GRAS ("generally recognized as safe") without evaluation of potential differences from genetically unmodified foods.

While the proponents of genetic modification of crops see it as a humanitarian product of cutting-edge science, their opponents see it as a technology based on limited and parochial assumptions,[9] deplorably naive about organisms, oblivious to ecology, economically motivated, and blind (in part, deceitfully) to the real causes of world hunger. Unlike the inclusive discussions that have taken place in countries like Cuba, GE crops and animals in the U.S. have been rushed into large-scale production without adequate scientific evaluation and public discussion. Why? The main reasons: belief systems and economics.

Bias and Conflicts of Interest

Proponents of GE food have consistently resisted engaging opposing perspectives. U.S. policy has not reflected the mass opposition to genetic engineering of foods, which crosses all political parties and ideologies. Regulators welcome favorable assessments, even if they are of poor quality, but give critical assessments a hard time even if they meet rigorous standards and are peer reviewed.[10] In addition, while there has been abundant funding for GE research, little money is available for context-sensitive agro-ecological approaches.[11] This bias appears to be driven by the extent to which GE research can be easily translated into patentable products and the promise of profit, while agroecology, though more sustainable, generally cannot. This means that public funds are biased toward supporting private interests. These biases have been incorporated into national policy; for example, international trade legislation includes funds for promoting agricultural biotechnology but not for seriously assessing it or developing agro-ecological, non-engineered approaches.[12]

Thus, it is important for the public to understand that we do not seek "equal time" or the operation of "unbiased science" in allocation of research resources. Serious conflicts of interest have dogged government- and industry-sponsored inquiry, with commissions considering the merits of GE food composed predominantly of members with industry ties and funding.[13]

Can There Be Open Debate in the Face of Incomprehension?

Proponents of GE foods do not appear willing to engage in transparent debate. It is not just that they fail to address the concerns of GE food critics, they actually are unable even to comprehend the criticisms. They frequently claim that they themselves are uniquely "scientific" and their critics are merely "emotional." Of course, this rhetorical strategy is a disingenuous public relations maneuver. But it also reflects genuine naivete. Arguments about GE food's threats to organismic, ecological, and cultural complexity and diversity seem incomprehensible to many GE enthusiasts, who see molecular genetics as the definitive universal code of life, whose encompassing truth must override all prior frameworks. GE presumes that genes are a universal alphabet that determines everything bottom-up. This universalism is abstracted from the full multi-scale complexity of organismic functioning and the belief that analogical (continuous) processes, including fluids, waves, organismic processes, lived experience, and more, can be digitized, without loss of nuance.[14]

One also sees an emboldened triumphalism, a sense of mission to improve the world on the basis of what are seen as "truths" revealed by molecular genetics. All human and other organic frailties are seen as susceptible to remediation by engineering or genetic recoding. Yet ironically this investment in the "universalism" of the genetic code has even interfered with application of genetic science itself. A growing number of studies have identified ways, some species specific, that non-coding DNA, as well as non-DNA proteins, modulate gene expression. Proponents of eliminating GE organisms assert that these findings cannot be comfortably incorporated into an ideological framework of genetic universalism and gene dominance, and are evidence for the need to consider the myriad of potential deleterious effects of GMOs on all levels of the biosphere.

Genetic Engineering as Technological Messianism

Technological messianism dovetails elegantly with the economic forces driving genetic modification. Inserting a specifically characterized gene sequence into an organism has been considered adequate justification for patenting the organism.[15] This patented seed offers numerous benefits to the patent-holding proprietor, further facilitating the extraction of profit from nature. For example, private ownership of Roundup Ready seeds assures a steady stream of corporate profit. Not only is the GE seed patented, but farmers, now disallowed from saving seeds from prior harvests, are forced to buy or rent new seeds every year from the manufacturer for each new crop. This newly juridically approved ownership of genetic complement allows a new kind of ownership of organisms the seeds grow into, and new ways of extracting profit from biological processes.

Another revenue stream ensured by the technology is that the farmers get locked into the need to purchase pesticides and herbicides, prominently glyphosate, from the same people who manufactured the GM seeds, in order for their crops to be viable. In this setting, predictable problems result in further profits for the manufacturers and further hardship for the farmer. For example, glyphosate resistance has, not surprisingly, arisen among targeted pests, locking farmers into purchasing and applying greater quantities of pesticide products to maintain farm production. All of this creates many challenges well-reviewed by other contributors to this volume.

The patenting of altered genes and seeds turns living beings into intellectual property and profit-generating biomachines. This occurs in an economic system where the overall goal is accumulation of profit. Getting rich means having more money, and "money" itself is an abstraction dissociated from the particular qualities of the commodities that are produced and sold. One can get rich from selling corn flakes or nerve gas—it doesn't really matter. The compulsion to implement a more efficient means of capital accumulation overwhelms all other considerations.

Thus, the mission to improve the world by redesigning it according to "genetic universals" complements the economic drive to control the market—and the world—according to the "universal money abstraction." Both the money abstraction and the genetic abstraction are divorced from any commonsense reality checks because they are divorced from any particular loyalties to specific context, whether it be place, species, person, or culture. Those who pursue the "money abstraction" and the "genetic abstraction" are, in terms of the "logic" of their activities, impervious to arguments coming from any domain of particularity that is outside their universalistic frame of reference. Such particular concerns may simply not register in the mind of anyone operating within this abstract universalist framework.

GE Foods as Literal Embodiments of a Belief System

The technological messianism of GE food advocates thus coexists poorly with other belief systems and by its inner logic runs over them. Certainly, this inability to coexist with other frames of reference characterizes fundamentalisms of many kinds. The problem is that the genetic engineering of food is more than a belief system; it is a technology, and moreover a technology that utilizes living organisms as its substrate, transforming them in unprecedented ways. GE foods do not merely *represent* a belief system; they *embody* it. GE foods incorporate the belief system that conceived them in their very tissues, their very flesh, indeed their genes, in a manner that goes beyond previous breeding techniques of industrial agriculture. They thus do not assert themselves merely as ideology or dogma, but even more as material—and organismic—force. And as a material living force enlisted in a messianic mission, they not only ideologically oppose but, even more, materially—and reproductively—displace non-genetically engineered organisms. Once an organism is genetically modified, there is no going back. And once genetically engineered organisms are in the environment, gene-sharing with non-genetically engineered wild species cannot be prevented or controlled.[16]

This aggressive, intrusive character of GE foods is not due just to the nature of the technology and its ecological risks; it also appears to be an explicit marketing strategy. In the words of one industry spokesperson: "The hope of the industry is that over time the market is so flooded [with GE food] that there's nothing you can do about it. You just sort of surrender."[17] A U.S. government official stated uncritically: "In four years, enough GE crops will have been planted in South Africa that the pollen will have contaminated the entire continent."[18] From this point of view, the biotech industry is likely to welcome the genetic contamination of maize by transgenic DNA in Oaxaca, Mexico, its very center of maize's origin.

Thus, the objection to GE as an intrusive and self-propagating biological colonialism cannot be refuted by recourse to scientific studies, because GE's proponents are not taken aback by such "revelation"—indeed, they are counting on that new level of colonialism as an integral part of its success! The assertion of one's right to food free of genetic engineering and, indeed, strong opposition to the supposed "right" of corporate proponents to pursue this questionable technology, are both critical bulwarks against the degradation of organismic and ecological integrity driven by a technology based on insensitivity, incomprehension and greed.

The Multilevel Objections to Genetically Engineered Food

Critics of GE food have not shared the conversion experience of the enthusiasts. GE food proponents may allege that ignorance is the reason for the critics' failure to see the genetic code as a comprehensive universalism, but GE critics see their opposition not as a failure or sign of ignorance, but as a *refusal* based on knowledge and a morality that counters the universalism. The reasons that GE opponents reject GE food are substantive and span multiple levels, from molecular genetics to ecology to culture, from simplest to most complex levels of organization.

Causation is not just genetic: At the level of genetics, there is abundant evidence that the genetic code is not uniquely determinative.[19] No one has ever created an organism out of raw DNA. Even if this creation should come to pass, which may be conceivable for very "simple" organisms but much more remote for multicellular organisms, other parts of the cell participate in reproduction and development, and significantly modulate the role of the DNA in ways that are not DNA controlled.[20]

Genes act in systems: There is also abundant evidence that genes do not act in isolation but in systems.[21] It is not unreasonable to think of a cell as a "little ecosystem." Insertion of foreign genetic sequences does not merely add a discrete and specific new function, nor does it leave the cell otherwise undisturbed. Instead, this genetic modification has the potential to create widespread alterations in gene expression patterns.[22] Mere knowledge of the genetic code, even with the tracking of the physical expression of individual gene insertions into the chromosome, does not even begin to give scientists the capacity to predict these types of broader systemic changes.[23] It is therefore the case that genetic modification has the potential to alter cellular metabolism in ways that we can neither understand, predict, nor control.[24] This unpredictability is not simply due to the complex interconnections both within the genome and between the genome and cell and tissue physiology. It is also due to the essentially random fashion in which genetic material has been introduced into the genomes of food organisms, e.g., plant pathogens via "gene guns," where the location within the strands of DNA is not controlled.[25] From this vantage point genetic engineering is not a precision technology so much as a gamble. The engineered organisms that do survive and make it to market are a small minority out of the many attempts to engineer organisms, most of which fail to produce viable outcomes. And the "successful" organisms brought to market often manifest significant problems that emerge during the organism's life course or after reproduction.

Context-sensitivity: The inter-species transfer of genes that are supposed to code for specific traits fails to account for the fact that genes and gene products are modified in ways that are specific not only to individual species, but also to particular tissue types within species.[26] Genes may play different roles when they are transferred into novel organisms than they play in the species from which they came. Thus, particularities of species and even tissues haunt and constrain genetic universalism. We can thus conclude that knowledge of the genetic code, while it provides new ways to manipulate organisms, does not go far enough in helping us understand how organisms are affected by these manipulations.

Undesired additional effects: This lack of knowledge, understanding, and control at the molecular and cellular level has ramifications when these techniques are applied to agricultural crops. Inserting a gene to add a desired characteristic—such as herbicide tolerance, frost tolerance, or salt tolerance—can lead to results other than the ones desired. First, the inserted genes may not function as intended, or may function optimally in only a narrow range of environmental conditions.[27] Beyond this, the organism may have unexpected additional metabolic alterations, some of which may lead to health risks such as allergenicity or toxicity. These possibilities have finally been acknowledged by the U.S. Food and Drug Administration,[28] after years of its insistence that GE foods were "substantially equivalent."[29] Some of these possibilities may be severe and obvious enough to be identified in industry's pre-market testing; but the narrow scope of testing required of these products, considered as GRAS or "substantially equivalent," will not be probing enough to identify other problems that emerge under ecologically variable stress in the field.

Cost implications of complications: The significant likelihood of these complications contributes greatly to the enormous cost of developing viable genetically engineered varieties, further belying public relations claims that genetic engineering of food is a practical, economical, people-oriented solution to world hunger.

Intrinsic instability: Another difference from traditionally bred organisms relates to gene silencing. The inserted genes may be modified or silenced by the organism. This can occur variably in different parts of the plant, and among different plants, and can worsen over the course of the growing season.[30] Despite former U.S. president Jimmy Carter's assurances to the contrary,[31] such erratic gene expression deviates strikingly from that of traditionally bred organisms and their native genes. It indicates a potentially serious intrinsic instability in genetically modified organisms. Such instability forebodes worrisome potential complications, particularly insofar as we allow our food supply to become dependent on these crops.

Who is testing? While some studies have demonstrated that these possibilities may occur, independent researchers are not generally funded to do these kinds of studies. Contrary to the complacent popular belief that our foods are well-regulated, GE organisms—as with pesticides and industrial chemicals—are generally only tested by the companies that produce them, and these tests are reviewed fairly uncritically by regulators. Should we entrust industry-sponsored or even industry-influenced science to seek evidence of such problems, let alone publicize such evidence if they find it?[32]

Magic bullets for complex systems? The recourse to GMOs to solve agricultural problems is an attempt to solve complex problems with a simple "magic bullet." Agriculture itself is a modification of growing patterns in the wild. In its current dominant "industrial" forms, it tends toward monoculture, or toward a reduced number of coexisting organisms.[33] Many traditional agricultural systems, as well as contemporary organic and agro-ecological methods,[34] address not only the characteristics of individual species but also the beneficial effects of intercropping on agricultural problems like pests and weeds. Industrial agriculture may attempt to fight infestations by applying pesticides or inserting pesticide-producing genes, as in BT corn and soy, but the efficacy is often modest, short lived, and rife with side effects, such as toxicity and the emergence of pesticide resistance. In contrast, organismic resourcefulness in getting around, adapting to, and defeating "magic bullets" via coevolution, is well established.[35] An agro-ecological approach to integrated pest management (IPM) which draws on intercropping and other inter-species interactions, can be safer, more effective, and more stable.[36] While some agricultural scientists see GMO's as one tool in a larger agro-ecological tool belt, genetic engineering techniques on their own are incapable of taking advantage of beneficial synergies in inter-species relationships—such synergies occur at multiple scales that are not considered by GE technologists. The inflexibility of GE technologies, has unfortunately led to the dismantling of agricultural research stations in the developing world that are not oriented to GE.

Ecological problems: Furthermore, GE creates ecological problems such as pollen flow to wild relatives, bio-invasion, and harm to other organisms through various direct and indirect pathways.[37] Regarding biodiversity, the way new GM organisms are developed tends to ignore rather than relate to local organism and ecology variants. Biotechnologists don't generally use scientific models that involve the interaction of organisms within specific ecological or cultural contexts. They tend to see biological features in a more general, context-independent way, rather than in relation to particular plants or animals that live in specific places with specific people. In addition, it is enormously expensive to produce GE food products, for one thing because it takes many thousands of laboratory failures before arriving at "viable" genetically modified varieties. There are thus multiple imperatives to widely market the seeds or animals that finally succeed in the lab in many greatly differing ecological and cultural locales. Locally adapted varieties are displaced, in favor of generic, yet patented, GE varieties.

Broader loss of diversity: If we broaden the context still further and consider the diversity and cognitive richness of local cultures, we find that GE and industrial agriculture are blind to their integrity and value.[38] For industrial agriculture, the imperative of production predominates, and considerations such as the stabilizing and nurturing effects of relationships, community, and traditions are ignored. But aside from the fact that genetic engineering's promise of improved yields is often not fulfilled,[39] there are further catastrophic impacts: farming communities are disrupted (particularly through the bankruptcy of smaller farms that cannot afford these technologies), and the accumulation of detailed local knowledge is lost.[40] Neither bountiful crops (when they occur) nor genetic manipulation can substitute for the cultures, communities, ecosystems and lives that are destroyed. What remains is a homogenized and degraded countryside, loss of biodiversity, cultural and material impoverishment, psychological devastation that passes from one generation to the next, and an abject debt-ridden dependency on multinational corporations for subsistence.

There Are Other Possibilities

Some kinds of science are capable of incorporating what is known generally into approaches that are grounded locally.[41] Genetic Engineering of food is not one of them. Thus, opposition to GE food is not anti-science. Instead, the relentless press of genetic modification, shielded from critics, is retarding genuine scientific progress. GE dominance throws good money after bad in an attempt to recoup an investment that should not have been made in the first place, and blocks many other more constructive approaches.

Asserting the right to non-GE food thus also maintains allegiance to an anti-colonial frame of reference that challenges the profit-extracting productivist mentality and the instrumentalist reductionism that genetic modification embodies. Productivism prioritizes quantity produced over other considerations in the belief that economic productivity and growth are the primary goals and measures of human activity; instrumentalism privileges usefulness over other considerations. Yet even in terms of their own rhetoric, the claims that GM food would lead to higher yields and lower inputs of pesticides (such as Roundup) have not been borne out by the evidence.[42] In fact, the opposite has been documented.[43]

Asserting the right to GM-free food is much more than a weak demand for small preserves or reservations of organic farming in the midst of vast spreads of GE crops, or a tame request for GE-free labels on our food and GE-free aisles in our supermarkets. Certainly, demands for protecting organic farming and for food labeling have tactical importance. But they are not enough, and in any case, pollen drifts over unanticipated distances, making it impossible to maintain crops that are organic and GE free in areas that include the cultivation of GM varieties.

Consequently, GE food—also known as "pesticide-saturated" food—is not the best way to feed the world's hungry people. What is needed is continued grassroots pressure, generating an emerging awareness of agroecology as a scientifically informed rational approach that is more sophisticated, healthier, sustainable, and eco-friendly than GE and industrial agriculture.

The question remains whether in the long run genetic modification of food crops will find a humbler role in a truly ecologically and culturally friendly agricultural strategy. I would argue that the current technologies are misconceived and intrinsically incapable of maturing in this fashion. While some idealistic scientific agronomists may wish to incorporate current GE technologies into sustainable agriculture, they are unlikely to have grappled with the full range of objections to GE, and also are probably quite naive about the economic imperatives driving the biotech industry's commitment to this approach—imperatives that will hijack the good will of those who see positive applications of biotechnology. Given the unlikelihood that current agricultural biotechnology will overcome its obsession with profits, production, and pesticides, the prudent thing to do at this time, therefore, is not just to strengthen our opposition, but to preserve intricate local knowledge along with biological and cultural diversity, and work toward a regenerative eco-cultural approach to food cultivation.

Notes

I would like to express my gratitude to Mitchel Cohen and Robin Falk Esser for their edits to this chapter, and the following people for their thoughtful and critical comments on various drafts of the 2005 version: Colin Gracey, George Scialabba, Chloe Silverman, Ruth Hubbard, Sheldon Krimsky, Peter Shorett, Abby Rockefeller, and Diana Cobbold.

18

Glyphosate Acting as a Glycine Analogue: Slow Insidious Toxicity

By Stephanie Seneff, PhD

Editor's Note: In a separate essay that had to be removed for lack of space but which will be posted to the website about this book, Dr. Seneff links the mass use of Monsanto's Roundup to an increase in autism. In this essay, she proposes a biochemical mechanism by which that can take place. She claims that glyphosate replaces the amino acid glycine in critical proteins throughout the human body. While her theory remains to be proved, mass spectroscopy could foreseeably distinguish the molecules involved to see if that is indeed the case.

Seneff rejects the simpler proposition that glyphosate may act to block protein synthesis by interfering with glycine incorporation or glycine transporter function and proposes a more difficult hypothesis: that glyphosate replaces glycine during protein synthesis. Exposure to Roundup, Seneff claims, has increased the incidence of autism through biochemical mechanisms.

The decision to include Seneff's hypothesis in "The Fight Against Monsanto's Roundup: The Politics of Pesticides" is mine and mine alone. I think she's onto something, difficult as it may be to prove to everyone's satisfaction. My inclusion of controversial essays here—my own and Seneff's in particular—in compilation with other activists and scientists who may or may not agree with our conclusions, should not be used to malign or detract from the work of the other brilliant contributors to this book. We are opening up space for thinking outside the spray truck.

—Mitchel Cohen

Glyphosate is the active ingredient in the pervasive herbicide Roundup, and its usage in agriculture has increased exponentially in the United States over the past two decades, in step with an alarming increase in the incidence of a long list of debilitating diseases and conditions, including diabetes, obesity, autism, Alzheimer's disease, Parkinson's disease, kidney disease, intestinal infection, pancreatic cancer, thyroid cancer, and many others.[1] While correlation does not always necessarily mean causation, there is no other chemical I can find used in agriculture that is so pervasive and has similarly increased dramatically in step with the dramatic increase in these diseases. That these diseases are becoming much more common is indisputable, and something is causing this disturbing pattern. I believe there is now sufficient evidence that glyphosate *could* cause these diseases to support a causal relation in the observational trends. I hope you will agree with me after reading this chapter.

The choice of Roundup over other herbicides is based on its status as a nonspecific herbicide (that is, it kills all plants except those that have been genetically engineered to resist it) and, especially, its reputation as being practically nontoxic to humans. Monsanto, glyphosate's original producer, was able to convince regulators back in the 1970s that it has very low toxicity to animals, in part using the argument that its main mechanism of toxicity to plants was based on its suppression of an enzyme in the shikimate pathway, ESPS synthase, which human cells never produce. This argument overlooks the fact that our gut microbes do possess the shikimate pathway, and they use this metabolic pathway to produce many vital nutrients that our own cells depend upon them to provide. The recent explosion in research papers on the gut microbiome attests to the fact that our gut microbiome today is chronically disrupted, and I think glyphosate plays a major role in this pathology.

A seminal paper by Seralini et al., first published in 2012 in the journal *Food and Chemical Toxicology*[2] and later republished in the journal *Environmental Sciences Europe*[3] after an ill-founded retraction by the original journal's editors, involved a long-term study of rats exposed to very small doses of glyphosate over their entire lifespan. One of the important realizations that comes out of this study is that glyphosate is a slow kill. After the rats had been exposed to glyphosate for three months, there were no significant differences in the health status between the exposed rats and the controls. The agrochemical industry has declared that toxicity studies don't need to go beyond three months, and nearly all of their reported studies are restricted to this maximum duration. Clearly, the toxicity of glyphosate could be missed altogether with such a strategy, convenient for keeping the chemical on the market.

Seralini's rats that were exposed to low doses of glyphosate eventually developed huge mammary tumors, kidney disease, liver disease, reproductive disorders, and early death. It can be supposed that disruption of the gut microbiome is a rather indirect mechanism of harm to the host and therefore takes more time to manifest as disease. However, I have come to believe that something much more insidious is going on, which over time results in an accumulation of glyphosate embedded in various proteins throughout the body.

The toxicologist Anthony Samsel and I have published a series of six papers together on glyphosate and disease. Remarkably, we had published four of these papers before we came to realize that glyphosate could be substituting for the coding amino acid glycine during protein synthesis by mistake. Glyphosate is in fact a glycine molecule, except that a hydrogen atom normally bound to the nitrogen atom has been displaced by a methyl phosphonyl group, as illustrated in Figure 1. Part of glyphosate's toxicity has

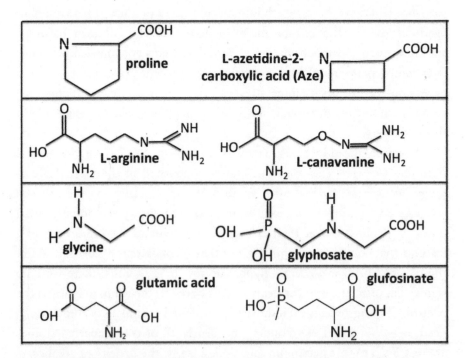

Figure 1: Molecular structure of four coding amino acids on the left and the corresponding non-coding amino acid analogues on the right. All of these molecules except glyphosate occur in nature.)

been presumed to be through its action as a glycine analogue. Glyphosate has been shown to excite NMDA (N-methyl-D-aspartate) receptors in rat hippocampus, and the argument is that it binds to the glycine-binding site in the receptor, pretending to be glycine.[4] In fact, a paper suggesting that glyphosate might be useful as a cancer drug has argued that its mechanism of toxicity could be through its action as a glycine analogue to suppress glycine synthesis in tumors.[5] Glyphosate disrupts the first step in the synthesis of chlorophyll, and this is probably in part because it displaces glycine as the substrate.[6] However, the enzyme that catalyzes this step has a highly conserved glycine-rich stretch that could be susceptible to glyphosate substitution, further contributing to this observed effect.[7]

Glycine is one of the basic building blocks of proteins, which are constructed through the famous DNA code. In December 2015, Anthony came up with the proposal that glyphosate could be substituting for glycine as a coding error during protein synthesis and disrupting protein function systemically. Proteins are synthesized as beads on a string, based on a four-letter code, where each unique three-letter subset codes for a specific amino acid. It turns out that protein synthesis is much more prone to error than DNA synthesis, so the code is inherently inaccurate. Interestingly, and perhaps surprisingly, an excess of glycine itself can lead to glycine substitution for its near-neighbor alanine. A methionine deficiency can lead to homocysteine substituting for methionine. Anthony was proposing was that, when the protein assembling machinery sees a code for glycine, it recognizes glyphosate by mistake, because it is in fact a glycine molecule, except for the extra material attached to the nitrogen atom.

I had previously considered and rejected this idea, because I mistakenly thought that the methyl phosphonyl side chain attached to the nitrogen atom would get in the way. Once I realized that the coding amino acid proline (shown in Figure 1) also has a side chain attached to its nitrogen atom, I became much more receptive and decided to explore the idea more deeply. While nearly all proteins have glycine residues in their peptide chain, relatively few of them have glycines that play an essential role in the protein's function. Glyphosate can probably substitute for glycine in many proteins without causing undue harm. But when glyphosate hits on one of these highly conserved, essential glycine residues, trouble arises.

I spent the next two months poring over the research literature, gathering a large collection of papers that concerned essential, highly conserved glycine residues in various proteins. We easily found dozens of human proteins with strong dependencies on highly conserved glycine residues. I would guess that there are hundreds, but not thousands, of such proteins. I was richly rewarded, because I began to see that essentially all of the diseases and conditions whose incidence in the U.S. population is rising dramatically in recent times can be explained by disruptions of glycine residues in specific proteins linked to those diseases.

As described by Samsel and Seneff below,[8] DuPont conducted a study on goats in 2007 in which it exposed the goats to radio-labeled N-acetyl-gyphosate and then extracted tissue samples from various organs, testing them for both radioactivity and glyphosate, and its metabolites.[9] Only 42 percent of the total radioactivity in extracted muscle tissue was identified through spectrophotometry, and so it became a mystery as to what had become of the glyphosate represented by the remaining 58 percent. The experimenters decided to apply digestive enzymes (pepsin and protease digests) in an attempt to free up radioactive molecules that might be protein-bound. This resulted in only negligible increased recovery, suggesting that glyphosate was bound to the proteins in such a way that it was extremely difficult to shake it loose. To me, this suggests, ominously, that once glyphosate is incorporated into a protein's peptide sequence, it becomes extremely difficult to break down the protein, and this is a perfect set-up for autoimmune disease. It also means that non-radiolabeled glyphosate in proteins might remain undetected. It turns out that proline, the coding amino acid with a side chain on its nitrogen atom, is also very difficult to break apart from the other amino acids linked to it in a peptide chain. A specialized enzyme called prolyl aminopeptidase is assigned the task of freeing up proline residues from peptide chains. Glyphosate's nitrogen side chain may similarly be causing glyphosate-containing proteins to resist proteolysis.

A partial list of glycine-dependent proteins and associated anticipated diseases if that protein is defective is provided in Table 1. When Anthony Samsel and I first started exploring glycine dependencies, it quickly became apparent that the strong correlations between the rise in glyphosate usage and the explosion in both diabetes and obesity could be explained by glycine dependencies in the insulin receptor and in lipase, the enzyme that digests fats.

Protein	Condition	Reference
insulin receptor	Diabetes	Bajaj *et al.*, 1987[10]
hormone-sensitive lipase	Obesity	Topf *et al.*, 2002[11]
trypsin	Celiac disease	Walter*et al.*, 1982[12]
folate receptor	Autism; neural tube defects	Chen *et al.*, 2013[13]
LDL receptor	High serum cholesterol	Koivisto *et al.*, 1995[14]
myosin	Chronic fatigue syndrome	Kinose *et al.*, 1996[15]
amyloid beta	Alzheimer's disease	Bucciantini *et al.*, 2002[16]
α-synuclein	Parkinson's disease	Du *et al.*, 2003[17]
TDP-43	ALS	Pesiridis *et al.*, 2009[18]
prion protein	Crutzfeld Jacob Syndrome	Harrison *et al.*, 2010[19]
aquaporin	Dehydration	Liu *et al.*, 2005[20]
chloride channels	Kidney failure	Tanuma *et al.*, 2007[21]
kinases	Cancer; Alzheimer's	Sternberg *et al.* 1984[22]
ubiquitin	impaired protein recycling	Zuin *et al.* 2014[23]

Table 1: Some examples of proteins with strong dependencies on glycine residues and predicted disease consequences if they are dysfunctional. Many of these diseases are rising dramatically in incidence in the population, in step with the rapid rise in glyphosate usage on core crops.

The insulin receptor depends crucially on a glycine residue in order to attach to the plasma membrane.[24] If it can't attach, it can't receive insulin and initiate glucose uptake into the cell. Fat cells contain a protein called hormone-sensitive lipase, which has two highly conserved glycine-containing motifs essential for its function—a GXSXG motif and an HGGG motif.[25] This protein allows the stored fat to be burned in response to hormonal stimulation. If it doesn't work, the fat can be expected to accumulate in the fat cells, unable to be released and utilized. Anthony Samsel tested porcine lipase for glyphosate contamination and found significant levels, as reported in Samsel and Seneff.[26] America's obesity epidemic arguably began in the mid-1970s, when glyphosate was first introduced into the food chain.

It turns out that receptors frequently have essential glycine residues, and we are still finding new susceptible receptors in our literature search. Two important ones beyond the insulin receptor are the LDL receptor and the folate receptor. A defective LDL receptor in the tissues can be expected to lead to high serum LDL, and, in fact, serum LDL levels have been going up in recent years, in step with the rise in glyphosate usage on core crops,[27] as shown in Figure 2. The popularity of statin drugs to lower cholesterol is a direct consequence, although the proper solution to the defect is not, in my opinion, to interfere with the liver's production of cholesterol, which is what statin drugs do. This will just further aggravate the systemic cholesterol deficiency problem in the tissues due to impaired cholesterol uptake.

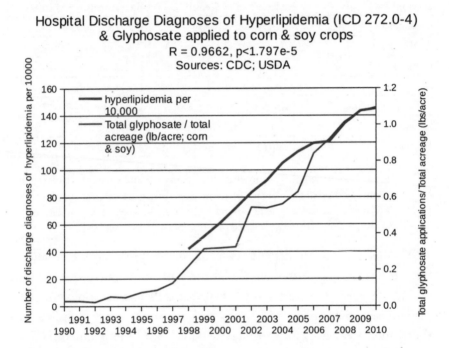

Figure 2: Time trends for hyperlipidemia in U.S. hospital discharge data compared with rates of glyphosate usage on core crops according to the USDA.

A defective folate receptor will cause a folate deficiency problem even in the presence of adequate dietary folate. This can lead to neural tube defects such as spina bifida, and glyphosate's disruption of folate uptake might be a factor in the regulatory decision to fortify wheat-based products with folic acid, first implemented in 1998, just as the genetically modified Roundup Ready crops were ramping up. Glyphosate's disruption of the shikimate pathway in gut microbes can be expected to interfere with their production of folate, which is derived from the shikimate pathway. It has been confirmed that gut microbes can synthesize folate to supply this critical nutrient to the host.[28,29] Folate deficiency due to glyphosate's adverse effect on gut microbes will work in concert with defective folate receptors to increase the risk to neural tube defects.

Myosin is a protein found in muscle fibers that is essential for muscle contraction. Remarkably, if the glycine residue at location 699 in myosin is replaced with alanine (one extra methyl group), the protein drops to only 1 percent of its capacity to contract.[30] It is possible that defective myosin due to glyphosate substituting for residue 699 is a factor in the epidemic we are seeing in chronic fatigue syndrome. Myosin also plays an essential role in the closure of the neural tube, as does folate, and these and multiple other factors[31] can explain observed links between glyphosate and anencephaly. The anencephaly epidemic (children born with no cerebral cortex) that occurred in Yakima Washington in 2011 and 2012 coincided with heavy use of glyphosate in the waterways to control noxious weeds.[32]

Amyloid beta, α-synuclein, TDP-43, and the prion proteins are all examples of proteins with essential, highly conserved glycine residues that have been targeted as the site of dysfunction associated with their misfolding attributed as a major factor in various debilitating neurological diseases, including Alzheimer's disease, Parkinson's disease, ALS, and Creutzfeldt Jacob disease.[33] Aquaporin is a really important protein in the plasma membrane that serves as a water channel to allow water to easily cross the membrane.[34] Impaired aquaporin function in the kidneys will lead to severe dehydration in the context of sweating due to vigorous exercise under the hot sun, because the kidneys will be unable to concentrate the urine, leading to excessive water loss.[35] This might be a contributing factor in the high rate of kidney failure among young agricultural workers in Central America who harvest sugar cane sprayed with glyphosate and then burned shortly before harvest.[36] Dehydration has been identified as a major factor in this mysterious disease, called Mesoamerican nephropathy.[37] Chloride channels also contain highly conserved essential glycine residues, and defective chloride channels will also contribute to kidney failure[38] but would in addition disrupt GABA receptor activity, which is linked to autism.[39]

There is a large class of enzymes called kinases, 99 percent of which contain an essential glycine residue that, if substituted by a bulkier negatively charged amino acid, results in *overactivity*.[40, 41] Glyphosate *is* a bulkier negatively charged amino acid. Thus, substituting glyphosate for this glycine residue should result in enzyme overexpression, leading to excessive phosphorylation of various substrates. Overphosphorylation of tau protein in the brain is a feature of Alzheimer's disease. Cyclin-dependent kinase 2 (CK2) is a tyrosine kinase that attaches a phosphate group to the amino acid tyrosine. Its overactivity has been linked to cancer, and a CK2 inhibitor is a common cancer drug. Antiphospholipid syndrome is a disease in which the body develops antibodies to phospholipids, which could be due to their overproduction by overactive kinases. It has a complex disease manifestation that includes venous thrombosis, pregnancy loss, preterm delivery, defective cardiac valves, haemolytic anemia, and cognitive impairment.[42]

Beyond the match-up between glycine-dependent proteins and glyphosate-correlated diseases, we also found other kinds of evidence of glyphosate's disruption of protein synthesis. We found papers that showed that, unlike DNA synthesis, protein synthesis is inherently a sloppy process.[43] Mistakes are commonly made, and many of them can be tolerated, but there is machinery that can detect that a protein has misfolded and orchestrate a program to rip it apart and reassemble it. However, even this process could be disrupted by glyphosate substitution for glycine, since ubiquitin, the protein that marks proteins for recycling, depends on a double glycine repeat at the very end of the protein to work properly[44, 45]

A study on protein expression in the rhizosphere, the microbial community living in the soil near the root zone of a plant, revealed that glyphosate exposure induced a sharp increase in proteins specifically involved in both protein synthesis and protein degradation.[46] This suggests that defective proteins are being produced that then have to be disassembled and reconstructed. A paper on cows exposed to glyphosate revealed a significant increase in serum urea, a breakdown product of amino acids, which could also be due to excessive protein turnover due to defects in the assembled proteins.[47]

Several *naturally produced* noncoding amino acids can substitute for coding amino acids that they resemble. One of these is the herbicide glufosinate, which derives its toxicity in part from its ability to substitute for glutamate during protein synthesis. Other examples are azetidine-2-carboxylic acid (Aze), which substitutes for proline and causes multiple sclerosis,[48] β-N-methylamino-L-alanine (BMAA), which substitutes for serine and causes an ALS-like disease,[49] and L-canavanine, which substitutes for arginine and causes metabolic failure and subsequent starvation.[50] L-canavanine is believed to be the toxin that ultimately killed Chris McCandless, the protagonist in the book and movie *Into the Wild*.[51] L-canavanine was produced by the wild potato whose seeds he consumed in large amounts toward the end of his life. Various organisms can synthesize these toxins under conditions of stress and as a defense against their enemies. The molecular structures of four coding amino acids and their respective noncoding analogues are shown in Figure 1.

Researchers have been puzzled to explain exactly how glyphosate works to disrupt the enzyme EPSP synthase in the shikimate pathway. A careful perusal of the research literature on the mechanism by which glyphosate disrupts this enzyme leads one to conclude the most plausible explanation is that glyphosate is substituting for a specific highly conserved glycine residue. This glycine residue is situated at the active site for binding to phosphoenol pyruvate, a substrate in the enzyme's catalytic reaction. Remarkably, multiple species of both microbes and plants have developed a natural resistance to glyphosate by changing the genetic code for EPSP synthase such that this essential glycine is converted to alanine.[52] Switching in alanine in place of the glycine residue reduces the protein's reactivity somewhat but completely stops glyphosate from disturbing its function. For instance, a strain of *E. coli* that replaced glycine with alanine at location 96 was shown to produce a version of EPSP synthase completely insensitive to glyphosate exposure even at extremely high levels.[53] Another microbe's EPSP synthase molecule with alanine substituted for glycine is the gene inserted into genetically modified crops to afford protection against glyphosate.[54]

Observations on Animals in Harm's Way

It is remarkable to me how many examples there are of stories of some species of animal experiencing some kind of health crisis with studies failing to reveal the cause. However, the studies almost never consider glyphosate as a possibility, even though it is often glaringly the most obvious exposure factor. The reason for such oversight must be that researchers believe the claims made repeatedly by Monsanto and by the regulators that glyphosate is safe.

One example that comes to mind is the epidemic we are currently witnessing in a chronic wasting disease in the deer population in the United States and Canada.[55] This is a prion disease like mad cow disease, which swept through Great Britain in the 1990s. The deer in the United States and Canada are being exposed to glyphosate used along the roadways but also directly in the forests, particularly in Canada, to kill deciduous trees in order to promote the growth of faster growing conifers. The prion protein has a unique palindromic sequence, AGAAAAGA, in its peptide that is believed to be the source of the misfolding pathology.[56] In particular, the two glycine residues in this sequence are singled out as being particularly problematic, even though they have been highly conserved in the protein over millennia and did not cause such problems in the past.

In December 2016, a massive die-off of marine life began to take place along the shores of the Bay of Fundy in Nova Scotia.[57] There were many news items about this alarming occurrence and its inherent mystery. Salinity and temperature in the waterway were found to be well within the normal range, and no infective agent was detected. However, nothing that I found in the literature linked it to what I think is the obvious cause. At the end of the preceding August, three separate counties in Nova Scotia abutting the Bay of Fundy legislated approval to use glyphosate to kill deciduous trees in the forests to provide clearance for planting faster-growing conifers.[58] The Bay of Fundy has a unique tidal bore that will push marine waters up into the rivers where there is an opportunity to pick up significant glyphosate residues from the forest spraying activities, carrying these residues back out to sea as the tide recedes.

There is much public awareness of the looming catastrophe we face because of bee colony collapse syndrome,[59] which threatens the survival of certain crops such as almonds that depend upon bees for cross pollination. Professor Don Huber, a retired professor from Purdue University, is among the very few who have proposed that glyphosate plays a key role in this disease.[60] While insecticides are surely also contributing, glyphosate disrupts the activity of cytochrome P450 enzymes in the bees, which are crucial for their ability to detox insecticides. So glyphosate works synergistically with the neonicotinoids to harm the bees. A study from Argentina found that field-realistic doses of glyphosate reduced the bees' appetite for nectar and impaired their memory and cognitive abilities.[61]

The monarch butterfly is another species that has come to symbolize the destruction of nature that seems to be characteristic of our times. The monarchs are famous for their long migration path down the middle of the United States to ultimately mate and reproduce in Mexico. Unfortunately for them, a key food source is milkweed, which is a problematic weed growing in the cornfields of Iowa and Kansas. The genetically engineered Roundup Ready corn that now encompasses over 90 percent of the crop allows farmers to indiscriminately spray Roundup on the crop to kill the milkweed. While it has been claimed that the loss of milkweed as a food source may be a key factor in the decimation of the monarchs, I believe the bigger problem is direct poisoning by the glyphosate contaminating the milkweed.

Since about 1998 there has been an epidemic in an unusual disorder among chickadees in the Great Lakes region and in Alaska.[62, 63] The disorder is associated with highly deformed beaks that eventually prevent the bird from obtaining adequate nutrition, leading to death by starvation. An extensive study investigating a possible role for trace metals and various environmental pollutants did not find anything obvious to explain the epidemic, and, notably, did not investigate glyphosate at all.[64] It was in 1998 that the genetically engineered Roundup Ready corn, soy, and sugar beet crops began to enjoy widespread popularity in the Great Lakes region. Chickadees are very much attracted to the bird feeders supplied with sunflower seeds sprayed with glyphosate right before harvest as a desiccant. The deformation in the beaks seems to be driven by overexpression of keratin, a durable, insoluble protein that is an important constituent of bird beaks, as well as hair, horns, nails, claws, and hooves. A protein that protects from overexpression of keratin synthesis is KEAP1, and it depends critically upon a terminal double glycine repeat domain to anchor it to the cytoskeleton, an important part of its protein expression.[65] Hence, disruption of KEAP1 due to glyphosate displacing either of these terminal glycine residues could easily explain the bird beak problem, and probably also contributes to founder (equine laminitis) that can cripple horses due to defective hooves. Laminitis is associated with metabolic syndrome in horses,[66] similar to metabolic syndrome in humans, which has become an epidemic in the past two decades. I attribute chronic glyphosate exposure to this epidemic in both horses and humans.

Hoy et al. correlated deterioration in the health of wild animals of Montana over the past two decades with similar disorders appearing over the same time period in humans, as determined from hospital discharge data.[67] They attributed both trends to glyphosate exposure. Several examples of severely damaged organs from various species of wildlife were illustrated, including thymus gland, lungs, heart, liver, eyes and skin. The parallel health issues among humans with time trends correlating significantly with the rise in glyphosate usage on core crops included newborn skin and eye disorders, lymph and blood disorders in children, metabolic disorders, genitourinary developmental defects, congenital heart conditions, enlarged right ventricle, liver cancer, and pulmonary disorders.

Human Diseases and Conditions

The United States faces a health care crisis today as it seems no solution is possible to pay for the escalating health care costs without bankrupting the country. Yet few seem to be inclined to ask why we need to pay so much for our health care—much, much more than any other country. I believe that chronic exposure to glyphosate from the food is the most important factor in causing an alarming increase in a huge number of debilitating diseases and conditions among people of all ages. And I propose that the underlying cause is an insidious slow accumulation of glyphosate embedded in diverse proteins throughout the body.

The United States has the highest rate of maternal mortality among industrialized nations, and mortality during childbirth has gone up fourfold in the past three decades.[68] The United States also has the highest rate of infant mortality on the first day of life. Our nation's children are suffering from many more chronic diseases every day, with alarming and growing rates of asthma, eczema, attention deficit hyperactivity disorder (ADHD), autism, teenage depression, adolescent rheumatoid arthritis, type 1 and type 2 diabetes, sleep disorder, gastroesophageal reflux disease (GERD), inflammatory bowel disease, and Celiac disease. Post-concussion syndrome has become an epidemic made famous by the diseased brains of football players,[69] and glyphosate can plausibly account for the increased risk to injury due to insufficient gelling of the fluids cushioning the brain.[70]

There is an epidemic today in nonalcoholic steatohepatitis (NASH), also known as fatty liver disease, in the United States, even showing up among children.[71] Many studies have shown that glyphosate causes liver disturbances,[72, 73, 74] and a recent study evaluating multiomic protein expression profiles in the liver in response to glyphosate showed that several enzymes related to fatty liver disease are overexpressed even under low exposure rates of glyphosate.[75] I believe that a major contribution to this effect of glyphosate is its impairment of fructose metabolism in the gut.[76] Multiple glycine-dependent proteins are involved in fructose metabolism, including EPSP synthase in the gut microbes, the enzyme targeted by glyphosate in killing plants. Unmetabolized fructose then makes its way to the liver via the hepatic portal vein, and the liver has to clear it before it gets into the general circulation, because fructose is a very powerful glycating agent. It is well established that the liver's processing of fructose, converting it into fat, is linked to fatty liver disease.[77, 78]

The elderly are facing an alarming increase in neurological diseases such as Parkinson's disease, ALS, and dementia. We have new diseases showing up among young and middle-aged adults that were previously practically nonexistent, such as AIDS, chronic fatigue syndrome, and eosinophilic esophagitis. Premenopausal breast cancer, along with pancreatic cancer and thyroid cancer, are all also alarmingly on the rise. Glyphosate is estrogenic, and exposure at minute amounts (measured in parts per trillion) caused estrogen-sensitive breast cancer cells grown in vitro to multiply.[79] Swanson *et al.*'s correlation data showed strong correlations between both pancreatic cancer and thyroid cancer and glyphosate usage on core crops.[80]

The p53 tumor suppressor gene is an important gene in cancer risk, and its inactivation leads to opportunistic growth of many cancers. Mutations in the p53 gene have been found in more than half of human tumors.[81] Specifically, mutations of the glycine residue at position 334 are the most common mutation of this gene. A substitution of valine for this glycine has been shown to produce a misfolded inactive peptide that forms amyloid fibrils.[82] A substitution of glyphosate for this glycine residue can be expected to have a similar effect.

There is an epidemic in polycystic ovary syndrome among young women, and this has been linked to nearly a threefold increased risk to diabetes[83] as well as an association with obesity. Researchers in Argentina who tested several cotton products for glyphosate found glyphosate residues in 85 percent of the feminine hygiene products they tested.[84] One might imagine that absorption through the skin of glyphosate in tampons could lead to damage to the reproductive organs. In fact, polycystic ovary syndrome is strongly associated with inhibition of aquaporin-9 (AQP-9) and overexpression of a protein kinase, phosphatidyl inositol 3-kinase (PI3K).[85] Both suppression of AQP-9 and overactivity of PI3K are predicted consequences of glyphosate substitution for highly conserved glycine residues in these proteins, as discussed earlier.

Men in America, but also in Europe and China, face an alarming loss in the quality and abundance of sperm, leading to an infertility crisis.[86, 87] Glyphosate severely suppressed the levels of testosterone in exposed male rats, and this was associated with a decrease in the diameter of seminiferous tubules, an increase in abnormal sperm morphology, and a decreased number of spermatids, all indicators of damage to the testes.[88] Others have also found direct damage to the male reproductive system in rats exposed to glyphosate.[89]

Food allergies have risen sharply since 2011, now affecting up to 8 percent of the children in the United States.[90] Peanut allergies are the most common cause of anaphylactic shock, accounting for over a quarter of the reported cases. I would suggest that peanut butter sandwiches play an important role in the rise in peanut allergy. Gluten intolerance is now so common that most grocery stores offer extensive selections of gluten-free products. Glyphosate applied to wheat as a desiccant right before harvest is a direct hit on glyphosate contamination in wheat-based products, most obviously bread. Allergies to casein in milk and to soy protein are also very common.

Glyphosate likely disrupts digestive enzymes such as trypsin and pepsin, and it sets up a leaky gut barrier allowing the undigested peptides to reach the general circulation. This is how immune cells become exposed to foreign proteins and develop antibodies to them. Through a process called molecular mimicry, the antibodies also bind to human proteins with peptide sequences that resemble those in the undigested proteins, and this causes an autoimmune attack on human tissues and organs. This can lead to Hashimoto's thyroiditis, type 1 diabetes, lupus, and neurological diseases like autism and multiple sclerosis.

We have an epidemic today in opioid drug addiction, and often a patient first gets hooked due to severe chronic pain (lower back pain, neck pain, shoulder pain, knee pain, foot pain, etc.). A recent paper published in the Annals of Internal Medicine states that 38 percent of U.S. adults were pre-scribed an opioid drug in 2015.[91] I suspect that glyphosate is playing a major role in this pathology, due to its rampant substitution for glycine in collagen molecules that make up an essential component of the joint. Collagen is the most abundant protein by far in the body, making up 25 percent of the body's protein mass. Collagen is also highly enriched with glycine, which constitutes over 20 percent of the collagen amino acid content. Collagen's triple-helix crystalline structure depends upon a G-X-Y repeat sequence. Randomly inserting glyphosate molecules into collagen will destroy its properties of water retention, tensile strength, and elasticity, leading to damage to the bones and joints.

A recent paper states that the prevalence of arthritis in the knee has doubled since the mid-twentieth century, and that this cannot be explained simply by the fact that we are living longer.[92] I suspect that glyphosate is play-ing a major role through its slow infiltration into the knee joint. Anthony Samsel tested bovine bone, ligaments, collagen, and gelatin, a food product derived from collagen, and found glyphosate contamination in all of them.[93] He also found glyphosate in vaccines such as MMR and the flu vaccine, where the manufacturing process involves growing the live virus on gelatin as a nutrient.

Impaired Mineral Homeostasis

One of the key mechanisms of toxicity of glyphosate is the adverse effect it has on the management of metals in the body. The trace minerals, so-called +2 cations such as iron, zinc, copper, cobalt, selenium, manganese and molybdenum, are exploited by the body to catalyze various enzymatic reactions essential for metabolism and protection from oxidative stress. These metals are, however, dangerous if they are present as free ions in the blood, because of their high reactivity. In consideration of this, the body has developed sophisticated mechanisms to hide many of these metals in transport proteins in the blood and also produces specific proteins to carry them across the cell wall for passage across the gut barrier and then later for uptake by a cell.

Ironically, with chronic glyphosate poisoning can come a situation of iron-based anemia simultaneous with iron-overload toxicity, in part because glyphosate tightly binds iron, preventing its uptake into the transport protein transferrin. Glyphosate then lets go of the metal it is carrying when it reaches a terminal watershed area of the blood, such as the kidneys or the brain stem, where the pH drops, causing glyphosate to lose binding capacity. The freed-up iron then becomes toxic to the tissues. Furthermore, if glyphosate can substitute for glycine during protein synthesis, it would severely disrupt the function of ceruloplasmin, an enzyme that oxidizes iron and delivers it to transferrin. Iron in this oxidized form (+3, ferric) is much less damaging as an agent of oxidative damage than iron in the +2 oxidation state (ferrous), which causes a great deal of cellular damage through the Fenton reaction. Ceruloplasmin also transports copper, which it conveniently uses as a catalyst in its action on iron. Ceruloplasmin has sixteen highly conserved essential glycine residues spread over all of its six sites where copper binds to the protein.[94] Defective ceruloplasmin due to glyphosate swaps for these glycines would result in impaired iron uptake across the gut wall and increased sensitivity to iron toxicity, as well as impaired copper transport. Furthermore, the transferrin receptor has an essential arginine-glycine-asparagine sequence at the site in the receptor where transferrin binds, prior to being taken up into the cell for the delivery of iron to the cell.[95] Autism is linked to deficiencies in both ceruloplasmin and transferrin.[96] This is particularly problematic for the brain, as it results in both iron and copper toxicity that can damage the delicate fatty acids in the plasma membranes of neurons.

Glyphosate and Parkinson's Disease

Parkinson's disease is one of the many diseases whose incidence is rising sharply in step with the sharp rise in glyphosate usage on core crops.[97] Both occupational[98] and accidental[99] exposure to glyphosate have led to Parkinson's disease in case studies. A Parkinson-like condition in worms has also been induced by glyphosate exposure.[100] Parkinson's disease is a neurological disease causing slow, stiff movement and difficulty walking, progressing eventually to death usually within fourteen years of initial diagnosis.

It is likely that impaired ceruloplasmin because of glyphosate contamination, resulting in toxicity of ferrous iron in the substantia nigra, is a factor in Parkinson's disease. The substantia nigra is the brain-stem nucleus that is the focal point of Parkinson pathology, where dopaminergic neurons are damaged, leading to a loss in dopamine supply, necessary to initiate movement. Glyphosate has been shown to suppress dopamine levels in rat brains, associated with impaired movement.[101] The activity of ceruloplasmin is significantly reduced in the cerebrospinal fluid in patients with Parkinson's disease, and this is associated with increased iron accumulation in the brain.[102]

α-Synuclein is a protein that accumulates in Lewy bodies in the brain of Parkinson's patients, and it is believed to be a factor in the disease process through a misfolding pathology. Recent research suggests that α-synuclein has prion-like properties: that misfolded α-synuclein can spread from affected to unaffected regions through self-propagating conformational changes induced by exposure of formerly unmodified α-synuclein molecules to misfolded seed proteins.[103]

What is extremely intriguing to me is that recent research suggests that misfolded α-synuclein may first appear *in the gut*, and then travels along the vagus nerve to the brain to infect the brain stem nuclei.[104] There is much greater opportunity for glyphosate exposure in the gut. A nine-residue peptide motif in α-synuclein, 66VGGAVVTGV74, containing three glycine residues, plays a crucial role in the fibrillization and cytotoxicity of α-synuclein.[105] If this sequence is deleted from the protein, it completely loses its toxicity. Parkinson's, like many other neurological diseases, is now believed to trace back to gut dysbiosis in its early stages.

α-Synuclein has a high affinity for iron, as well as manganese and copper, and all three of these metals have been implicated in Parkinson's disease.[106] Glyphosate substitution for glycine in the peptide sequence 66VGGAV-VTGV74 can be expected to cause α-synuclein's ability to bind to these metals to be enhanced, given glyphosate's strong metal-binding properties. It could be that glyphosate substitution for glycine in α-synuclein induces the attraction to metal ions that leads to the misfolding that ultimately causes neuronal damage.

Glyphosate and Toxic Metals

Besides the essential minerals needed in trace amounts, another set of metals are highly toxic and not used by the body at all, or very rarely, during metabolic processes. These include mercury, aluminum, lead, arsenic, chromium, and nickel, among others. A paper investigating an epidemic in kidney failure among agricultural workers in Sri Lanka proposed that glyphosate was working synergistically with arsenic in the soil, by binding arsenic and then delivering it to the kidneys, where the arsenic was then released in the acidic environment of the kidneys.[107] Both the arsenic and the glyphosate then become toxic to the kidneys.

There has been a great deal of concern in the recent past about the idea that aluminum toxicity may be a major contributor to multiple health issues, particularly neurological diseases.[108, 109] Even though aluminum has a +3 charge rather than the +2 charge characteristic of most metals that glyphosate binds, glyphosate can bind aluminum in a configuration where two glyphosate molecules surround the aluminum atom,[110] and this complex can be expected to get past the gut barrier much more easily than free aluminum, because aluminum's highly concentrated positive charge is canceled out by the bound glyphosate molecules. Glyphosate also creates a leaky gut barrier that allows small molecules easy access to the main circulation. A theoretical paper has argued that glyphosate can carry aluminum to the brain stem nuclei and cause toxicity to the pineal gland, perhaps explaining the epidemic of sleep disorder in the United States.[111] The mechanism would be similar to what goes on with arsenic and the kidneys. As far as I am aware, no studies exist that have examined whether glyphosate enhances the toxicity of lead or mercury, but it can be argued that it would, given the known effects it has on other +2 cations like copper, zinc, manganese, and iron. In fact, one has to wonder whether glyphosate was a major contributor to the high levels of lead found in the drinking water in Flint, Michigan. Flint could be viewed as the sugar beet capital of the world, as the city is surrounded by genetically modified Roundup Ready sugar beets. Glyphosate was first patented as a pipe cleaner due to its uncanny ability to strip metal off of metal pipes. Glyphosate contamination in water flowing through lead pipes would easily cause an enhanced load of lead in the drinking water drawn from those pipes.

What is much more ominous to consider is the possibility of glyphosate getting embedded into certain proteins in place of glycine and then binding to a metal such as aluminum or manganese or copper and causing the protein to misfold, leading to neurological diseases such as the prion diseases or Alzheimer's disease. It has been argued that manganese may have played an important role in the mad cow epidemic in the United Kingdom in the 1990s.[112] The manganese would have gained access to the nerves in the spinal column when the pesticide phosmet was applied to the backs of the cows, to treat for ringworm. Manganese is unusual in its ability to travel along nerve fibers, thus allowing it to reach the brain directly from the spinal column. My suspicion is that glyphosate is substituting for one or both of the glycine residues in the palindromic sequence AGAAAAGA in the prion protein and then causing a much greater likelihood of the protein binding to manganese and then misfolding. Anthony Samsel and I published our third paper together on the topic of glyphosate and manganese, describing how glyphosate disrupts manganese homeostasis, causing both manganese toxicity and manganese deficiency.[113]

Together with several colleagues, I published a paper on glyphosate and Amyotrophic Lateral Sclerosis (ALS) in 2016, in which we carefully examined the genetic mutations linked to familial ALS, and found a remarkably strong link to glycine residues in multiple proteins.[114] Familial ALS accounts for only about 5–10 percent of the total cases of ALS. Two proteins that have genetic links are copper-zinc superoxide dismutase (SOD1) and TDP-43. Specific glycine mutations within SOD1 as well as multiple mutations, many involving glycine residues, within a glycine-rich region of TDP-43 are linked to familial ALS. Glyphosate substitution for glycine residues in these same proteins could explain cases of ALS missing the genetic links. We proposed that manganese toxicity along with copper deficiency plays a role here, as it does in the prion diseases, and that glyphosate enhances the toxicity of manganese. Similar parallel studies associate toxic metals, misfolded proteins, and highly conserved glycine residues in amyloid beta in Alzheimer's and in α-synuclein in Parkinson's disease, as previously discussed.[115]

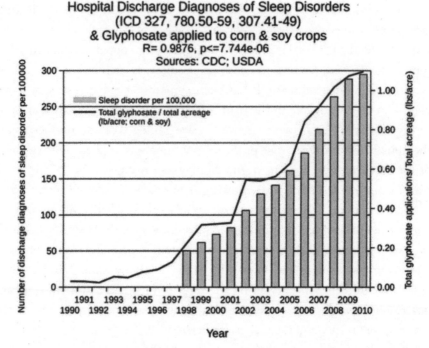

Figure 3: Time trends in sleep disorder according to the U.S. Centers for Disease Control (CDC) Hospital Discharge Data and in glyphosate usage on core crops, according to the U.S. Department of Agriculture

How to Test This Idea

While many chemistry experts are skeptical that glyphosate could insinuate itself into proteins by mistake in place of glycine, it has not yet been proven that this does *not* happen. And, in my opinion, no compelling reasons preclude the possibility. If you assume it does happen, suddenly it becomes clear why we are facing an epidemic in so many autoimmune and neurological diseases today. What is particularly troublesome is that glyphosate can be predicted to directly hit hard on digestive enzymes that break down dietary proteins, like trypsin, pepsin and prolyl aminopeptidase (which has three regions where highly conserved glycines are in the amino acid sequence).[116] Once trypsin is disrupted, the processes that restore the gut barrier to normalcy following a reaction to a toxic disturbance are impaired, and the leaky gut barrier is sustained much longer than it should be, resulting in the dispersal of undigested immunogenic peptides throughout the body, and even into the brain once a leaky brain barrier is in place. The immune system goes on fire, developing antibodies to these peptide fragments, which then, through molecular mimicry, start attacking the body's own tissues, including the collagen in the joints and the myelin sheath surrounding nerve fibers. A complex and unpredictable response leads to a long list of debilitating diseases people are confronting today.

What is the definitive experiment that needs to be done to prove or disprove my hypothesis? The experiment I briefly described earlier, conducted by DuPont on goats, was a good start. For an even more convincing test, I would propose growing in a cultured petri dish *E. coli* microbes engineered to produce large amounts of a protein that is naturally rich in glycine residues. Or, if feasible, human cells that naturally produce a glycine-rich protein could be grown in vitro. For example, fibroblasts produce significant amounts of collagen, and nearly one out of every four amino acids in collagen is a glycine residue. Various cultures of fibroblasts could be exposed to varying amounts of radiolabelled glyphosate, and then the collagen they produce could be purified and then tested for levels of radioactivity before being specifically tested for glyphosate contamination through spectrophotometry. This test would be conducted both before and after subjecting the extracted collagen to extensive proteolysis. I predict that, without proteolysis, the amount of radiolabel would be much higher than the amount of detected glyphosate. The numbers would converge after protein digestion, but probably still there would be glyphosate molecules stubbornly bound to the collagen peptide sequence. Because some of the individual glyphosate molecules would be freed up by proteolysis, these would now be visible through glyphosate and glyphosate metabolite signatures in the spectrophotometry spectrum.

An epidemic in an ALS-like condition in Guam following World War II has been traced to BMAA, a naturally produced noncoding amino acid analogue of serine. In a paper published in 2005, it was observed that inconsistent results are often obtained when measuring levels of BMAA in tissue samples and that it is imperative to apply extensive proteolysis prior to analysis in order to free up the protein-bound BMAA. These authors wrote, "When the insoluble, protein-containing fraction following TCA (trichloroacetic acid) extraction is further hydrolysed to release BMAA from protein, there is a further pool of protein-bound BMAA that is present in a ratio of between 60:1 and 120:1 compared with the pool of free BMAA."[117] In other words, following extensive proteolysis there was at least sixty times as much BMAA in the sample as was detected without extensive proteolysis. I believe this problem is compounded in the case of glyphosate due to its side chain on the nitrogen atom that causes the protein to resist proteolysis.

In my opinion, the evidence is overwhelming that glyphosate is wreaking havoc on the earth's ecosystem, with multiple species beyond our own being adversely affected by this insidious, pervasive, toxic chemical. My sincere hope is that regulators will finally wake up and realize the devastation this chemical is causing and will begin to shut down the factories where it is being produced, heavily fining anyone who is caught using it in any capacity to control weeds or to desiccate a crop at the harvest time. Barring this, individual consumers can play a powerful role in curtailing its usage by refusing to buy any foods that are not certified organic. For you, personally, and your family, switching to a 100 percent certified organic diet will have enormous benefits in terms of a reduction in your health care costs, an increased sense of well-being and wholesomeness, and an increased life span, for all your family members.

19

Respecting the Underground Ecosystem and Gut Microbiome

By Jay Feldman

I wanted to give the not quite final word to my colleague Jay Feldman, executive director of Beyond Pesticides, editor of its spectacular magazine Pesticides and You, *and a coplaintiff in the lawsuit against New York City filed by the No Spray Coalition way back in 2000. Jay wrote this essay as an introduction to the Summer 2017 issue of that magazine, but it can just as well serve, with a certain degree of symmetry, as a coda for this book. (You can find that magazine and the essays Jay refers to at www.BeyondPesticides.org.)*
—Mitchel Cohen

In his talk at Beyond Pesticides' 35th National Pesticide Forum in April, David Montgomery, PhD, captured the essence of the conversations critically needed in all our communities and action that must be taken for a sustainable future. In many ways, the talk was a personal story of revelation, rethinking of scholastic thinking, understanding relationships in nature, and appreciation of the power and fragility of the natural world.

Underground Ecosystem

Dr. Montgomery, a professor of geology at University of Washington, MacArthur Fellow, and author of three books on soil health, human health, and taking action, explains the steps that his co-author and wife, biologist Anne Biklé, took to convert their garden soil, which contained a mere 1 percent organic matter, to a healthy ecosystem. He said, "We were cycling organic matter into this underground ecosystem in ways that led us to learn things that frankly quite surprised us and started us on this view of a completely different relationship of the natural world to human societies." He clearly explains the contribution that soil microbes bring to soil and plant health and the effect that the management of the land has on our bodies and particularly our gut biome.

Similarly, Don Huber, PhD, professor emeritus of plant pathology at Purdue University, gave us a complete picture of the adverse impact that pesticides have on the ecosystem and our health. We hold our national conference every year to keep ourselves updated on the underlying science that must drive change, to share strategies from around the country on transitioning to organic practices that respond to our increasing scientific understanding and to bring back to the policy debate in our communities appropriate land and building management practices that protect and nurture life. Bring Drs. Montgomery's and Huber's words to your campaign to align community practices with sustained health. As Dr. Montgomery said, "[W]e need to think about our microbial crew, or the microbiomes of plants and people, in terms of protecting, restoring, and cultivating the beneficial microorganisms that are key elements of those communities."

Respecting Complexity

Rachel Carson warned us in her book *Silent Spring* in 1962 that when we use pesticides, we are adversely affecting complex biological communities. And people understood the value of the microbial community (sometimes referred to as ecosystem services) when Sir Albert Howard constructed the definition of organic in *The Soil and Health: The Study of Organic Agriculture* (1940), and *An Agricultural Testament* (1947).

After building the case for nurturing the underground ecosystem, Dr. Montgomery concludes, "[I]f we use a broad spectrum biocide [pesticide], we are taking out all the beneficial organisms."

Holistic Solutions

There is a tendency to try to simplify problems and then look for simple solutions. In truth, the problems of environmental degradation and health threats induced by toxic chemical exposure require holistic solutions with changes in systems that establish our much-ignored relationship with nature.

Indiscriminate Effects to Microbiota

We must remember that when the U.S. Environmental Protection Agency (EPA) registered the neonicotinoid insecticides, which are clearly tied to elevated rates of decline in bee and pollinator populations, it did not have a field study to evaluate the chemicals' overall impact on ecosystem health, let alone impacts on individual species.

Whether we are talking about the soil or aquatic food web, the agency did not do the analysis. But, it really is not difficult to see that systemic pesticides that enter the vascular system of a plant and express themselves through pollen, nectar, and guttation droplets are going to have a wide range of nontarget effects.

The same is true for plants genetically engineered to contain a pesticide gene. What is the overall impact on the soil microbiota when growing a plant in the ground that exudes pesticides indiscriminately? And, with an eye to economic impact, indiscriminate pesticide use is causing insect and weed resistance, which adversely affects productivity and keeps those on the pesticide treadmill looking for the next best chemistry to throw at the ecosystem.

We celebrate the victories in communities across the country that have adopted and are working to adopt organic land management, with practices that build soil health. It is critical that we enrich our understanding and effectiveness to meet the challenges ahead in our communities and states.

AMONG THE REDWOODS

for Judi Bari

Hello old friends, it's been many years
Ere I walked among ye, towering

Duffy mulch absorbs each step
Woods burl with life emerging

Here, warless warriors born before the Crusades
Bear regal witness to the graceless masquerade

Of new and newer holocausts. Omipotent hum
Of chains saw your wisdom

Establish greed's dominion
Profits über alles; Let me touch your skin

And not possess it, kiss your ancient lips, spill
My memories skyborne into your arms

Hear your silence slap the buzzing air
Fingers blazing with contempt

And join your final stand. At last
To soar among ye, towering.

Mitchel Cohen editing this book at Albert Einstein's home,
Princeton New Jersey, May 2018

ACKNOWLEDGMENTS

I want to especially thank Marilyn Berkon, Cathryn Swan, Isis Feral, Carolina Cositore, Jack Shalom, Marcy Gordon, Jennifer Jager, Stuart Newman, Brian Tokar, Robin Esser, Beth Youhn, Howie Cohen, and Hector Carosso for suggestions, editing, and tracking down references. And thanks also to toxicologist Dr. Robert Simon, attorney Joel Kupferman and Karl Coplan, Anti-pesticides activists Rev. Billy Talen, Savitri D., Kimberly Flynn, Meg Feeley, Robert Lederman, Aton Edwards, Howard Brandstein, Afrime Ottaway, Maris Abelson, Bob Kane, Eva Yaa Asantewaa, Arnie Gore, Tom Smith, Lori Evans, Steve Greenspan, Curtis Cost, Jim West, Barbara Deutsch, Karl Grossman, Melissa Ennen, Chris Kinder, Meg Sears, Rhio, David Schwartzman, Maggie Zhou, Carl Lawrence, Joel Meyers, Deeadra Brown, Robet Gold, Don Hickock, Gary Davidson, Karen Ingenthron, Connie Lesold, Neysa Linzer, Kellie Gasink, Eric Daillie, Donna Reilly, Donna Romo Gianell, Roy Doremus, Bob Lesko, and those many newly inspired environmental activists working with the No Spray Coalition against pesticides for staying the course and shaping a radical vision for a different kind of society; and to the numerous No Spray activists and supporters who died in part from exposure to pesticides in the course of this struggle. And, of course, thanks go to Rachel Carson for starting us down this path.

BIOGRAPHIES

Haideen Anderson is an artist living in San Francisco, and member of "The Missile Dick Chicks" performance troupe.

Mitchel Cohen coordinates the No Spray Coalition in New York City, which successfully sued the City government over its indiscriminate spraying of toxic pesticides. He graduated from Stuyvesant High School in Manhattan and received his B.A. from SUNY Stony Brook, where he co-founded the Red Balloon Collective in 1969—a radical-thinking and creatively acting group that put their bodies and art on the line against the U.S. war machine and in defense of social and environmental justice.

Mitchel ran for Mayor of NYC as one of five Green Party candidates. That primary election fell on Sept. 11, 2001, when life as we know it was changed forever. He was editor of the national newspaper, "Green Politix," for one faction of the national Green Party (before being purged), as well as the NY State Green Party newspaper; chaired the non-commercial, listener-sponsored WBAI radio (99.5 FM) Local Board, and hosted a weekly show, "Steal This Radio," for a different internet-based station. His writings include: *The Social Construction of Neurosis*, and other pamphlets; *What is Direct Action?* a book that draws on personal experiences as well as lessons from Occupy Wall Street; two books of poetry—*One-Eyed Cat Takes Flight* and *The Permanent Carnival*—and scores of pamphlets, which he sells on the streets and subways of his beloved Brooklyn. His forthcoming book of poetry and short stories is called, *The Rubber Stamp Man* and will be available in 2019.

Carolina Cositore is a writer, social worker, and New York-born activist, who lived and worked in Cuba for ten years with the news agency *Prensa Latina*. She is author of *Caonao and Jagua* (a Cuban adventure legend), and two mystery books, *That Time in Havana* and *That Time in New York*.

Robin T. Falk Esser, PhD, earned her Ph.D. in Physiology & Biophysics (cardiac electrophysiology, neurophysiology, and pharmacology) at SUNY

Stony Brook Health Sciences Center; published papers in *Science* and *Nature*, and co-authored scientific papers published in *The Journal of Physiology, The Journal of General Physiology,* and *The Biophysical Journal.* She was a New York City high school science teacher for 30 years, teaching Advanced Placement Environmental Science, AP Bio, and Chemistry, and was a recipient of the Siemens AP Science award for excellence in teaching (subject of *NY Times* articles and Tom Brokaw NBC Nightly National News interview on exemplary teaching). She is a life-long political activist; working as a teenager in the 1965 anti-apartheid mobilization, 1968 Poor People's Campaign with Dr. Martin Luther King, Jr., against the Vietnam War and in the draft resistance movement, and with the Red Balloon Collective at SUNY Stony Brook.

Jay Feldman is a cofounder and executive director of Beyond Pesticides. Jay dedicated himself to finding solutions to pesticide problems after working with farmworkers and small farmers through an EPA grant in 1978 to the national advocacy organization Rural America (1977–1981). Since that time, Jay has helped to build Beyond Pesticides' capacity to assist local groups and impact national pesticide policy. He has tracked specific chemical effects, regulatory actions, and pesticide law. Jay has a Master's in urban and regional planning with a focus on health policy from Virginia Polytechnic Institute and State University (1977), and a B.A. from Grinnell College (1975) in political science. In September 2009, U.S. Department of Agriculture Secretary Tom Vilsack appointed Jay to the National Organic Standards Board (NOSB), where he completed a five-year term in January 2015.

Martha Reed Herbert, PhD, MD, is a pediatric neurologist, neuroscientist, systems thinker, and writer who trained in medicine at Columbia University, Pediatrics at Cornell-New York Hospital and Neurology at the Massachusetts General Hospital, Harvard Medical School. She founded the collaborative multidisciplinary TRANSCEND Research Program, which takes a whole-body-brain approach to challenged brain development. She received her PhD in the History of Consciousness at the University of California, Santa Cruz. In the spirit of integrating her radical ecofeminist critique of science with a commitment to regenerative biocultural praxis, she is the scientific director of the Documenting Hope Project and principal investigator of its FLIGHTT (Facilitated Longitudinal Intensive Investigation of Genuine Health Transformation) Research Program on recovery from pediatric chronic illness. She founded the Higher Synthesis Foundation in Cambridge, Massachusetts, whose mission is to study and inspire successful regenerative approaches to complex health and environmental problems, including through registry building to capture data from individualized, nuanced practices, particularly those that fall under the umbrella of its "Gifted Clinicians, Transformative Treatments, Sensitive Measures" Project and her Body-Brain-World thematic.

John Jonik was born and still lives in Philadelphia. He won top awards for painting at Pennsylvania Academy of the Fine Arts as a student and afterwards, and has been a magazine cartoonist since early 1970s for *National Lampoon, New Yorker, Playboy, Cosmopolitan, Gourmet, Natural History*, etc., with political cartoons added since GHW Bush began bombing Iraq. He is active in areas concerning pesticides, environment, death penalty, media and private corporate influence in public government.

Sheldon Krimsky, PhD, is Lenore Stern Professor of Humanities and Social Sciences in the Department of Urban & Environmental Policy & Planning in the School of Arts & Sciences and Adjunct Professor in Public Health and Community Medicine in the School of Medicine at Tufts University. www.tufts.edu/~skrimsky He received his BS and MS in physics from Brooklyn College, CUNY and Purdue University respectively, and an MA and PhD in philosophy at Boston University. His research focuses on the linkages between science/technology, ethics/values and public policy. He has authored, co-authored or edited 16 books and published over 200 papers and reviews that have appeared in: *JAMA, Nature, Nature Genetics, Nature Biotechnology, Nature Medicine, NEJM, American Journal of Bioethics, The Lancet, Plos One,* and *Plos Biology* among others.

Jonathan R Latham, PhD, is co-founder and Executive Director of the Bioscience Resource Project, which publishes ground-breaking and topical analyses of the science underlying food and agriculture systems in the peer-reviewed academic literature and elsewhere. Jonathan is also the Editor of *Independent Science News*, and is the Director of the Poison Papers project which publicizes documents of the chemical industry and its regulators. Dr. Latham holds a Master's degree in Crop Genetics and a PhD in Virology. He was subsequently a postdoctoral research associate in the Department of Genetics, University of Wisconsin, Madison. He has published scientific papers in disciplines as diverse as plant ecology, plant virology, human genetics and plant genetic engineering. Dr Latham talks frequently at international events and scientific and regulatory conferences on the research conducted by the Project. He has written for *Truthout*, *MIT Technology Review*, the *Guardian*, *Resilience*, and many other magazines and websites.

Stacy Malkan is co-founder and co-director of U.S. Right to Know, a non-profit public interest group standing up for truth and transparency in our food system. She is also a co-founder of the Campaign for Safe Cosmetics and author of the award-winning book, *Not Just a Pretty Face: The Ugly Side of the Beauty Industry*.

Stephanie Seneff, PhD, is a Senior Research Scientist at MIT's Computer Science and Artificial Intelligence Laboratory in Cambridge, Massachusetts. She has a BS degree from MIT in biology and MS, EE and PhD degrees from MIT in electrical engineering and computer science. She has published over 200 peer-reviewed papers in scientific journals and conference proceedings. Her recent interests have focused on the role of toxic chemicals and micronutrient deficiencies in health and disease, with a special emphasis on the pervasive herbicide, Roundup, and the mineral, sulfur. In 2012, Dr. Seneff was elected Fellow of the International Speech and Communication Association (ISCA).

Vandana Shiva, PhD, is an Indian scholar, physicist, environmental activist, food sovereignty advocate, and editor and author of numerous books. The founder of Navdanya, a movement for biodiversity conservation and farmers' rights, and founder and director of the Research Foundation for Science, Technology and Natural Resource Policy, Dr. Shiva received the Right Livelihood Award, commonly known as the "Alternative Nobel Prize." Other awards include the Order of the Golden Ark, Global 500 Award of the UN, Earth Day International Award, the Lennon Ono Grant for Peace, and the Sydney Peace Prize 2010. *Time* magazine identified

Shiva as an "environmental hero," and *Asia Week* has called her one of the five most powerful communicators of Asia. Her writings include *Who Really Feeds the World?: The Failures of Agribusiness and the Promise of Agroecology*; *Staying Alive: Women, Ecology, and Development*; *The Violence of the Green Revolution: Third World Agriculture, Ecology, and Politics*; *Biopiracy: The Plunder of Nature and Knowledge*; *Monocultures of the Mind*; *Water Wars: Privatization, Pollution, and Profit*; and *Stolen Harvest: The Hijacking of the Global Food Supply*, as well as over 300 papers in leading scientific and technical journals.

Cathryn Swan helps coordinate the work of the No Spray Coalition against pesticides. For ten years she has edited the *WashingtonSquareParkBlog.com*, documenting the ins and outs of the famed park and opposing privatization of public space in New York City. She's the author of *Tales from Washington Square Park*, and studied aromatherapy, which she puts to practical use through her B-Girl company.

Brian Tokar is an activist and author, a Lecturer in Environmental Studies at the University of Vermont, and an active board member of 350Vermont and the Institute for Social Ecology. He is the author of *The Green Alternative* (1987, Revised 1992), *Earth for Sale* (1997), and *Toward Climate Justice: Perspectives on the Climate Crisis and Social Change* (2010, Revised 2014). He co-edited the 2010 book, *Agriculture and Food in Crisis*, and also edited two collections on biotechnology issues: *Redesigning Life?* (2010) and *Gene Traders* (2004). Recently he has contributed to several international collections, including Routledge Handbooks on the Climate Change Movement and on Climate Justice, *Climate Justice and the Economy*, *The Global Food System: Issues and Solutions*, *Social Ecology and Social Change*, and *Pluriverse: The Post-Development Reader*. He received a 1999 Project Censored Award for the original version of his investigative history of Monsanto, first published in *The Ecologist*.

Steve Tvedten was a pest-control operator for over fifty years. He authored *The Best Control 2: Encyclopedia of Integrated Pest Management*, a compilation of safer alternatives to cancer-causing pesticides, and offers it at no charge via the internet at www.thebestcontrol2.com.

Patricia (Patti) Wood is founder and executive director of Grassroots Environmental Education, a not-for-profit organization that uses science-driven arguments for clean air, clean water, and a safe food supply, and for stricter regulation of chemical toxins. A Visiting Scholar at Adelphi University, Ms. Wood lectures on the environment and related health issues in the College of Nursing and Public Health.

ENDNOTES

Foreword

1 (Vandana Shiva, *Origins: The corporate war against nature and culture*).
2 https://www.ncbi.nlm.nih.gov/pmc/articles/PMC5044953/
3 http://www.ipsnews.net/2016/07/biodiversity-gmos-gene-drives-and-the
   -militarised-mind/
4 https://peoplesassembly.net/monsanto-tribunal-and-peoples-assembly-report/

Preface

1 Jordan Schactel, "The Revolving Door: All 3 FDA-authorized COVID shot companies now employ former FDA commissioners," in Dossier.substack.com.
2 Mitchel Cohen, "*Genetic Engineering, Pesticides, and Resistance to the New Colonialism,*" Chapter 15, for many more details about the "revolving door". Also, Brian Tokar, "Monsanto: Origins of an Agribusiness Behemoth," Chapter 4.
3 Danny Hakim, "Monsanto Weed Killer Roundup Faces New Doubts on Safety in Unsealed Documents," *The New York Times*, March 14, 2017. The documents themselves are available at www.poisonpapers.org. Also, "The Poison Papers Expose Decades of Collusion between Industry and Regulators over Hazardous Pesticides and Other Chemicals," *Bioscience Resource Project*, July 26, 2017.
4 Stefan Oelrich, speech to the World Health Summit on Capitalizing on the Momentum of Innovation from Covid-19, October 24, 2021. https://youtu.be/-oI9vGKW0_Y. Unfortunately, YouTube has removed the vido of Oelrich's talk, stating: "This video has been removed for violating YouTube's Community Guidelines". But I transcribed Oelrich's words before YouTube censored them, and I wrote to Bayer for clarification: Are they planning to introduce "gene therapies" that permanently change an individual's genetic code? Bayer has not responded.
5 "Big Pharma exec: COVID shots are 'gene therapy'", Cause/Action, originally published I *WND*, November 15, 2021. https://cqrcengage.com/causeaction/app/document/36532017.
6 "Through the discovery and engineering of novel CRISPR systems, the company is enabling the full potential of its platform to read and write the code of life. Mammoth aims to develop permanent genetic cures through best-in-class in vivo and ex vivo therapies." "Bayer and Mammoth Biosciences to Collaborate

on Novel Gene Editing Technology," Bayer Global, January 10, 2022. https://media.bayer.com/baynews/baynews.nsf/id/Bayer-and-Mammoth-Biosciences-to-collaborate-on-novel-gene-editing-technology.

7   Lorenz Borsche,1,* Bernd Glauner,2 and Julian von Mendel3, "Covid 19 Mortality Risk Correlates Inversely with Vitamin D Status, and a Mortality Rate Close to Zero Could Theoretically Be Achieved at 50 ng/ml 25(OH)D3: Results of a Systematic Review and Meta-Analysis," *Nutrients*, October 2021, 13(10):3596 —1Independent Researcher, D-69117 Heidelberg, Germany; 2Independent Researcher, D-72076 Tübingen, Germany; ed.eborpllec@lgb; 3Artificial Intelligence, IU International University of Applied Sciences, D-99084 Erfurt, Germany; ten.nailujred@bup; *Correspondence: ed.xmg@ehcsrob

8   Hietanen, E.; Linnainmaa, K.; Vainio, H. Effects of phenoxy herbicides and glyphosate on the hepatic and intestinal biotransformation activities in the rat. *Acta. Pharmacol. Toxicol.* 1983, 53, 103-112.

9   Of many reports, see "The Efficacy of Vitamin D Supplementation in Patients With Severe and Extremely Severe COVID-19 (COVID-VIT)," which appears to be one of the most interesting recent quantitative studies, conducted by the Federal Research Clinical Center of Federal Medical & Biological Agency, Russia, to be completed in January 2023.

10 Susan Harris, "Vitamin D and African Americans," J Nutr. 2006 Apr;136(4):1126-9. doi: 10.1093/jn/136.4.1126. PMID: 16549493.

11 William F. Marshall, III M.D., "Coronavirus infection by race: What's behind the health disparities?" Mayo Clinic, August 2020, https://www.mayoclinic.org/diseases-conditions/coronavirus/expert-answers/coronavirus-infection-by-race/faq-20488802; also Maritza Vasquez Reyes, MA, LCSW, CCM, "The Disproportional Impact of COVID-19 on African Americans", Health and Human Rights Journal, Dec. 2020. https://www.ncbi.nlm.nih.gov/pmc/articles/PMC7762908/.

12 Harris, *ibid.*

13 Alison Caldwell, PhD, "Study suggests high vitamin D levels may protect against COVID-19, especially for Black people," *U Chicago Medicine*, March 19, 2021; See also Borsche, Lorenz, Bernd Glauner, and Julian v. Mendel 2021. "COVID-19 Mortality Risk Correlates Inversely with Vitamin D3 Status, and a Mortality Rate Close to Zero Could Theoretically Be Achieved at 50 ng/mL 25(OH)D3: Results of a Systematic Review and Meta-Analysis" *Nutrients* 13, no. 10: 3596. https://doi.org/10.3390/nu13103596; Samsel, Anthony, and Stephanie Seneff. 2013. "Glyphosate's Suppression of Cytochrome P450 Enzymes and Amino Acid Biosynthesis by the Gut Microbiome: Pathways to

Modern Diseases" *Entropy* 15, no. 4: 1416-1463. https://doi.org/10.3390/e15041416

14 "Vitamin D and Its Potential Benefit for the COVID-19 Pandemic," Nipith Charoenngam, Arash Shirvani, Michael F Holick, March 17, 2021: PMID: 33744444, PMCID: PMC7965847, DOI: 10.1016/j.eprac.2021.03.006. See also, Dr. Roger Sehuelt, MD, "The Role of Vitamin D in Covid-19," https://www.youtube.com/watch?v=ha2mLz-Xdpg

15 April Glaser, "McDonald's french fries, carrots, onions: all of the foods that come from Bill Gates farmland," NBC News, June 9, 2021.

16 In June 2020, the Prince of Wales and the head of the annual Davos summit launched an initiative calling for the pandemic to be seen as a chance for what they called a Great Reset of the global economy. "We have an incredible opportunity to create entirely new sustainable industries," Prince Charles said. "The time to act is now." Klaus Schwab is the Founder and Executive Chairman of the World Economic Forum. The WEF holds annual gatherings in Davos, Switzerland for billionaires and the world's decision-makers, underlines the role of biotechnology in consolidating what the WEF calls "The Great Reset": "The third and final priority of a Great Reset agenda is to harness the innovations of the Fourth Industrial Revolution to support the public good, especially by addressing health and social challenges. During the COVID-19 crisis, companies, universities, and others have joined forces to develop diagnostics, therapeutics, and possible vaccines; establish testing centers; create mechanisms for tracing infections; and deliver telemedicine. Imagine what could be possible if similar concerted efforts were made in every sector. ... the pandemic represents a rare but narrow window of opportunity to reflect, reimagine, and reset our world to create a healthier, more equitable, and more prosperous future." In other words, a world re-structured and arranged by billionaire bio- and social-engineers. While some portray this as "socialism," it is anything but. It's actually a new global form of feudalism, in which corporate monopoly control of the global and genetically engineered food supply is a necessary and critical facet.

17 BBC News, June 24, 2021, "What is the Great Reset - and how did it get hijacked by conspiracy theories?" Of course, the BBC report—chock-full of good information and crucial links, forgets to mention that the "Great Reset" was written and strategized a decade before the "opportunity" presented by the Pandemic as described in the previous footnote. https://www.bbc.com/news/blogs-trending-57532368.

18 Im Li Ching, "GM Mosquitoes: Flying through the Regulatory Gaps?," in Sheldon Krimsky and Jeremy Gruber, *The GMO Deception: What you need to*

know about the Food, Corporations, and Government Agencies Putting Our Families and Our Environment at Risk, (Skyhorse Publishing: 2014).

Introduction

1   So powerful are Marc Antony's words as written in Shakespeare's Julius Caesar, that I took the liberty of freeing their meaning from their original context, in order to marshal them in the fight against pesticides. I'm hoping that William Shakespeare would not strenuously object.

2   Some ecologists believe that all organisms, whether "beneficial" to human purposes or not, have an intrinsic right to exist. Thus, the use of the judgment "beneficial" is considered by deep ecologists, for example, to be anathema to ecological vision.

3   A Monsanto representative had this to say about Neil Young's song: "Many of us at Monsanto have been and are fans of Neil Young. Unfortunately, for some of us, his current album may fail to reflect our strong beliefs in what we do every day to help make agriculture more sustainable. We recognize there is a lot of misinformation about who we are and what we do—and unfortunately several of those myths seem to be captured in these lyrics."

4   Sam Levin, "Revealed: how Monsanto's 'intelligence center' targeted journalists and activists," *The Guardian*, Aug. 8, 2019. https://www.theguardian.com/business/2019/aug/07/monsanto-fusion-center-journalists-roundup-neil-young. Carey Gillam's exposés can be found on the website "US Right to Know," https://usrtk.org. Her books include *Whitewash: The Story of a Weed Killer, Cancer and the Corruption of Science* (2017) and *The Monsanto Papers - Deadly Secrets, Corporate Corruption, and One Man's Search for Justice* (2021).

5   Paul Elias, *Associated Press*, October 23, 2018.

6   Elias, *ibid.*

7   Stephanie K. Baer, news report for *BuzzFeed News*, Aug. 10, 2018, "A Jury Has Awarded Nearly $290 Million To A Man Who Says A Popular Weed Killer Caused His Cancer." https://www.buzzfeednews.com/article/skbaer/weed-killer-cancer-jury-verdict

8   In October 2021, Monsanto—which had been losing case after case—scored a partial win in court against a parent whose child developed cancer as a result of repeated exposures to Roundup. The child's attorney said the jury doubted that a few exposures to Roundup could have been enough to cause cancer. However, he said the jury did not address the larger question of the alleged carcinogenicity of Roundup overall, and the appeal is currently underway.

9   Court of Appeal of the State of California, First Appellate District, affirming the lower court's decision, August 9, 2021. https://usrtk.org/wp-content /uploads/2021/08/Pilliod-Opinion.pdf

10 So-called "medical experiments", often amounting to torture, have been done systematically on women in Puerto Rico and elsewhere, prisoners in the US, American Indians, soldiers, Black Americans (Tuskegee being just one example of many, as discussed in later chapters in this book.

11 © Copyright 1960 (renewed) and 1963 (renewed) by Woody Guthrie Publications, Inc. & TRO-Ludlow Music, Inc. (BMI). For the full lyrics, see https:// www.woodyguthrie.org/Lyrics/Pastures_Of_Plenty.htm.

12 Ariel Wittenberg, "EPA pesticide ban overlooks some farmworkers," *The Green-Wire*, Sept. 14, 2021.

13 David Dorado Romo, "Jan. 28, 1917: The Bath Riots," *Zinn Education Project*, https://www.zinnedproject.org/news/tdih/bath-riots. The other quotes following in this section are taken from that same article.

14 Ibid.

15 Microbiology Society, "Typhus in World War I", https://microbiologysociety. org/publication/past-issues/world-war-i/article/typhus-in-world-war-i.html.

16 The United States Memorial Holocaust Museum, 100 Raoul Wallenberg Place, SW, Washington, DC 20024-2126.

17 Holocaust Museum, ibid.

18 Holocaust Museum, ibid.

19 Max Blau and Lylla Younes, of *ProPublica*, has issued an extraordinary report concerning BASF, IG Farben and its Nazi history, "The Dirty Secret of America's Clean Dishes," December 2021. https://www.propublica.org/article /the-dirty-secret-of-americas-clean-dishes

20 Holocause Museum, Ibid.

21 Beth Youhn, "Which 'Cide Are You On?", *The Fight Against Monsanto's Roundup: The Politics of Pesticides* (2019).

22 Richard Gale and Gary Null, "Monsanto knew all along! Secret studies reveal the truth of Roundup toxicity," *The Ecologist*, Sept. 18, 2015. Documentation of Roundup's hazards goes back to the 1990s, when RoundupReady crops were first released. The full "Monsanto Files" filled the issue of the *Ecologist*, Sep/Oct 1998. Joe Mendelson, then of the Center for Food Safety, summarized there what was known at the time—and it was plenty!

23 Beth Youhn, *Ibid.*

24 Amanda Mills, "Federal Appeals Court orders EPA to ban toxic pesticide Chlorpyrifos," *Nation of Change*, April 30, 2021.

25 Center for Food Safety, "Farmworkers and Conservationists Sue EPA for Re-Approving Monsanto Cancer-Causing Pesticide," March 20, 2020. https://www.centerforfoodsafety.org/press-releases/5965/farmworkers-and-conservationists-sue-epa-for-re-approving-monsanto-cancer-causing-pesticide.

26 Alva and Alberta Pilliod of Livermore, California, were awarded $2 billion. (Andrew Blankstein and Adiel Kaplan, "California jury hits Monsanto with $2 billion judgment in cancer lawsuit," *NBC News*, May 13, 2019.) Another jury in 2019 awarded cancer victim Edwin Hardeman $5.3 million in compensation for his illness and $75 million in punitive damages, which are intended to punish a defendant and deter future misconduct. The jury found that punitive damages were required because Monsanto had failed to warn users about its product. A judge later reduced the punitive award to $20 million. (Maura Dolan, "Appeals court upholds $25-million verdict against maker of Roundup," *Los Angeles Times*, May 14, 2021).

27 *BigBuds magazine*, "GMO Cannabis is the Likely Goal": "Michael Straumietis, founder and owner of hydroponics nutrients company Advanced Nutrients, has constantly warned the marijuana community about Monsanto, Scotts Miracle-Gro, GMO marijuana, and corporate takeover of the marijuana industry. "Monsanto and Bayer share information about genetically modifying crops," Straumietis notes. "Bayer partners with GW Pharmaceuticals, which grows its own proprietary marijuana genetics. It's logical to conclude that Monsanto and Bayer want to create GMO marijuana."

28 Benedikt Kammel and Joel Rosenblatt, "Bayer Request to Settle Future Roundup Claims Is Denied by Judge," *Bloomberg News*, May 27, 2021.

29 "Bayer Confirms End of Sale of Glyphosate-Based Herbicides for U.S. Lawn & Garden Market," *Sustainable Pulse*, July 29, 2021.

30 Andrew Kimbrell, executive director of the Center for Food Safety (CFS).

31 Kenny Stancil, "Historic Victory: Bayer to End U.S. Residential Sales of Glyphosate-Based Herbicides," *Natural Blaze*, July 29, 2021. https://www.commondreams.org/news/2021/07/29/historic-victory-bayer-end-us-residential-sales-glyphosate-based-herbicides. "Timely review" by EPA? The words just melt in your mouth, don't they?

32 Joe Martino, "Bayer To Stop Selling Glyphosate Products In US, Including RoundUp," *The Pulse*, August 2, 2021. https://thepulse.one/2021/08/02/bayer-to-stop-selling-glyphosate-products-in-us-including-roundup/

33 Jef Feeley, Tim Bross, and Bloomberg, "Bayer is facing a new wave of herbicide lawsuits—and this time it's not over Monsanto's Roundup," *Fortune*, February 17, 2020.

34 "Biden EPA Reveals Prior Approval of Monsanto's Roundup Failed to Account for Risks to Monarch Butterflies and Other Endangered Species, Drift Harm to Farmers," Center for Food Safety, May 19, 2021. https://www .centerforfoodsafety.org/press-releases/6367/biden-epa-reveals-prior -approval-of-monsantos-roundup-failed-to-account-for-risks-to-monarch -butterflies-and-other-endangered-species-drift-harm-to-farmers,

35 The Poison Papers Expose Decades of Collusion between Industry and Regulators over Hazardous Pesticides and Other Chemicals," Bioscience Resource Project, July 26, 2017.

36 "Trump Signs Executive Order to Further Gut Federal GMO Oversight," Center for Food Safety, June 13, 2019.

37 Madeline Knight, "Biden Chooses Tom "Mr. Monsanto" Vilsack as Agriculture Secretary," *Left Voice*, December 23, 2020.

38 Mitchel Cohen, "*Genetic Engineering, Pesticides, and Resistance to the New Colonialism,*" Chapter 15, for many more details about the "revolving door". See also Jordan Schachtel, "The Revolving Door: All 3 FDA-authorized COVID shot companies now employ former FDA commissioners," in Dossier.substack.com.

39 Danny Hakim, "Monsanto Weed Killer Roundup Faces New Doubts on Safety in Unsealed Documents," *The New York Times*, March 14, 2017. The documents themselves are available at www.poisonpapers.org. Also, "The Poison Papers Expose Decades of Collusion between Industry and Regulators over Hazardous Pesticides and Other Chemicals," *Bioscience Resource Project*, July 26, 2017.

40 Joanna Walters, *The Guardian*, "Trump administration could be sued over pesticide threat to orca and salmon," January 21, 2018.

41 *Ibid.*

42 A study by Moms Across America in 2014 found glyphosate in breast milk, which was especially alarming (https://www.momsacrossamerica.com/glyphosate _testing_results). But even some activists in the non-profit world dispute the validity of the conclusions made by the Moms Across America study. (See, for example, Jennifer R. Schroeder, "Pesticides found in mothers' breast milk—so what?" *The Conversation*, May 13, 2014.) Still, even amidst all her dodging the question (why should *any* quantities of glyphosate be in mother's milk?), the author admits that glyphosate was indeed detected in breast milk (https://the-conversation.com/pesticideFs-found-in-mothers-breast-milk-so-what-26427). In fact, it's also accumulating in soybeans. (See T. Bøhn, et al., "Compositional differences in soybeans on the market: Glyphosate Accumulates in Roundup Ready GM Soybeans," *Food Chemistry* 153 (June 2014): 207–15.) Why accept the EPA's claim that there is any safe level of glyphosate?

43 https://wikileaks.org/plusd/cables/07PARIS4723_a.html

44 *Ibid.*

45 Peter Kornbluh, "The Pinochet File: *A Declassified Dossier on Atrocity and Accountability*," in a radio segment "Make the Economy Scream": Secret Documents Show Nixon, Kissinger Role Backing 1973 Chile Coup," *Democracy Now!*, September 20, 2013, interviewed by Aaron Maté and Amy Goodman.

46 Dave Murphy, Food Democracy Now, http://www.fooddemocracynow.org /campaign/hillary-s-monsanto-how-clinton-state-department-became-global -marketing-arm-monsanto.

47 Tom Philpott, "Taxpayer Dollars Are Helping Monsanto Sell Seeds Abroad," *Mother Jones*, May 18, 2013.

48 "U.S. Taxpayer Money Used to Help Promote Monsanto GMO Products Overseas, Documents Reveal," *New York Daily News*, May 14, 2013.

49 Anita Katial, Senior Director Europe Operations at USDA Foreign Agricultural Service (FAS), is named as the responsible officer for the pro-biotech propaganda effort on behalf of the U.S. government. https://wikileaks.org/plusd/cables/09HONGKONG128_a.html

50 https://wikileaks.org/plusd/cables/07CHIANGMAI155_a.html

51 Mitchel Cohen, "*Genetic Engineering, Pesticides, and Resistance to the New Colonialism,*" Chapter 15 for many more details about the "revolving door" between industry and government regulatory agencies. Also, see Jordan Schachtel, "The Revolving Door: All 3 FDA-authorized COVID shot companies now employ former FDA commissioners," in Dossier.substack.com. The revolving door between the FDA "regulators" and the pharmaceutical companies—and this holds as well for the World Health Organization officials—serves to promote Monsanto's Roundup and many other agri-industrial products that are toxic to human health and the environment.

52 https://wikileaks.org/tpp-final/

53 Julian Borger, "CIA officials under Trump discussed assassinating Julian Assange—report: Mike Pompeo and officials requested 'options' for killing Assange following WikiLeaks' publication of CIA hacking tools, report says." *The Guardian*, Sept. 27, 2021.

54 Many thanks to Patricia Dahl, an organizer with Stand with Assange NY, for outlining some of the secret involvements of the U.S. government with Monsanto and other corporate polluters that were first brought to light by Julian Assange and Wikileaks. See Michael Ratner, *Moving the Bar: My Life as a Radical Lawyer*, (OR Books: 2021) for an extensive first-hand review of the Assange legal case by his chief attorney.

55 A similar question resonates in this one: How significant is the fight, say, for workers in one shop to fight for and achieve higher wages since capitalism

keeps offering new low-waged jobs that desperate people around the world are willing to take, under the coercion of the frequently unbearable costs of daily life? (Karl Marx was one of the first to address this question in a small pamphlet, "Value, Price and Profit".)

56 Sheldon Krimsky, *Science and the Private Interest*. With a foreword by Ralph Nader (2003: Roman & Littlefield Pubs); also, Krimsky, *Conflicts of Interest In Science: How Corporate-Funded Academic Research Can Threaten Public Health*, (Hot Books, 2019);

57 The Standing Rock Sioux and environmental advocates won a temporary victory in July 2020 when a judge ordered that the Dakota Access Pipeline must shut down by August 5, 2020 pending a court-ordered environmental review. Microsoft News called this "a major defeat for the Trump administration and the oil companies that have been on the wrong side of history for years." https://www.msn.com/en-us/news/us/victory-for-standing-rock-the-dakota-access-pipeline-must-shut-down-by-august-5/ar-BB16oYiF. However, by May of 2021 the environmental review process had still not been completed, and a judge ruled that oil could flow through the pipeline while such review is pending. Meanwhile, in June of 2021 eco-activists in Hubbard County, Minnesota, chained themselves to a semi truck carrying drilling equipment Monday in an attempt to stop construction of Line 3, a $9.3 billion pipeline meant to transport some of the most climate-destructive oil in the world into the states. (Samir Ferdowsi, "Protesters Chained Themselves to a Semi Truck to Stop the Next Big U.S. Oil Pipeline," "At least 500 water protectors were arrested protesting the Line 3 pipeline, which will carry toxic tar sands oil from Canada into the U.S.," *Vice News*, June 17, 2021). Earlier that month (June 2021), the Biden administration revoked the permit for yet another pipeline subject to mass protests, the Keystone XL pipeline.

Chapter 1: Roundup the Usual Suspects

1 State of California, Office of Environmental Health Hazard Assessment (OEHHA), July 7, 2017. www.oehha.ca.gov/proposition-65/crnr/glyphosate-listed-effective-july-7-2017-known-state-california-cause-cancer#_ftnref3. The corporation has appealed California's listing of glyphosate as a probable carcinogen.

2 Sustainable Pulse, April 20, 2018. www.sustainablepulse.com/2018/04/20/california-defeats-monsanto-in-court-to-list-gyphosate-as-probable-carcinogen/.

3   Danny Hakim, "Monsanto Weed Killer Roundup Faces New Doubts on Safety in Unsealed Documents," *The New York Times*, March 14, 2017. The documents themselves are available at www.poisonpapers.org/.

4   "The Poison Papers Expose Decades of Collusion between Industry and Regulators over Hazardous Pesticides and Other Chemicals," *Bioscience Resource Project*, July 26, 2017.

5   Ibid.

6   Ibid.

7   Kathianne Boniello, "Your Oatmeal May Be Killing You," *New York Post*, April 30, 2016. See also Wills Robinson, "Is Your Oatmeal Killing You? Quaker Oats Is Sued for $5 Million Following Claims Weed Killer Is Used in Production," *Daily Mail*, May 1, 2016. "In April [2016], a new series of tests by the Alliance for Natural Health-USA has revealed popular breakfast foods including eggs, bagels, wholewheat bread and coffee creamers include 'alarming' levels of a widely-used agricultural herbicide."

8   Organic Consumers Association, "Peace, Love . . . and Monsanto's Weedkiller," www.organicconsumers.org/sites/default/files/downloads/ben_and_jerrys _leaflet_color.pdf. See also https://www.organicconsumers.org/news/peace -love-and-glyphosate-your-ice-cream."

9   Zen Honeycutt, "Glyphosate Found in All 5 Major Orange Juice Brands," Moms Across America, October 11, 2017, www.momsacros samerica.com/all _top_5_orange_juice_brands_positive_for_weedkiller. The following brands were found to contain glyphosate: Tropicana, Minute Maid, Stater Bros, Signature Farms, Kirkland.

10 Tony Mitra, "Vaccine-glyphosate link exposed by Anthony Samsel," *Farm Wars*, September 3, 2018. http://farmwars.info/?p=15100

11 "Monsanto's Weedkiller Is Contaminating Popular Wines and Beers," Raw Story via Alternet, March 27, 2018, www.rawstory.com/2018/03/monsantos -weedkiller-contaminating-popular-wines-beers/. Even many organic beers and wines showed some contamination with glyphosate, and the nonorganic drinks were also found to contain higher levels of arsenic as well as the glyphosate.

12 Carey Gillam, "Just Released Docs Show Monsanto 'Executives Colluding With Corrupted EPA Officials to Manipulate Scientific Data,'" EcoWatch, www.ecowatch.com/monsanto-papers-2467891575.html. Gillam is the author of *Whitewash: The Story of a Weed Killer, Cancer, and the Corruption of Science*. "For me," writes Gillam, "*Whitewash* is more than an exposé about the hazards of one chemical or the actions of one company. It's also a call to remember the lessons of Rachel Carson and *Silent Spring* as evidence mounts that we are in a

very precarious point as the push for pesticide dependence and the drive for corporate profits take precedence over people's lives and our environment."

13 Ibid.

14 Ibid.

15 Robert F. Kennedy, Jr., *The World Mercury Project,* August 1, 2017. https://www .facebook.com/WorldMercuryProject/posts/1949188845328527

16 Gilliam, www.ecowatch.com/monsanto-papers-2467891575.html.

17 Dr. Stephanie Seneff, "Glyphosate and Autism," Also reference "Glyphosate Acting as a Glycine Analogue: Slow Insidious Toxicity" in this book. See Dr. Joseph Mercola, "Monsanto's Roundup Herbicide May Be Most Important Factor in Development of Autism and Other Chronic Disease," June 9, 2013, www.articles.mercola.com/sites/articles/archive/2013/06/09/monsanto -roundup-herbicide.aspx. And see Zoë Schlanger, "Study Finds 25 Percent Higher Rate of Autism Where Mosquito Killer Is Sprayed from Planes," *News-week,* June 30, 2016.

18 See Rev. Billy Talen, "Monsanto is the Devil (the Earth also Rises)," unpublished submission to this book, and from speech at Monsanto's headquarters in 2017 in St. Louis, Missouri.

19 Environmental Working Group, "2,4-D Herbicide & GMO Crops," June 22, 2014, www.ewg.org/research/24D. The herbicide 2,4-D was a component of the notorious Vietnam-era herbicide Agent Orange and long known to be toxic to people and the environment. Invented in 1946, it was used in some products that contained dioxin impurities until the mid-1990s. Also see U.S. government link that gives history of pesticide use in the United States: www.ncbi.nlm.nih .gov/books/NBK236351/

For further reference, see "The Risks of the Herbicide 2,4-D," Lars Neumeister, *Genewatch,* January 2014. Despite assurances to the contrary, Neumeister reports that 2,4-D pesticide products have been found that contain measurable quantities of dioxin.

The chemical structure of 2,4-D itself contains a benzene ring, which makes it carcinogenic, with or without the dioxin:

2,4-D dichlorophenoyacetic acid

See H. Chen and D. A. Eastmond, "Topoisomerase Inhibition by Phenolic Metabolites: A Potential Mechanism for Benzene's Clastogenic Effects," *Carcinogenesis* 16, no. 10 (1995): 2301–2307. See also S. Rappaport et al., "Human Benzene Metabolism Following Occupational and Environmental Exposures," *Chemico-Biological Interactions* 184, no. 1–2 (2010): 189–195.

20 "Colombia to Use Glyphosate in Cocaine Fight Again," *The Guardian*, April 19, 2016. "The defense minister, Luis Carlos Villegas, said instead of dumping glyphosate from American-piloted crop dusters, as Colombia did for two decades, the herbicide will now be applied manually by eradication crews on the ground.... A better eradication strategy, the experts insist, is the one already in place and which the government has been promising to scale up. In that approach, work crews pull up coca bushes by the roots, thus ensuring plants can't grow back as happens after exposure to glyphosate."

21 Javiera Rulli, ed., *United Soya Republics: The Truth About Soya Production in South America,* GRR Grupo de Reflexión Rural, Argentina: 2007, www.lasojamata .net/files/soy_republic/Chapt01IntroductionSoyModel.pdf.

22 Private lawns simply did not exist for the United States working class until the 1950s, and even as late as 1987 thru 1990 there were 362 deaths associated with riding mowers, many because they tipped over and eviscerated the lawn care specialist. (See Neil Genzlinger, "Can't We All Get a Lawn?" *New York Times Sunday Book Review,* June 18, 2006.) Lawns around one's home were solely a manufactured desire, taking off from the aristocracy in England and Scotland. Hollywood and American television promulgated the well-scrubbed version of the American dream (writer Henry Miller blasted it as the air-conditioned nightmare), which was to be able to buy one's own family home in the suburbs. As thousands of soldiers returned from World War II, they and their families needed housing, and between 1948 and 1952 Abraham Levitt and his sons William and Alfred on Long Island built what would be six thousand houses, with signature unfenced lawns. "This was the first American suburb to include lawns already in place when the first tenants took possession. The Levitts, who also build subdivisions in New Jersey, Pennsylvania, Cape Cod, and Puerto Rico (several of them also called Levittown), pioneered the established lawn, which residents were required to keep up but forbidden to fence in. The importance of a neat, weed-free, closely-shorn lawn was promoted intensely in the newsletters that went out to all homeowners in these subdivisions, along with lawn-care advice on how to reach this ideal." These soldiers were "trained in neatness and obedience, and these were the conformist fifties, when everyone was on the watch for signs of Communism and crabgrass. At times, the two seemed morally equivalent." (See "Lawn History," *Planet Natural Research Center: Answers*

& Advice for Organic Gardeners, www.planetnatural.com.) Well, this might have been a white suburbanite's dream, but there were plenty of people of color as well as working-class whites for whom a suburban house and lawn were not only beyond financial reach but also ideologically absurd.

23 Neil Genzlinger, "Can't We All Get a Lawn?" *New York Times Sunday Book Review,* June 18, 2006.

24 Monsanto is currently championing another herbicide, dicamba, because many "weeds" have grown resistant to Roundup. The company is now engineering a new wave of crops to be resistant to dicamba, in much the same way as it engineered and marketed Roundup for Roundup-Ready Soy and Corn.

25 See, for one, Richard Levins, "The Struggle for Ecological Agriculture in Cuba," Red Balloon publications, 1991.

26 Glyphosate is sprayed on wheat just before harvest. Residual glyphosate is consumed by humans in wheat products, where it inhibits the microflora in the human gut, particularly the bacteria cytochrome P450 (CYP), which is needed to sustain normal gut homeostasis. A major proximal outcome is celiac disease, along with other disorders, including cancer. See David Haines and Stephanie C. Fox, et al., "Chemical Sensitivity, Identifying and Removing Threats," in *Attacking Illness At Its Roots: Biotherapeutic Strategies in Precision Medicine* David Haines and Stephanie C. Fox, Hoboken NJ, Wiley Life Science Books (Work in-progress).

27 Theo Colborn, "Pesticide Use in the U.S. and Policy Implications: A Focus on Herbicides," *Toxicology and Human Health,* February 1, 1999.

28 Donald Atwood and Claire Paisley-Jones, *Pesticides Industry Sales and Usage: 2008–2012 Market Estimates.* Biological and Economic Analysis Division Office of Pesticide Programs Office of Chemical Safety and Pollution Prevention, U.S. Environmental Protection Agency, 2017.

29 Organophosphates include such chemical pesticides as Malathion, which in 1999 was sprayed all over New York City and surroundings to kill mosquitoes said to be carrying the West Nile virus.

30 "Not ready for Roundup: Glyphosate Fact-Sheet," Greenpeace, 1997. This document has been apparently removed from Greenpeace's website, but it can still be found online through the Wayback machine at web.archive.org/web /20040111060519/www.greenpeaceusa.org/media/factsheets/glyphosatetext .htm.

31 Ibid.

32 Ibid.

33 Ibid. Also, see pubchem.ncbi.nlm.nih.gov/compound/1_4-dioxane#section =Top, where dioxane is listed as being a probable carcinogen.

34 "Common Weed Killer (Roundup) Shows Evidence of Environmental and Health Problems," *Organic Gardening,* July 2000.

35 "DNA Damage?" The Detox Project, www.detoxproject.org/glyphosate/dna -damage/.

36 "The Quality of Our Nation's Waters, Pesticides in the Nation's Streams and Ground Water, 1992–2001," U.S. Geological Survey, pubs.usgs.gov/circ /2005/1291/.

37 *Third National Report on Human Exposure to Environmental Chemicals*, Centers for Disease Control, 2005.

38 See particularly studies done by Philip Landrigan and others at Mt. Sinai Hospital on the effects of pesticide exposure on children. For example, Dr. Philippe Grandjean and Philip J. Landrigan, MD, "Neurobehavioural Effects of Developmental Toxicity," *The Lancet* 13, no. 3 (March 2014).

39 www.nospray.org/wp-content/uploads/2015/07/JudgeDanielsDecision -memorandum-opinion-and-order-no-spray-vs-nyc.pdf.

40 Monika Krüger, Philipp Schledorn, Wieland Schrödl, Hans-Wolfgang Hoppe, Walburga Lutz, and Awad A. Shehata, "Detection of Glyphosate Residues in Animals and Humans," *Journal of Environmental & Analytical Toxicology* 4, no. 2 (2014).

41 Emily Marquez, "Study Finds Glyphosate in Pregnant Women," *Pesticide Action Network*, April 5, 2018.

Chapter 2: Better Active Today than Radioactive Tomorrow

1 "From its origins, chemical agriculture has been a form of warfare—it is a war against the soil, against our reserves of fresh water, and against all the microbes and insects that are necessary for the growing of healthy food. Since the earliest origins of modern industrial agriculture, agribusiness has been at war against all life on earth, including ourselves." Brian Tokar, "Agribusiness, Biotechnology and War," *Z Magazine*, September 2002.

2 John Bellamy Foster and Brett Clark, "Rachel Carson's Ecological Critique," *Monthly Review*, February 2008.

3 Meir Rinde, "Richard Nixon and the Rise of American Environ-mentalism," *Distillations: Science+Culture+History*, Spring 2017, www.sciencehistory.org /distillations/magazine/richard-nixon-and-the-rise-of-american -environmentalism.

4 Ibid.

5 "Some students confessed . . . their worries that 'leaders in the political and industrial establishment are deliberately pushing the environment issue 'to take some of the force out of the anti-war, anti-racism, anti-poverty issues.'" Edmund Muskie of Maine, one of the Senate's prominent environmentalists and the unsuccessful Democratic vice-presidential candidate in 1968, warned that a

green movement should not become a "'smoke screen' obscuring the 'challenge of equal opportunity.'" Zoë Carpenter, "In 1970, Environmentalism Was Poised to 'Bring Us All Together.' What Happened? Today, the Environment Is a Controversial Issue Divided Along Partisan Lines—But It Wasn't Always That Way." *The Nation,* April 20, 2015.

6 An Dien and Jon Dillingham, "Da Nang Agent Orange Cleanup a First Step, But Questions Abound," *Thanh Nien News,* August 17, 2012. http://www.thanhniennews.com/politics/da-nang-agent-orange-cleanup-a-first-step-but-questions-abound-5632.html.

7 An Dien and Jon Dillingham, "US Chemical Companies Concealed Effects of Dioxin, Say Advocates," Centre for Research on Globalization, August 6, 2009, www.globalresearch.ca/vietnam-chemical-companies-us-authorities-knew-the-dangers-of-agent-orange/14720.

8 The United States, however, never signed onto the International Criminal Tribunal, to prevent US officials from being tried for crimes against humanity. Despite the US refusal to sign the agreement, it has pressured other countries' officials be brought before the Court, such as Slobodan Milosevic, to be tried for war crimes and "crimes against humanity."

9 Jon Dillingham, "Vietnam: Chemical Companies, U.S. Authorities Knew the Dangers of Agent Orange," *Europe Solidaire,* August 10, 2009, www.europe-solidaire.org/spip.php?article14706.

10 Ibid.

11 Ibid.

12 See, for one of many, American Cancer Society's *Cancer Action Network,* "ACS CAN President Says It's Time to Put an End to Big Tobacco's Lies," December 19, 2017. https://www.acscan.org/news/acs-can-president-says-it's-time-put-end-big-tobacco's-lies.

13 Mark Hertsgaard and Mark Dowie, "How Big Wireless Made Us Think That Cell Phones Are Safe: A Special Investigation," Nation Magazine, April 23, 2018, www.thenation.com/article/how-big-wireless-made-us-think-that-cell-phones-are-safe-a-special-investigation/.

14 Dillingham, "Vietnam." Dillingham quotes Dr. Wayne Dwernychuk, a retired senior advisor at Hatfield Consultants and researcher studying contamination from dioxin herbicides in Vietnam, who debunked Young's reports: "Young is paid by the chemical companies," Dwernychuk told *Thanh Nien Daily* in 2009. "I don't believe a word he says."

15 Ibid.

16 Ibid.

17 Loana Hoylman, "Agent Blue: Arsenic-Laced Rainbow," *The VVA Veteran,* May/June 2015, www.vvaveteran.org/35–3/35-3_agentblue.html.

18 Dillingham, "Vietnam."

19 National Research Council, *Arsenic in Drinking Water: 2001 Update*. (Washington, DC: The National Academies Press, 2001). https://www8 .nationalacademies.org/onpinews/newsitem.aspx; Also, Katharine Q. Seelye, "Arsenic Standard for Water Is Too Lax, Study Concludes," *The New York Times*, Sept. 11, 2001. https://www.nytimes.com/2001/09/11/us/arsenic -standard-for-water-is-too-lax-study-concludes.html.

20 "Bush U-Turn on Arsenic Rule," *CBS News*, Oct. 31, 2001, www.cbsnews.com/ news/bush-u-turn-on-arsenic-rule/.

21 "New Evidence Confirms Cancer Risk from Arsenic in Drinking Water," *National Academy of Sciences, Engineering, and Medicine*, Sept. 11, 2001. https:// www8.nationalacademies.org/onpinews/newsitem.aspx?RecordID=10194.

22 Dr. Robert Simon, letter to the No Spray Coalition against pesticides, May 2018.

23 David Heath, "Politics of Poison: How Politics Derailed EPA Science on Arsenic, Endangering Public Health. Delay Keeps Pesticides with Arsenic on the Market," The Center for Public Integrity, June 28, 2014, updated January 26, 2015, www.publicintegrity.org/2014/06/28/15000/how-politics-derailed-epa -science-arsenic-endangering-public-health.

24 Ibid. Quoting from Heath's report:

Evidence from the Center's investigation pointed to one congressman: Mike Simpson of Idaho. Simpson was one of the Republicans who signed the letter to the EPA administrator complaining about the missing 300 studies. He was the chairman of the subcommittee that controlled funding for the EPA, where the language first appeared. He was also a member of another committee where the language surfaced again in a different report. He even asked the EPA administrator about arsenic at a subcommittee hearing.

Simpson, who worked as a dentist and state legislator before entering Congress, is a frequent critic of the EPA. But in the 2012 and 2014 election campaigns, he has been portrayed as too liberal by Tea Party candidates funded by the right-wing Club for Growth.

In a brief interview outside his Capitol Hill office, Simpson accepted credit for instructing the EPA to stop work on its arsenic assessment.

"I'm worried about drinking water and small communities trying to meet standards that they can't meet," he said. "So we want the Academy of Science to look at how they come up with their science."

Simpson said he didn't know that his actions kept a weed killer containing arsenic on the market. He denied that the pesticide companies lobbied him for the delay.

But lobbyist Grizzle offered a different account.

"I was part of a group that met with the congressman and his staff a number of years ago on our concerns," Grizzle said, adding that there were four or five other lobbyists in that meeting but he couldn't remember who they were.

25 N. Defarge, J. Spiroux de Vendômois, and G. E. Séralini, *Toxicology Reports* 5 (2018): 156–163, www.sciencedirect.com/science/article/pii/S221475001730149X. See also www.ncbi.nlm.nih.gov/pmc/articles/PMC5756058/.

26 Ibid.

27 Henry Rowlands, "Shocking Study Shows Glyphosate Herbicides Contain Toxic Levels of Arsenic," *Sustainable Pulse* (Jan. 8, 2018). https://sustainablepulse.com/2018/01/08/shocking-study-shows-glyphosate-herbicides-contain-toxic-levels-of-arsenic/#,W6G_HvkpCig.

Chapter 3: The Future Ain't What It used to Be

1 See, for instance, James Weinstein, *The Corporate Ideal in the Liberal State: 1900–1918*, (Boston: Beacon Press, 1968). Also, see Mitchel Cohen, *Big Science, Fragmentation of Work, & the Left's Curious Notion of Progress*, (New York: Red Balloon Collective, 2017).

2 Meir Rinde, "Richard Nixon and the Rise of American Environmentalism," *Distillations Magazine*, Science History Institute, Spring 2017, www.sciencehistory.org/distillations/magazine/richard-nixon-and-the-rise-of-american-environmentalism. In Brooklyn, NY, a large municipal dump across the Belt Parkway from Starrett City public housing was covered over. Starrett City was for many years associated with stagnant water and dirty air. When the buildings went private, in order to sell the apartments to unsuspecting buyers the entire area, including the former dump, were re-christened "Spring Creek".

3 Ibid.

4 www.businessinsider.com/bayer-monsanto-merger-has-farmers-worried-2018-4.

5 Zeynep Tufekci, "Facebook's Surveillance Machine," *New York Times*, March 19, 2018.

6 Henry Krinkle, "Federal Contract Site Reveals US Reliance on Monsanto and Israeli Firm for White Phosphorus Supply," *Current Events Inquiry*, February 7, 2013, www. ceinquiry.wordpress.com/2013/02/07/white-phosphorus/.

7 Department of the Army, "Justification and Approval for Other Than Full and Open Competition: Control No.: 12–118," January 29, 2013., www.fbo.gov/utils/view?id=aa766891fe32cf9ae7f87f3c7d3611a3. Signed by Amy VanSickle, Special Competition Advocate, on Oct. 24, 2012.

8 The "Convention on the Prohibition of the Development, Production, Stockpiling and Use of Chemical Weapons and on their Destruction" is administered by the Organisation for the Prohibition of Chemical Weapons (OPCW), an intergovernmental organization based in The Hague, the Netherlands.

9 Kurtis Bright, *Natural Blaze*, "Monsanto's Dirty Links With Government Over Deadly White Phosphorus Manufacturing," Nov. 28, 2016. www.naturalblaze .com/2016/11/monsantos-dirty-links-with-government-over-deadly-white -phosphorus-manufacturing.html.

10 See, for example, Henry David Thoreau, *Walden*, 1854.

Chapter 4: Monsanto: Origins of an Agribusiness Behemoth

1 Ronnie Cummins, "Activists in 16 Nations Carry Out Successful Global Days of Action," *Food Bytes*, no. 3, November 4, 1997, Little Marais, Minnesota: Pure Food Campaign. See B. Tokar "Resisting the Engineering of Life," www.social -ecology.org/wp/2001/05/resisting-the-engineering-of-life/.

2 Greenpeace U.S.A., "Greenpeace Quarantines Genetically-Altered Monsanto 'X-field' in Iowa," press release, October 10, 1996; "Greenpeace stops genetically engineered soybeans destined for Europe on Mississippi River," press release, November 19, 1996.

3 Paul Brown, "Printers Pulp Monsanto Edition of Ecologist," *The Guardian*, September 27, 1998. Downloaded August 14, 2017, from organicconsumers .org.

4 "Chemical Producers: Dow Chemical, DuPont, Monsanto and Union Carbide have ranked among Top 10 biggest chemical makers since 1940," *Chemical and Engineering News*, January 12, 1998, 193.

5 Marc S. Reisch, "From Coal Tar to Crafting a Wealth of Diversity," *Chemical and Engineering News*, January 12, 1998, 90.

6 Pamela Peck, "Vermont's Polystyrene (Styrofoam) Boycott," Vermonters Organized for Cleanup, (Barre, Vermont), 1989.

7 Theo Colborn, Dianne Dumanoski, and John Peterson Myers, *Our Stolen Future* (New York: Penguin Books, 1996), 90.

8 Michelle Allsopp, Pat Costner, and Paul Johnson, *Body of Evidence: The Effects of Chlorine on Human Health* (Exeter, England: University of Exeter, Greenpeace Research Laboratories, May 1995); see also Joseph E. Cummins, "PCBs—Can the World's Sea Mammals Survive Them?" *The Ecologist* 28, no. 5 (September/ October 1998): 262–263.

9 Colborn et. al, *Our Stolen Future*, 101–104.

10 Jonathan Kozol, *Savage Inequalities: Children in America's Schools* (New York: Crown Publishers, 1991), 7, 20.

11 "Death of Animals Laid to Chemical," *New York Times*, August 28, 1974, 36. See also Colborn et. al, *Our Stolen Future*, 116.

12 "Citizen Inquiry Uncovers Blatant Violation of Environmental Law Surrounding the Proposed Times Beach Incinerator," Times Beach Action Group, (St. Louis, Missouri), November 1995.

13 Philip Shabecoff, *A Fierce Green Fire: The American Environmental Movement* (New York: Hill and Wang, 1993), 210–212; Brian Tokar, *Earth for Sale: Reclaiming Ecology in the Age of Corporate Greenwash* (Boston: South End Press, 1997), 59–60; Times Beach Action Group, ibid.

14 Lisa Martino-Taylor, "Legacy of Doubt," *Three River Confluence*, no. 7/8 (Fall 1997): 27.

15 Steve Taylor, personal communication with author, August 5, 1998.

16 Peter Downs, "Is the Pentagon Involved?" *St. Louis Journalism Review*, June 1998.

17 Hugh Warwick, "Agent Orange: The Poisoning of Vietnam," *The Ecologist* 28, no. 5 (September/October 1998), 264–265. Peter H. Schuck, *Agent Orange on Trial: Mass Toxic Disasters in the Courts*, (Cambridge, MA: Harvard University Press, 1987), 86–87, 155–164. Monsanto's share of Agent Orange production was 29.5 percent, compared to Dow's market share of 28.6 percent; however, some batches of Agent Orange contained more than 47 times more dioxin than Dow's. The other defendants in the case were Hercules Chemical, Diamond Shamrock, T. H. Agriculture and Nutrition, Thompson Chemicals, and Uniroyal.

18 Cate Jenkins, "Criminal Investigation of Monsanto Corporation—Cover-Up of Dioxin Contamination in Products—Falsification of Dioxin Health Studies," USEPA Regulatory Development Branch, November 1990. Jed Greer and Kenny Bruno, "Monsanto Corporation: A Case Study in Greenwash Science," in *Greenwash: The Reality Behind Corporate Environmentalism*, (Penang, Malaysia: Third World Network, 1996), 141.

19 Jock Ferguson, "Chemical Company Accused of Hiding Presence of Dioxins," (Toronto), *Globe and Mail*, February 19, 1990, A9. The punitive damages in Kemner vs. Monsanto were overturned on appeal two years later.

20 Cate Jenkins, "Criminal Investigation of Monsanto Corporation."

21 Samuel Fromartz, "Monsanto's sordid historical legacy?" Food & Environment Reporting Network, January 11, 2022. https://mailchi.mp/thefern.org/back-forty-monsanto-historical-legacy?e=5cffaaf04b]

22 Kenny Bruno, "Say it Ain't Soy, Monsanto," *Multinational Monitor* 18, no. 1–2, (January/February 1997). The quote is from stock analyst Dain Bosworth; Mark Arax and Jeanne Brokaw, "No Way Around Roundup," *Mother Jones*, January/February 1997.

23 Testimony of Champion Paper Company, Vermont Forest Resources Advisory Council, Island Pond, Vermont, June 26, 1996.

24 Carolyn Cox, "Glyphosate Fact Sheet," *Journal of Pesticide Reform* 11, no. 2 (Spring 1991).

25 Ibid. See also Joseph Mendelson, "Roundup: The World's Biggest-Selling Herbicide," *The Ecologist* 28, no. 5 (September/October 1998), 270–275.

26 Carolyn Cox, "Glyphosate, Part 2: Human Exposure and Ecological Effects," *Journal of Pesticide Reform* 15, no. 4 (Fall 1995).

27 Sylvia Knight, "Glyphosate, Roundup and Other Herbicides—An Annotated Bibliography," Vermont Citizens' Forest Roundtable, January 1996.

28 Pesticide Action Network North America, "Monsanto Agrees to Change Ads and EPA Fines Northrup King," January 10, 1997.

29 "Case of Mislabeled Herbicide Results in $225,000 Penalty," *Wall Street Journal*, March 25, 1998.

30 J. Greer and K. Bruno,. "Monsanto Corporation," 145–46.

31 Sarah Anderson and John Cavanagh, "The Top 10 List," *The Nation* December 8, 1997, 8.

32 Larry Rohter, "To Colombians, Drug War Is a Toxic Foe," *New York Times*, May 1, 2000.

33 Jennifer Ferrara, "Revolving Doors: Monsanto and the Regulators," *The Ecologist* 28, no. 5 (September/October 1998), 280–286.

34 Craig Canine, "Hear No Evil," *Eating Well*, July/August 1991, 41–47; Brian Tokar, "The False Promise of Biotechnology, *Z Magazine*, February 1992, 27–32; Debbie Brighton, "Cow Safety, BGH and Burroughs," *Organic Farmer*, Spring 1990, 21.

35 Andrew Christiansen, *Recombinant Bovine Growth Hormone: Alarming Tests, Unfounded Approval*, Rural Vermont, (Montpelier, Vermont), July 1995; see also Tokar, "The False Promise," 28–29.

36 Christiansen, *Recombinant Bovine Growth Hormone*, 10, 17; "FDA's Review of Recombinant Bovine Growth Hormone," U.S. General Accounting Office, August 6, 1992 (GAO/PEMD-92-96).

37 Mark Kastel, *Down on the Farm: The Real BGH Story*, Rural Vermont, Fall 1995.

38 Brian Tokar, "Biotechnology: The Debate Heats Up," *Z Magazine*, June 1995, 49–55; Diane Gershon, "Monsanto Sues over BST," *Nature* 368, March 31, 1994, 384. The Vermont state labeling law was defended by the state on the grounds of consumer preference, rather than public health, and was ultimately struck down by a federal judge, who ruled that mandatory rBGH labeling was a violation of the companies' alleged constitutional right to refuse to speak.

39   D.S. Kronfeld, "Health Management of Dairy Herds Treated with Bovine Somatotropin," *Journal of the American Veterinary Medical Association* 204, no. 1 (January 1994), 116–130; Samuel S. Epstein, "Unlabeled Milk from Cows Treated with Biosynthetic Growth Hormones: A Case of Regulatory Abdication," *International Journal of Health Services* 26, no. 1 (1996), 173–185.

40   Sonja Schmitz, "Cloning Profits: The Revolution in Agricultural Biotechnology," in Brian Tokar, ed., *Redesigning Life?—The Worldwide Challenge* to *Genetic Engineering* (London: Zed Books), 2001.

41   Bruno, "Say it Ain't Soy, Monsanto."

42   Monsanto Company, *1997 Annual Report*, 16, 37.

43   Charles M. Benbrook, "Impacts of Genetically Engineered Crops on Pesticide Use in the U.S.—The First Sixteen Years," *Environmental Sciences Europe* 24, no. 24 (2012); Scott Kilman, "Superweeds Hit Farm Belt, Triggering New Arms Race," *Wall St. Journal*, June 4, 2010.

44   Lorraine Chow, "Arkansas Could Become First State to Ban Dicamba," *EcoWatch* August 28, 2017.

45   Hope Shand, "*Bacillus Thuringiensis*: Industry Frenzy and a Host of Issues," *Journal of Pesticide Reform* 9, no. 1 (Spring 1989), 18–21; Ricarda A. Steinbrecher, "From Green to Gene Revolution: The Environmental Risks of Genetically Engineered Crops," *The Ecologist* 26, no. 6 (November/December 1996), 273–281; Brian Tokar, "Biotechnology vs. Biodiversity," *Wild Earth* 6, no. 1 (Spring 1996), 50–55.

46   Union of Concerned Scientists (1998) "EPA Requires Large Refuges," *The Gene Exchange* (Volume: SummerEdition), no. 1; Union of Concerned Scientists (1998) , "Transgenic Insect-Resistant Crops Harm Beneficial Insects," *The Gene Exchange* (Volume: Summer Edition), no. 4; Union of Concerned Scientists, "Managing Resistance to Bt," *The Gene Exchange* 6, no. 2–3 (December 1995), 4–7.

47   Brian Tokar, "Agribusiness, Biotechnology and War," *Z Magazine*, September 2002.

48   Rural Advancement Foundation International, *The Life Industry 1997: The Global Enterprises that Dominate Commercial Agriculture, Food and Health*, November/December 1997. The comment about Asgrow was quoted by Brewster Kneen in *The Ram's Horn*, (Ottawa, Canada), no. 160, June 1998, 2.

49   Monsanto Company, *1997 Annual Report*, 17; Rural Advancement Foundation International, *The Life Industry*; Union of Concerned Scientists, "Expanding in New Dimensions: Monsanto and the Food System," *The Gene Exchange*, December 1996, no. 11.

50   Barnaby J. Feder, "Monsanto Says it Won't Market Infertile Seeds," *New York Times*, October 5, 1999, 1; John Vidal, "How Monsanto's Mind Was Changed,"

The Guardian, October 9, 1999, 15; Greenpeace Business Conference transcript, posted to electronic list, biotech_activists@iatp.org, October 7, 1999.

51 Bob Burton, "Advice on Making Nice," *PR Watch*, 6, no. 1 (First Quarter 1999), 1–6.

52 "Monsanto v. U.S. Farmers 2012 Update," Washington: Center for Food Safety, November 2012.

53 K. Makin and A. Dunfield, "Monsanto Wins Key Biotech Ruling," *Globe and Mail*, May. 21, 2004; "Percy Schmeiser Claims Moral and Personal Victory in Supreme Court Decision,"; "Monsanto Canada Inc. v. Schmeiser," May 21, 2004, 2004 SCC 34, No. 29437.

54 Andrew Pollack, "Monsanto Buying Leader in Fruit and Vegetable Seeds," *New York Times*, January 25, 2005.

55 ETC Group, "Breaking Bad: Big Ag Mega-Mergers in Play" December 2015, www.etcgroup.org.

56 Beth Burrows, "Government Workers Go Biotech," Edmonds Institute, May 19, 1997.

57 Genentech, Inc., "Genentech Names Moore New Head of Government Affairs Office Based in Washington, D.C.," April 13, 1998. See also Senator Al Gore, "Planning a New Biotechnology Policy," *Harvard Journal of Law and Technology* 5 (Fall 1991), 19–30.

58 James Turner, "The Aspartame/NutraSweet Timeline," Washington: National Institute of Science, Law, and Public Policy, www.rense.com.

59 Biotechnology Industry Organization, September 20, 2001, accessed November 11, 2008, www.bio.org.

60 Michael Pollan, "Big Food Strikes Back: Why Did the Obamas Fail to Take On Corporate Agriculture?" *New York Times Magazine*, October 5, 2016.

61 Center for Food Safety, "Part 340 Rules Sign On Letter," email message to author, May 25, 2017.

62 Robert Langreth and Nikhil Deogun, "Investors Cool to Pharmacia Merger Plan," *Wall Street Journal*, December 21, 1999.

63 "Appendix 2: GMOs are Dead," in *Ag Biotech: Thanks, But No Thanks?*, Deutsche Bank Alex. Brown investor's report on DuPont Chemical, July 12, 1999, 18. The Appendix was reportedly released by Deutsche Bank as an independent report to investors on May 21, 1999; Scott Kilman and Thomas M. Burton, "Monsanto Feels Pressure From the Street," *Wall Street Journal*, October 21, 1999.

64 Andrew Pollack, "Widely Used Crop Herbicide Is Losing Weed Resistance," *New York Times*, January 14, 2003.

65 ETC Group, "Breaking Bad"; Jack Kaskey, "Bayer-Monsanto Combination Likely Too Big in U.S. Cottonseed," *Wall St. Journal*, September 14, 2016;

"Joining Forces: Market shares resulting from proposed mergers," September 15, 2015.

66 Eric Lipton, "Food Industry Enlisted Academics in G.M.O. Lobbying War, Emails Show," *New York Times*, September 6, 2015.

67 Carey Gillam, "Newly Released 'Monsanto Papers' Add To Questions Of Regulatory Collusion," Huffington Post, August 2, 2017. The quoted attorney is Brent Wisner. See also Jacob Bunge, "Monsanto Marshals Scientists in Herbicide Suit," *Wall Street Journal*, August 3, 2017.

68 Danielle Ivory and Robert Faturechi, "The Deep Industry Ties of Trump's Deregulation Teams," *New York Times*, July 11, 2017.

Chapter 5: Poisoning the Big Apple

1 Dan Halper, "Summer of Spray: Pesticide Spraying in New York City 1999–2000. Necessary Mosquito Control Efforts or Dilute Chemical Warfare?" NY Environmental Law & Justice Project, Fall 2000.

2 Rathone, H.S. & Nollet, M.L. "Pesticides: Evaluation of Environmental Pollution," CRC Press 2012 p. 321. Also, Wayne Riddle, "Nerve Gases and Pesticides: Links are Close," *New York Times*, March 30, 1984.

3 In addition to the No Spray Coalition, plaintiffs included the National Coalition Against Misuse of Pesticides (now, Beyond Pesticides), Save Organic Standards-NY, Disabled in Action, and individuals Mitchel Cohen, Valerie Sheppard, Robert Lederman, Eva Yaa Asantewaa, and Howard Brandstein.

4 Cheminova—the manufacturer of the Malathion used in New York City spraying—had been ordered by the EPA in the mid-1990s to add the warning against spraying over water to its label. But five years later when the spraying began, the label still did not display such warning. A few months later, it was added after much protest and legal challenges by the New York Environmental Law and Justice Project.

5 Some areas like Houston installed bat houses in swampy areas to deal with mosquitoes naturally, without spraying. There are numerous alternatives to the mass application of pesticides, some of which are documented at www .NoSpray.org/alternatives. See also https://www.reference.com/pets-animals /many-mosquitos-bats-eat-night-d1e92a5f7b73fac2.

6 Actually, West Nile virus was not unknown to scientists in the United States. It was first discovered by western scientists from Rockefeller University in Uganda in 1937, and many experiments with it were performed in the late 1950s at Sloan Kettering, Rockefeller University, Yale, and at the U.S. Army biological warfare weapons lab at Fort Detrick, Maryland.

7 New York City kept no records of the numbers of people who were sickened or who died as a result of the pesticide spraying, so it is not possible to compare deaths from the spraying with deaths from encephalitis caused by exposure to the virus. The No Spray Coalition set up its own hotline, and it was flooded with calls from individuals who were sickened by the spraying. Eight core members of the Coalition died from illnesses caused or exacerbated by exposure to the pesticides over the next few years. What we do have in the way of hard data are visits by eight spray truck drivers, New York Environmental Law and Justice Project (NYELJP) represented, "who were not given adequate safety training or protective gear, and who consequently suffered from pesticide poisoning." (See www.nyenvirolaw.org/legal-actions/no-spray-coalition/.) The drivers were diagnosed at the occupational health and safety clinic at Mount Sinai Hospital. See, *NoSpray et al. v. The city of NY et al.*, 2000, www.nospray.org /wp-content/uploads/2015/05/plaintiffs-reply-no-spray-lawsuit-2001.pdf.

8 Like Malathion, massively sprayed in 1999 to kill mosquitoes, glyphosate belongs to the same family of organophosphates. The following is from Philip J. Chenier, *Survey of Industrial Chemistry*, 3rd ed. (New York: Springer Science+Business Media), 384: "Glyphosate is an aminophosphonic analogue of the natural amino acid glycine, and like all amino acids, exists in different ionic states depending on pH. Both the phosphonic acid and carboxylic acid moieties can be ionised and the amine group can be protonated and the substance exists as a series of zwitterions. Glyphosate is soluble in water to 12 g/l at room temperature. The original synthetic approach to glyphosate involved the reaction of phosphorus trichloride with formaldehyde followed by hydrolysis to yield a phosphonate. Glycine is then reacted with this phosphonate to yield glyphosate, and its name is taken as a contraction of the compounds used in this synthesis—viz. glycine and a phosphonate."

9 When No Spray Coalition researcher Jim West and I attended a talk in December 2000 at the New York Academy of Medicine titled "Challenges of Emerging Illness in Urban Environments" we questioned Dr. Marcelle Layton, who played a key role in defining the WNV "epidemic." Layton admitted that no actual virus had been found in the autopsied brains of any of those the CDC claimed had died from West Nile virus. What had been detected upon autopsy were antibodies to St. Louis Encephalitis (https://www.cdc.gov/mmwr /preview/mmwrhtml/mm4838a1.htm), which resembles genetically West Nile virus, but no viral particles were found. (Some of those said to have died from West Nile viral encephalitis were taking chemotherapy for existing cancers, and all had pre-existing conditions that compromised their immune systems.) Jim West's research is here: www.harvoa.org/wnv. However, see Deborah S. Asnis, Rick Conetta, Alex A. Teixeira, Glenn Waldman, Barbara A. Sampson, "The

West Nile Virus Outbreak of 1999 in New York: The Flushing Hospital Experience," *Clinical Infectious Diseases*, Volume 30, Issue 3, 1 March 2000, Pages 413–418, https://doi.org/10.1086/313737 which states that autopsies on most of those humans who died showed that they had microglial nodules in their brain—white blood cells—indicating infection from some cause, rather than poisoning, air pollution, oil refinery emissions, or pesticides exposure, according to The National Center for Biotechnology Information, which is part of the United States National Library of Medicine, a branch of the National Institutes of Health. https://www.ncbi.nlm.nih.gov/pmc/articles/PMC523558/

10 Paul H.B. Shin, "Discover West Nile Pathway into Brain," *NY Daily News*, November 22, 2004.

11 "Pathologically, the virus is found in the brain stem in humans, but in birds 'it's found everywhere. It's much easier to find,' said Dr. [Marcelle] Layton." See Pippa Wysong, "East Nile, West Nile, All Around the World," *Medscape*, www.medscape.org/viewarticle/418826.

12 Muray Weiss, "No Proof of Saddam Ties to Bug Virus—Sources." *New York Post*, October 11, 1999.

13 Felix Grün and Bruce Blumberg, "Minireview: The Case for Obesogens," *Molecular Endocrinology* 23, no. 8 (August 2009), 1127–1134, www.ncbi.nlm.nih.gov/pmc/articles/PMC2718750/; Bhan et al., 2014: www.sciencedirect.com/science/article/pii/S0960076014000314; Prins, 2014: press.endocrine.org/doi/abs/10.1210/en.2013–1955. Pyrethroids have a mode of action similar to chlorinated pesticides such as cyclodienes (chlordane, aldrin, etc), which were banned in the United States in the 1980s due to their dangerous impact on human heath and the environment. In addition, over the last decade, pesticides have contributed to the collapse of bee colonies in New York and throughout the United States, and spray drift has forced reclassification of produce from now-ruined organic farms.

14 *NoSpray et al. v. The city of NY et al.*, 2000, http://nospray.org/wp-content/uploads/2015/05/plaintiffs-reply-no-spray-lawsuit-2001.pdf.

15 Frank Lombardi and Martin Mbugua, "More Mosquito War Protests Greet New Plan for Spraying," *New York Daily News*, Saturday, September 18, 1999.

16 Bugged by Spraying," *Newsday*, October 10, 1999.

17 *Newsday*, September 14, 1999.

18 Jodi Wilgoren, "Spraying Expands in New York Encephalitis Fight," *New York Times*, Sept. 14. 1999.

19 "As Mosquito Spraying Continues, Officials Stress Its Safety" by Andrew C. Revkin, *NY Times*, September 14, 1999. https://www.nytimes.com/1999/09/14/nyregion/as-mosquito-spraying-continues-officials-stress-its-safety.html.

20 People who are exposed to paint, glue, or degreaser fumes at work may experience memory and thinking problems in retirement, decades after their exposure, according to a new study. See www.sciencedaily.com/releases/2014/05/140512213734.htm and E. L. Sabbath, L. A. Gutierrez, C. A. Okechukwu, A. Singh-Manoux, H. Amieva, M. Goldberg, M. Zins, and C. Berr, "Time May Not Fully Attenuate Solvent-Associated Cognitive Deficits in Highly Exposed Workers," *Neurology* 82 no. 19(May 2014), 1716. https://www.nytimes.com/1999/09/14/nyregion/as-mosquito-spraying-continues-officials-stress-its-safety.html.

21 Piperonyl butoxide (PBO), which is generally mixed with pyrethroid insecticides, is also a suspected reproductive toxicant. J. Jankovic, "A Screening Method for Occupational Reproductive Health Risk," *American Industrial Hygiene Association Journal*, 57, no. 7 (July 1996), 641–49. 1996. Another test that indicates PBO may be carcinogenic is reported by a California environmental products company, Safe2Use, which cited a study by Environmental Chemistry, Inc., a Texas environmental laboratory that primarily serves the chemical industry.

22 Among these "inert" ingredients are polyethylbenzene (PEB), also known as heavy aromatic solvent naptha (petroleum), which is widely used in pesticides. PEB is listed on the EPA Office of Pesticide Programs' Inert Pesticide Ingredients List No 2, which is a list of 64 substances the EPA "believes are potentially toxic and should be assessed for effects of concern. Many of these inert ingredients are structurally similar to chemicals known to be toxic; some have data suggesting a basis for concern about the toxicity of chemical." PEB is related to ethylbenzene, which is listed as a suspected reproductive toxicant and a suspected respiratory toxicant by the EPA. The white mineral oil, also known as hydrotreated light paraffinic petroleum distillate, is also listed on the EPA's Inert Pesticide Ingredients List No 2 of potentially toxic chemicals. According to Cornell's Pesticide Management Education Program, hydrocarbons used as solvents in spray products are likely to result in coughing, fever, and chest pain (hydrocarbon pneumonitis) when these liquid mists are breathed in.

23 H. Chen and D. A. Eastmond, "Topoisomerase Inhibition by Phenolic Metabolites: A Potential Mechanism for Benzene's Clastogenic Effects." *Carcinogenesis* 16, no. 10 (1995), 2301–2307. And see S. Rappaport et al., "Human Benzene Metabolism Following Occupational and Environmental Exposures," *Chemico-Biological Interactions* 184, no. 1–2 (2010), 189–195.

24 Cat Lazaroff, "Brain Damage Found in U.S. Gulf War Syndrome Victims," *Environmental News Service*, May 25, 2000, www.ens-newswire.com/ens/may2000/2000-05-25-07.asp. When we combine these vectors with the administration of experimental, genetically engineered vaccines and a field of

radiation from Uranium 238 weapons, the assault on the immune system is heightened far beyond even the sum of each of those causes taken separately. (This is known as a "synergistic" effect.)

25 Ibid. The study outlined three interrelated but separate causes for brain deterioration found in many Gulf War veterans. Some soldiers in the Gulf War wore flea collars meant for pets, exposing them to toxic levels of pesticides. In 1997, Dr. Robert Haley, UT Southwestern chief of epidemiology and lead author of the study, and his colleagues defined three Gulf War syndromes. Syndrome 1, commonly found in veterans who wore pesticide containing flea collars, is characterized by impaired cognition. Syndrome 2, called confusion ataxia, the most severe and debilitating of the syndromes, is found among veterans who said they were exposed to low-level nerve gas and experienced side effects from anti-nerve gas, or pyridostigmine (PB), tablets. Syndrome 3, characterized by central pain, is found in veterans who wore insect repellent with high concentrations of DEET, a common ingredient in many mosquito and tick repellents. Veterans with Syndrome 3 also experienced severe side effects from PB tablets. Haley RW, Marshall WW, McDonald GG, Daugherty M, Petty F, Fleckenstein JL. Brain abnormalities in Gulf War syndrome: Evaluation by 1H magnetic resonance spectroscopy. *Radiology* 2000; 215: 807–817.

26 Ibid. These brain-cell losses are similar to those found in patients with brain diseases like amyotrophic lateral sclerosis (ALS, or Lou Gehrig's disease) and multiple sclerosis, as well as dementia and other degenerative neurological disorders, although the affected areas of the brain are different.

27 "Gulf War syndrome is believed to be caused by exposure to a class of chemicals known as anticholinesterases." Encyclopedia Britannica. https://www.britannica.com/science/Gulf-War-syndrome.

28 2011 Health Advisory #17: Pesticide Spraying Notification to Reduce Mosquito Activity and Control West Nile virus in Queens, August 19, 2011.

29 "Surveillance for Acute Insecticide-Related Illness Associated with Mosquito-Control Efforts—Nine States, 1999–2002," Center for Disease Control, www.cdc.gov/mmwr/preview/mmwrhtml/mm5227a1.htm.

30 General Accounting Office, "Information on Pesticide Illness and Reporting Systems," GAO-01-501T.

31 J. Vera Go et al., "Estrogenic Potential of Certain Pyrethroid Compounds in the MCF-7 Human Breast Carcinoma Cell Line," *Environmental Health Perspectives* 107, no. 3 (1999); M. C. R. Alavanja et al., "Use of Agricultural Pesticides and Prostate Cancer Risk in the Agricultural Health Study Cohort," *American Journal of Epidemiology* 157 (2003), 800–814.

32 Beyond Pesticides, "Daily News Archive," August 31, 2005.

33 Heidi Singer, "Malathion Played Role in Death of Fish," *Staten Island Advance,* January 22, 2000.

34 Studies done by Dr. Mary Wolff and others at Mt. Sinai Hospital causally linked Sumithrin to the proliferation of breast cancer cells in women and low sperm counts in men.

35 Dr. Samuel Epstein, MD, and Dr. Quentin Young, MD, as quoted in *Pesticides and You* 22, no. 2 (Summer 2002).

36 Increased pesticides exposure—along with the impact of diet sodas (aspartame), radiation from nuclear-weapons tests, power plants, and cellphone towers—correlates with and may be the cause of the vast increase in multiple sclerosis, Parkinson's disease, and other neurological and immune-compromising diseases.

37 *Third National Report on Human Exposure to Environmental Chemicals,* Centers for Disease Control, 2005.

38 *The Quality of Our Nation's Waters, Pesticides in the Nation's Streams and Ground Water,* 1992–2001, United States Geological Survey, pubs.usgs.gov/circ/2005 /1291/.

39 M. T. Salam et al., "Early-Life Environmental Risk Factors for Asthma: Findings from the Children's Health Study," *Environmental Health Perspectives* 112 no. 6 (May 2004), 760–765.

40 Journal of the Am Mosquito Control Assoc, Dec; 13(4): 315–25, 1997 Howard JJ, Oliver New York State Department of Health, SUNY-College ESF, Syracuse 13210, USA.

41 Physicians and Scientists for a Healthy World, "The Multigenerational, Cumulative and Destructive Impacts of Pesticides on Human Health, Especially on the Physical, Emotional and Mental Development of Children and Future Generations. A Submission to The House of Commons Standing Committee on Environment and Sustainable Development,"February 2000; Elizabeth Guillette et al., "Anthropological Approach to the Evaluation of Pre-School Children Exposed to Pesticides in Mexico." *Environmental Health Perspective* 106, no.6 (June 1998); Jonathan Kaplan et al., *Failing Health. Pesticides Use in California Schools. Report by Californians for Pesticide Reform, 2002,* American Academy of Pediatrics, Committee on Environmental Health; *Ambient Air Pollution: Respiratory Hazards to Children,* Pediatrics 91, 1993, www.pediatrics .aappublications.org/content/pediatrics/91/6/1210.full.pdf.

42 George Haas, "West Nile Virus, Spraying Pesticides the Wrong Response, *American Bird Conservancy,* October 23, 2000.

43 Michael R. Reddy, "Efficacy of Resmethrin Aerosols Applied from the Road for Suppressing Culex Vectors of West Nile Virus," *Vector-Borne and Zoonotic Diseases* 6, no. 2 (June 2006).

44 Up until that moment, the City had refused to negotiate. Here is a typical example of the City's mindset: its counsel had written, "The New York City Department of Health and Mental Hygiene, unfortunately, cannot accept the terms of settlement proposed by plaintiffs and does not think that further settlement discussions would be productive at this time."

45 Haliburton, "KBR Awarded U.S. Department of Homeland Security Contingency Support Project for Emergency Support Services," January 24, 2006, www.halliburton.com/public/news/pubsdata/press_release/2006/kbrnws _012406.html. KBR is the engineering and construction subsidiary of Halliburton. The contract, for $385 million, "provides for establishing temporary detention and processing capabilities to augment existing ICE Detention and Removal Operations (DRO) Program facilities in the event of an emergency influx of immigrants into the U.S.," facilities that can and will readily be used for other emergency purposes.

46 Carol Brouillet, "Opposing the Emergency Health Powers Act," *IndyBay* (April 2, 2002), https://www.indybay.org/newsitems/2002/04/02/1203561 .php.

47 See transcript of testimony of Dr. Adrienne Buffaloe on the effects of pesticides—particularly malathion and pyrethroids—on people, based on her own evaluations of her patients, before hearings held by a congressional panel chaired by Representative Gerald Ackerman of Queens, New York. http:// nospray.org/dr-adrienne-buffaloe-on-pesticides-at-hearing-held-by-rep -ackerman/.

48 Joanne Wasserman and Michael R. Blood, "Officials Confirm Skeeter Workers' Illnesses," New *York Daily News,* January 25, 2001.

49 No Spray Coalition researcher Jim West's articles on air pollution and diseases, including West Nile, can be found on his website, www.harpub.tk/noxot. Also, in *Townsend Letter for Doctors & Patients,* July 2002, and on ABCNews.com (through the reporting of Nicholas Regush). Jim West found that the source of the Monroe, Louisiana, WNV epidemic of 2002 could likely have been the 130 oil refineries near southern Louisiana, which has long been the center for St. Louis encephalitis in the United States. The second largest concentration of oil refineries is in New Jersey. Staten Island, a main locale in NYC for West Nile, is downwind from three of those refineries.

50 Many municipalities passed resolutions against spraying. See http://nospray .org Issue 26 July 3rd 2001.

51 Pesticide delivery systems come in many forms, but the contents are equally toxic. Where areas are being saturated twenty-four hours a day for months with the lures in traps or the twist ties, this continual low-dose can be more toxic

for many people than a one-time exposure, even of a high-dose. See www
.dontspraycalifornia.org.

52 See www.nospray.org/2017/08/12/spray-vs-no-spray-14-cities-comparative
-analysis-pesticide-spraying-west-nile-virus/.

53 Ibid.

54 Maria Sause and Rio Davidson, "Lincoln County (Oregon) Bans Aerial Pesticide Spraying. Voters Vote YES on Measure 21–177 in Narrow Victory over
Pesticide Companies: First Electoral Ban of Pesticides in the Country!" Citizens for a Healthy County, May 30, 2017.

55 www1.nyc.gov/site/doh/health/health-topics/zika-virus.page.

56 Chlorinated water drastically affects malathion, turning it into malaoxon, which
is seventy-seven times more deadly than malathion. This applies not only to
chlorinated swimming pools but also to cisterns into which some pour Cholox
as an antibacterial agent.

57 Anvil 10+10 is comprised of four ingredients: sumithrin; piperonyl butoxide;
polyethylbenzene, also known as heavy aromatic solvent naptha (petroleum);
and white mineral oil, also known as hydrotreated light paraffinic petroleum
distillate. It is toxic to bees and fish and kills natural predators of the mosquito,
including dragonflies, bats, frogs, and birds. Sumithrin is a suspected gastroin-
testinal, kidney, and liver toxicant and a suspected neurotoxicant. Piperonyl
butoxide is a suspected carcinogen. These are just a few of the known health
issues related to these pesticides.

58 Grün and Blumberg, "Minireview."

59 Local Law 37 authorizes the commissioner of the Department of Health the
power to grant city agencies a temporary waiver of the law's prohibitions only
after consideration of whether the prohibitions, in the absence of the waiver,
would be unreasonable with respect to (i) the magnitude of the infestation,
(ii) the threat to public health, (iii) the availability of effective alternatives, and
(iv) the likelihood of exposure of humans to the pesticide. (See *§17–1206*,
Local Law 37.)

The NYC Department of Health got around Local Law 37 by authorizing
to itself pro forma waivers of that law's prohibitions against broadcast spraying
of the pesticide Anvil 10+10, even though it acknowledged that at least one of
the chemicals it sought to spray is categorized by the U.S. Environment Protection Agency as a possible carcinogen. The DOH failed to establish the magnitude of infestation and the threat to public health; it failed to investigate the
availability of effective alternatives or consider as significant the prodigious evidence disputing the City's minimizing of the likelihood of human exposure to
the pesticide, all of which are required by law. The issuance of waivers in such

circumstances undermines Local Law 37's protection *from the dangers of pesticides* to public health and the environment.

The NYC DOH made two claims for why it approved waivers in 2011: First was the desire to gain "control of adult mosquitoes in the Rockaways where the severity of infestation has created a public health nuisance. In these communities adjacent to the Jamaica Bay, mosquitoes force residents indoors during the summer months, negatively affecting the residents' quality of life and reducing healthy outdoor activity. The spraying of adulticide provides a knockdown of the populations in the area and gives the residents a reprieve from the nuisance of these mosquitoes"; ("Decision on Local Law 37 Waiver Number DOH11-0002," May 18, 2011). In other words, the required establishment of a "threat to public health" was glossed over and turned into "reducing healthy outdoor activity." Seond, the Deputy Commissioner affirmed—without any proof, substantiation or further documentation—that Anvil 10+10 was "approved for the control of adult mosquitoes in areas where monitoring has indicated a risk to the public of West Nile Virus transmission" (*Decision on Local Law 37 Waiver Number DOH11-000,* May 18, 2011.).. No substantiation was offered as to what constitutes an "indicated risk to the public of West Nile Virus transmission," nor were any concerns expressed over the pesticides' effects on human health or of alternatives to spraying carcinogenic pesticides, as required under Local Law 37. Consequently, neither of the rationales presented by the NYC DOH meets Local Law 37's four criteria for approval and receipt of a waiver. In granting both waivers to itself, the NYC Department of Health stands in violation of Local Law 37 for failing to even address, let alone substantiate, *any* of the requirements and concerns explicitly listed in Local Law 37 for the granting of waivers.

Local Law 37 outlines the process whereby a city agency may request a waiver of the restrictions established pursuant to §17–1203, and limits such waiver to be in effect for no longer than one year. The provisions in Local Law 37 went into full effect in 2006. Since that time, the Health Department has granted to itself a waiver for adulticide spraying for mosquitoes *every single year,* like clockwork. Each individual waiver, taken by itself, provides four or five months of "temporary" relief from the prohibitions of Local 37. But as part of a larger pattern, the steady string of waivers for application of Anvil 10+10 between 2006 and 2011 has meant that Local Law 37 has *never* provided any protection from the admittedly carcinogenic chemical included in this pesticide. The Health Department's authority to grant *temporary* waivers to City agencies was not intended as an ongoing or permanent mechanism and must not be permitted to become a vehicle for circumventing prohibition of the *seasonal* use of prohibited, carcinogenic pesticides in perpetuity.

60 According to the Mayo Clinic, fewer than 1 percent of people who are infected become severely ill (www.mayoclinic.org/diseases-conditions/west-nile-virus /symptoms-causes/syc-20350320). About 70 to 80 percent of people infected will never display symptoms, and many others experience only mild flu-like symptoms (Centers for Disease Control, www.cdc.gov/westnile/faq/genquestions .html). In addition, the average person's risk of contracting West Nile is extremely low; even in areas where the virus is present, only a very small number of mosquitoes carry the virus. (NYC Department of Health and Mental Hygiene, www.nyc.gov/html/doh/html/environmental/mosquitos.shtml). As of September 29, 2015, there had been just seventeen reported cases of West Nile in the state of New York and one death out of a population of about 20 million people (Centers for Disease Control, www.cdc.gov/westnile/statsmaps /preliminarymapsdata/histatedate.html). Nationally, there have been 1,028 cases and 54 deaths out of more than 320 million people (CDC, ibid.).

61 The audio interview can be heard at my website: www.MitchelCohen.com.

62 A video of the spraying is available here: www.nospray.org/2015/09/07/watch -how-nyc-sprays-neighborhoods-by-truck-pesticides-you-wont-believe-it/.

63 The book is available for no charge at www.thebestcontrol2.com. (Tvedten's story, "Why I Stopped Using Pesticide Poisons," is a later chapter in this book.)

64 Email to a CDC staffer, October 21, 1999.

65 Karen Charman, "Pesticide Wars: The Troubling Story of Dr. Omar Shafey," November 16, 2001, www.TomPaine.com. Two years later, Shafey was fired by the Florida Department of Health for an alleged overcharge of a mere $12.50 on a travel-reimbursement voucher. The inspector general said Shafey should have claimed reimbursement for three-quarters of a day's per diem instead of a full day when he returned to Tallahassee, a charge Shafey disputes.

66 *New York Daily News*, January 7, 2000, 2.

67 Richard J. Ochs, "Government by Anthrax," June 9, 2002. www.free fromterror .net.

68 HIV/AIDS played an early role here as the government experimented with how to orchestrate hysteria to control the beliefs and activities of a target population. Fortunately, the direct action approach of ACT UP thwarted their plans, forcing government scientists to regroup and try their maneuvering elsewhere.

69 Rob Wallace, "Big Farms Make Big Flu: Dispatches on the Infectious Disease, Agribusiness, and the Nature of Science," Monthly Review Press, New York: 2016.

70 https://www.independent.co.uk/news/world/americas/donald-rumsfeld -makes-5m-killing-on-bird-flu-drug-6106843.html

71 Maureen Groppe, "Frist's new Senate role could bring help for Lilly. The majority leader, a doctor, wrote bill that shields vaccine makers from preservative suits." *Indianapolis Star,* Dec. 24, 2002.

72 Judith Miller, Stephen Engelberg and William J. Broad: "U.S. Germ Warfare Research Pushes Treaty Limits," *The New York Times,* September 4, 2001.

73 Joining Hauer at Kroll Associates was former NY Police Commissioner Bratton, now head of the Los Angeles Police Department. Before going on to LA, Bratton stopped over in Venezuela, where he was a special adviser to the right-wing Mayor of Caracas during the U.S.-sponsored coup attempt, which was repelled by a mass uprising of the working class and poor in defense of their enormously popular, elected president Hugo Chavez. Questions as to Bratton's role in Venezuela have yet to be asked.

74 Contrary to public knowledge, Rockefeller University had been experimenting with the supposedly "unknown" WNV for decades.

75 OraVax, owned by Peptide Therapeutic of Cambridge, England, was having problems getting beyond research and bringing products to market. By 1996 and 1997, its survival at stake, OraVax tried to win part of the Pentagon's expanding germ work as a subcontractor to make smallpox and other vaccines. By early 1998, that work had failed to materialize and the company's stock price was down 90 percent from $10 a share in the initial public offering. "The way to handle it is to be open, so people understand that I may have a potential bias," Monath said.[NY Times, 8/7/98]

76 See William Blum, "Rogue State: A Guide to the World's Only Superpower," Zed Books 2001–2002. In 1981, an epidemic of dengue hemorrhagic fever (DHF) swept across the island of Cuba. Transmitted by blood-eating insects, usually mosquitos, the disease produces severe flu-like symptoms and incapacitating bone pain. Between May and October 1981, over 300,000 cases were reported in Cuba with 158 fatalities, 101 of which were children under 15. (Bill Schaap, "The 1981 Cuba Dengue Epidemic," *Covert Action Information Bulletin* (Washington, DC), No. 17, Summer 1982, p.28–31) The Center for Disease Control later reported that the appearance in Cuba of this particular strain of dengue, DEN-2 from Southeast Asia, had caused the first major epidemic of DHF ever in the Americas. (Reported on their website: http://www.cdc.gov /ncidod/dvbid/dengue.htm)

In 1956 and 1958, declassified documents have revealed, the US Army loosed swarms of specially bred mosquitos in Georgia and Florida to see whether disease-carrying insects could be weapons in a biological war. The mosquitos bred for the tests were of the Aedes aegypti type, the precise carrier of dengue fever as well as other diseases. (*San Francisco Chronicle*, October 29, 1980, p.15)

In 1967 it was reported by *Science* magazine that at the US government center in Fort Detrick, Maryland, dengue fever was amongst those "diseases that are at least the objects of considerable research and that appear to be among those regarded as potential BW [biological warfare] agents." (*Science* (American Association for the Advancement of Science, Washington, DC), January 13,1967, p.176) On a clear day, October 21, 1996, a Cuban pilot flying over Matanzas province observed a plane releasing a mist of some substance about seven times. It turned out to be an American crop-duster plane operated by the US State Department, which had permission to fly over Cuba on a trip to Colombia via Grand Cayman Island. Responding to the Cuban pilot's report, the Cuban air controller asked the US pilot if he was having any problem. The answer was "no". Two months later, Cuba observed the first signs of a plague of *Thrips palmi*, a plant-eating insect never before detected in Cuba. It severely damages practically all crops and is resistant to a number of pesticides. Cuba asked the US for clarification of the October 21 incident. Seven weeks passed before the US replied that the State Department pilot had emitted only smoke, in order to indicate his location to the Cuban pilot. (For further details of the State Department's side of the issue, see *New York Times*, May 7, 1997, p.9) By this time, the Thrips palmi had spread rapidly, affecting corn, beans, squash, cucumbers and other crops. In response to a query, the Federal Aviation Administration stated that emitting smoke to indicate location is "not an FAA practice" and that it knew of "no regulation calling for this practice." In April 1997, Cuba presented a report to the United Nations which charged the US with "biological aggression" and provided a detailed description of the 1996 incident and the subsequent controversy. In August, signatories of the Biological Weapons Convention convened in Geneva to consider Cuba's charges and Washington's response. In December, the committee reported that due to the "technical complexity" of the matter, it had not proved possible to reach a definitive conclusion.

77 Patricia Doyle, "Deadly West Nile virus for Profit Vaccine Award Announced," *NoSpray Newz*, August 2000. Also, http://www.rense.com/general3/profit.htm.

78 See especially NoSpray Newz #17, at http://www.NoSpray.org.

79 The Office of Public Health Emergency Preparedness is now called the Office of the Assistant Secretary for Preparedness and Response (ASPR).

80 Ceci Connolly, "Smallpox Vaccination for Medical Workers Proposed," *Washington Post*, Sept. 4, 2002.

81 Lawrence K. Altman and Sheryl Gay Stolberg, "Smallpox Vaccine Backed for Public," *NY Times*, Oct. 5, 2002. The article offers an inside glimpse into a split in the thinking of the Bush administration, noting that "Vice President Dick Cheney favors a mass vaccination approach, while Mr. Bush favors a more moderate approach." Strangely, Jerome Hauer is said to have been removed from his position as head of the Office of Public Health Emergency Preparedness "primarily for conflicts he had with Scooter Libby over whether the risks of smallpox vaccination were worth the benefit. Hauer charged that the Office of the Vice President was pushing for the universal vaccination despite the vaccine's health risks, primarily exaggerate the risk of biological terrorism." This, despite the long record of Hauer's aggressive advocacy of mandatory vaccination and exaggeration of the risk of biological terrorism.

82 Of the many books and websites that report this story, see Barron's "How to prepare for the AP environmental science advanced placement exam," written by Gary Thorpe. Also, Paul Hawken, Amory Lovins, and L. Hunter Lovins, "Natural Capitalism," 2010 edition.

83 For fascinating sleuthing in tracking down the origins of the Borneo cat story and controversies challenging its validity, see Patrick T. O'Shaughnessy, "Parachuting Cats and Crushed Eggs: The Controversy Over the Use of DDT to Control Malaria," Jan. 2008, http://www.ncbi.nlm.nih.gov/pmc/articles /PMC2636426/#bib19.

84 Homer Bigart., "A DDT Tale Aids Reds in Vietnam," *New York Times*, February 2, 1962: 3.

85 The covert U.S. biological and chemical warfare program, much of it developed at Fort Detrick, Maryland, has historically tested its weapons on U.S. soldiers, American Indian reservations, ghetto populations, colonies (like Puerto Rico), and prisoners—in other words, on controlled and bounded populations. Widescale testing on others is now becoming increasingly frequent and aggressive—albeit shrouded in secrecy and disinformation.

Chapter 6: Children & Pesticides

1 Statement of American Academy of Pediatrics, *Pediatrics* 130, no. 6 (December 2012), 1757–63.

2 V. Garry, et al., "Pesticide Appliers, Biocides, and Birth Defects in Rural Minnesota," *Environmental Health Perspectives* 104, no. 4 (April 1996), 394–99.

3 M. Krüger M, A. A. Shehata, W. Schrödl, and A. Rodloff, "Glyphosate Suppresses the Antagonistic Effect of Enterococcus spp. on Clostridium botulinum," Anaerobe 20 (April 2013) 74–8, doi: 10.1016/j.anaerobe.2013.01.005 . Also see www.ncbi.nlm.nih.gov/pubmed/23396248.

4 "Could Common Insecticides Be Tied to Behavior Issues in Kids?" HealthDay News, March 2, 2017, www.medicinenet.com/script/main/art.asp ?articlekey=202021.

5 Roni Caryn Rabin, "A Strong Case Against a Pesticide Does Not Faze E.P.A. Under Trump," *New York Times,* May 15, 2017, www.nytimes.com/2017/05/15 /health/pesticides-epa-chlorpyrifos-scott-pruitt.html.

6 Grandjean and Landrigan. "Developmental Neurotoxicity of Industrial Chemicals," *The Lancet* 368 (November 2006).

7 L. J. Akinbami, J. E. Moorman, and X. Lui, "Asthma Prevalence, Health Care Use, and Mortality: United States, 2005–2009, National Center for Health Statistics, *National Health Statistics Reports* 32, January 2011.

8 M. T. Salam et al., "Early Life Environmental Risk Factors for Asthma: Findings from the Children's Health Study." *Environmental Health Perspectives* 112, no. 6 (May; 2004), 760–65.

9 Chen et al., "Residential Exposure to Pesticide During Childhood and Childhood Cancers: A Meta-Analysis," *Pediatrics* 136, no. 4 (October 2015).

10 J. Rudant, et al., "Household Exposure to Pesticides and Risk of Childhood Hematopoietic Malignancies: The ESCALE Study (SFCE)," *Environmental Health Perspectives* 115 (December 2007), 1787–1793.

Chapter 7: It's Not That Anyone *Wants* to Kill Butterflies

1 "The Environmental Impacts of Glyphosate," Friends of The Earth, 2013.

2 Warren Cornwall, "The Missing Monarchs," *Slate,* January 29, 2014, www.slate .com/articles/health_and_science/science/2014/01/monarch_butterfly_decline _monsanto_s_roundup_is_killing_milkweed.html.

3 Ibid.

4 J. A. Springett and R. A. J. Gray, "Effect of Repeated Low Doses of Biocides on the Earthworm *Aporrectodea caliginosa* in Laboratory Culture," *Soil Biology and Biochemistry* 24, no. 12 (1992), 1739–1744. Repeated applications of glyphosate significantly affect the growth and survival of earthworms.

5 S. A. Hassan et al., 1988. "Results of the Fourth Joint Pesticide Testing Programme Carried out by the IOBC/WPRS-Working Group 'Pesticides and Beneficial Organisms,'" *Journal of Applied Entomology* 105, no. 1–5 (1988), 321–329.

6 L. C. Folmar, H. O. Sanders, and A. M. Julin, "Toxicity of the Herbicides Glyphosate and Several of its Formulations to Fish and Aquatic Invertebrates," *Archives of Environmental Contamination and Toxicology* 8, no. 3 (1979), 269–278. Environmental factors, such as high sedimentation and increases in

temperature and pH levels, increase the toxicity of Roundup, especially to young fish.

7 Impacts of Pesticides on Wildlife (2017) Beyond Pesticides, https://beyondpesticides.org/programs/wildlife.

8 www.theguardian.com/environment/2017/oct/21/insects-giant -ecosystem-collapsing-human-activity-catastrophe.

9 Northwest Coalition for Alternatives to Pesticides (NCAP), "Herbicide Fact-sheet: Glyphosate (Roundup)." *Journal of Pesticide Reform* (1998).

10 Umberto Quattrocchi, *CRC World Dictionary of Plant Names: Common Names, Scientific Names, Eponyms,* (Boca Raton, FL: CRC Press), 1999.

11 C. Robertson, "Insect Relations of Certain Asclepiads," *I. Botanical Gazette* 12 (1887), 207–216, as summarized at www.wikipedia.org/wiki/Asclepias.

12 "Milkweed Poisoning of Horses," New Mexico State University, College of Agricultural, Consumer and Environmental Sciences (ACES), October 2017, aces.nmsu.edu/pubs/_b/B709/.

13 www.pollinatorparadise.com/what_is_pollination.htm.

14 http://monarchwatch.org/blog/2014/01/24/importance-of-monarch -conservation/.

15 http://metatexte.net/ezine/pdf/Huber.pdf.

16 http://xerces.org/neonicotinoids-and-bees/.

17 Simon Marks, "Watchdog Links Pesticide to Bee Decline," Politico, December 5, 2017, www.politico.eu/article/food-safety-watchdog-links-pesticide-to-bee -decline/.

18 Tami Canal, "EPA Finally Admits What Has Been Killing Bees For Decades," *March Against Monsanto,* January 10, 2016, www.march-against-monsanto .com/epa-finally-admits-what-has-been-killing-bees-for-decades/.

19 Rosemary Mosco, "14 Darling Facts About Ladybugs," *Mental Floss,* May 24, 2016.

20 Alan Watts, *Cloud-Hidden, Whereabouts Unknown* (New York: Vintage Books), 1974.

Chapter 9: Why I Stopped using Pesticide Poisons

1 www.toxipedia.org/display/toxipedia/The+Ideal+Pesticide.

2 www.thebestcontrol2.com.

Chapter 10: Where and How Is It *Still* Possible to Eat *Relatively* Safely?

1 Proterra Investment Partners predicted at a conference of agriculture investors that the compound annual growth rate for non-GMO farmland from 2017-2022 would be 16.2 percent.

Chapter 11: Consequences of Glyphosate's Effects on Animal Cells, Animals, and Ecosystems

1   Monsanto produces glyphosate herbicide mixes, or GlyBH.

2   "Registration Review - Preliminary Ecological Risk Assessment for Glyphosate and its Salts,"; p. 2, https://www.epa.gov/ingredients-used-pesticide-products/draft-human-health-and-ecological-risk-assessments-glyphosate. Registration Review - Preliminary Ecological Risk Assessment for Glyphosate and its Salts; p. 2

3   Glyphosate's herbicidal efficacy has been based on its inhibition of the shikimate pathway in plant cells (Boocock, M. & Coggins, J., 1983).

4   The internal concentration is greater than that found in the organisms' surroundings.

5   The internal concentration is greater, often by 10 times or more, at each consecutive feeding level.

6   Although glyphosate is often referred to as an *organophosphate*, it is technically an *organic phosphorus compound*, since it is missing the ester link, which must be present in order for a substance to be classified as an organophosphate, in the strict chemical definition of the term.

7   Benthic macroinvertebrates are tiny "bugs" or "worms"(often larva of arthropods) which inhabit muddy leaf-littered bottoms of streams. They are used as index species, to characterize the health of the particular ecosystem.

8   Nanoparticles maybe added by the pesticide manufacturer to increase glyphosate's persistence in the environment and its potency, or to reduce the percentage of glyphosate needed in the mix. The environmental and health effects of nanoparticles, in GlyBH, as well as food, clothing, cosmetics, fertilizers products, appear to be extensive and are now being evaluated. Dr. Mengshi Lin's group at the University of Missouri, School of Agriculture is one starting point for researching experiments investigating the safety of nanoparticles.

9   Discussed in Chapter 3 of this book (Cohen, M.)

10   However, *Burlington Northern & Santa Fe Ry. v. United States (Burlington Northern)*, 556 U.S. 599 (2009) has put that in jeopardy, although joint and several liability is still the rule, rather than the exception. (Kilbert, K. 2012).

11   Estimate is of the United Nations Environmental Program.

12   See earlier in page 9 of this chapter for discussion on neonicotinoids and the chapter in this book by J. Latham.

13   ThisWhich could be one example of Chaos Theory in action.

Chapter 12: Unsafe at any Dose? Glyphosate in the Context of Multiple Chemical Safety Failures

1   Nagel, S and Bromfield, J. (2013) Bisphenol A: A Model Endocrine Disrupting Chemical With a New Potential Mechanism of Action; Endocrinology. 2013 Jun; 154(6): 1962–1964. https://www.ncbi.nlm.nih.gov/pmc/articles /PMC3740487/.

2   https://www.thestreet.com/story/10471527/1/sunoco-restricts-sales-of -chemical-used-in-bottles.html.

3   http://www.nydailynews.com/life-style/health/france-bans-contested -chemical-bpa-food-packaging-article-1.1219611.

4   http://www.bbc.com/news/world-europe-11843820.

5   Jenna Bilbrey, "BPA-Free Plastic Containers May Be Just as Hazardous," *Scientific American,* Aug. 11, 2014. Biphenol-S may be more hazardous than BPA and is now widely used since BPA was banned.

6   Worldwide Integrated Assessment of the Impact of Systemic Pesticides on Biodiversity and Ecosystems(2015) Environmental Science and Pollution Research Volume 22, Issue 1, January 2015 http://link.springer.com/journal /11356/22/1/page/1, www.tfsp.info/, www.iucn.org/news/secretariat/201709 /severe-threats-biodiversity-neonicotinoid-pesticides-revealed-latest-scientific -review, and www.iucn.org/content/systemic-pesticides-pose-global-threat -biodiversity-and-ecosystem-services.

7   http://www.ncbi.nlm.nih.gov/pmc/articles/PMC2718750/.

8   http://www.sciencedirect.com/science/article/pii/S0960076014000314.

9   http://press.endocrine.org/doi/abs/10.1210/en.2013–1955.

10 http://infoscience.epfl.ch/record/169263/files/Perinatal Exposure to Bisphenol A, Ayyakkannu, Laribi.pdf.

11 http://presse.inra.fr/en/Resources/Press-releases/Bisphenol-A-and-food -intolerance-a-link-established-for-the-first-time.

12 http://www.sciencedirect.com/science/article/pii/S0890623814000203.

13 http://www.researchgate.net/publication/51743683_Impact_of_early-life _bisphenol_A_exposure_on_behavior_and_executive_function_in_children /file/72e7e524ec8adb78d9.pdf.

14 http://europepmc.org/articles/PMC3440080.

15 http://www.sciencedirect.com/science/article/pii/S0161813X14001715.

16 http://press.endocrine.org/doi/full/10.1210/er.2011–1050.

17 http://www.efsa.europa.eu/en/efsajournal/pub/3978.htm.

18 http://en.wikipedia.org/wiki/Terbuthylazine.

19 http://www.treehugger.com/sustainable-agriculture/tyrone-hayes-mis fortune -frogs-crooked-science-and-why-we-should-shun-gmos.html.

20 https://thepumphandle.wordpress.com/popcorndiacetyl/.

21 http://blogs.edf.org/health/2010/01/12/won't-we-ever-stop-playing-whack-a-mole-with-regrettable-chemical-substitutions. Diacetyl was a chemical added to the butter dribbled on popcorn to enhance its buttery flavor. The lung's bronchioles, the smallest airways, are the ones most damaged in popcorn lung. See https://www.medicalnewstoday.com/articles/318260.php.

22 http://www.deq.state.va.us/Portals/0/DEQ/Land/RemediationPrograms/Brownfields/Weaver1-195-1-PB-8r.pdf.

23 http://greensciencepolicy.org/hbcd-is-on-the-way-out-but-use-of-questionable-alternatives-will-persist/.

24 https://www.independentsciencenews.org/news/many-european-pesticide-approvals-are-unlawful-says-eu-ombudsman/.

25 http://generationgreen.org/2010/03/the-toxies-an-award-show-for-bad-actor-chemicals/.

26 http://www.sciencedaily.com/releases/2014/02/140227134843.htm.

27 Sarah A. Vogel. (2009) The Politics of Plastics: The Making and Unmaking of Bisphenol A "Safety" Am J Public Health. 2009 November; 99(Suppl 3): S559–S566; https://www.ncbi.nlm.nih.gov/pmc/articles/PMC2774166/.

28 https://www.fda.gov/Food/IngredientsPackagingLabeling/FoodAdditivesIngredients/ucm355155.htm.

29 http://www.sciencedirect.com/science/article/pii/S0013935114002473.

30 https://student.societyforscience.org/article/corals-dine-microplastics.

31 http://www.pvc.org/en/p/cadmium-stabilisers.

32 http://www.ncbi.nlm.nih.gov/pmc/articles/PMC3734497/.

33 https://www.independentsciencenews.org/health/the-failing-animal-research-paradigm-for-human-disease/

34 https://www.ncbi.nlm.nih.gov/pmc/articles/PMC3587220/.

35 https://www.independentsciencenews.org/news/the-experiment-is-on-us-animal-toxicology-testing-science/

36 http://www.sciencedirect.com/science/article/pii/S0890623814000203.

37 http://www.sciencedirect.com/science/article/pii/0041008X81901903.

38 http://toxsci.oxfordjournals.org/content/122/1/1.full.

39 http://www.sciencedirect.com/science/article/pii/S0091674995700730.

40 http://carcin.oxfordjournals.org/content/36/Suppl_1/S254.full.pdf+html.

41 http://www.sciencedirect.com/science/article/pii/S1471489214000988

42 http://citeseerx.ist.psu.edu/viewdoc/download?doi=10.1.1.276.7132&rep=rep1&type=pdf.

43 http://press.endocrine.org/doi/full/10.1210/er.2011-1050.

44 http://www.altex.ch/resources/altex_2014_2_157_176_Charukeshi1.pdf.

45 http://webarchive.nationalarchives.gov.uk/20090505194948/http://www
.bseinquiry.gov.uk/report/volume1/toc.htm.

46 http://www.cogem.net/index.cfm/en/publications/publicatie/can-interactions
-between-bt-proteins-be-predicted.

47 http://www.figo.org/sites/default/files/uploads/News/Final%20PDF_8462
.pdf.

48 http://journals.plos.org/plosmedicine/article?id=10.1371/journal.pmed
.0040005.

49 http://www.cogem.net/index.cfm/en/publications/publicatie/research-report
-ecological-and-experimental-constraints-for-field-trials-to-study-potential
-effects-of-transgenic-bt-crops-on-non-target-insects-and-spiders.

50 http://en.wikipedia.org/wiki/Industrial_Bio-Test_Laboratories.

51 http://nepis.epa.gov/.

52 https://en.wikipedia.org/wiki/Industrial_Bio-Test_Laboratories.

53 http://www.webcitation.org/69A19G61r.

54 https://en.wikipedia.org/wiki/Industrial_Bio-Test_Laboratories.

55 http://faculty.haas.berkeley.edu/dalbo/Regulatory_Capture_Published.pdf.

56 http://www.oecd.org/chemicalsafety/testing/oecdguidelinesforthetestingof
chemicalsandrelateddocuments.htm.

57 http://digibug.ugr.es/bitstream/10481/24821/1/ehp-117-309.pdf.

58 http://en.wikipedia.org/wiki/Good_laboratory_practice.

59 http://www.researchgate.net/publication/8331752_Equal_treatment_for
_regulatory_science_extending_the_controls_governing_the_quality_of
_public_research_to_private_research/file/3deec525c2f5d0c924.pdf.

60 http://digibug.ugr.es/bitstream/10481/24821/1/ehp-117-309.pdf.

61 http://home.comcast.net/~jurason/main/bio4.htm.

62 https://www.independentsciencenews.org/health/designed-to-fail-why
-regulatory-agencies-dont-work/.

63 https://www.independentsciencenews.org/news/eu-safety-institutions-caught
-plotting-an-industry-escape-route-around-looming-pesticide-ban/.

64 https://www.independentsciencenews.org/health/designed-to-fail-why
-regulatory-agencies-dont-work/.

65 http://www.vallianatos.com/.

66 http://www.amazon.com/Poison-Spring-Secret-History-Pollution/dp
/1608199142/ref=cm_cr_pr_product_top/188-1966921-5851069.

67 http://www.whistleblowers.org/index.php?option=com_content&task=view
&id=74.

68 https://www.independentsciencenews.org/health/how-epa-faked-the-entire
-science-of-sewage-sludge-safety-a-whistleblowers-story/.

69 *McElmurray v. United States Department of Agriculture*, United States District Court, Southern District of Georgia. Case No. CV105-159. Order issued Feb. 25, 2008.

70 http://www.desmogblog.com/2015/03/02/internal-documents-reveal-extensive-industry-influence-over-epa-s-national-study-fracking.

71 https://www.documentcloud.org/documents/1678647-greenpeace-foia-returns.html#document/p332/a205594

72 http://www.peer.org/

73 http://www.peer.org/news/news-releases/2015/04/21/egregious-epa-misconduct-delivers-whistleblower-win/

74 http://www.boulderweekly.com/article-12640-muzzled-by-monsanto.html.

75 http://inthesetimes.com/article/18504/epa_government_scientists_and_chemical_industry_links_influence_regulations.

76 http://www.sciencedirect.com/science/article/pii/0041008X7990471X.

77 http://www.washingtonpost.com/wp-dyn/content/article/2010/11/29/AR2010112903764.html.

78 http://www.amazon.com/Pandoras-Poison-Chlorine-Environmental-Strategy/dp/0262700840.

Chapter 13: Glyphosate on Trial: The Search for Toxicological Truth

1 International Agency for Research on Cancer (IARC), IARC Working Group on the Evaluation of Carcinogenic Risks to Humans, Some Organophosphate Insecticides and Herbicides, IARC Monographs on the Evaluation of Carcinogenic Risks to Humans, monograph 112, (Lyons, France, 2017).

2 European Food Safety Authority, "Conclusion on the Peer Review of the Pesticide Risk Assessment of the Active Substance Glyphosate," EFSA Journal 13, no. 11 (2015): 4302.

3 See letter from Allan Hirsch, Chief Deputy Director, OEHHA to Philip W. Miller, vice president Global Corporate Affairs, Monsanto Company, June 26, 2017, https://oehha.ca.gov/media/downloads/proposition-65/crnr/comments/letterphilipmillerandiarcrespondstoreuters.pdf.

4 David Michaels, Doubt Is Their Product (Oxford: Oxford University Press, 2008).

5 Gerald Markowitz and David Rosner, The Lead Wars (Oakland: University of California Press, 2014).

6 Baum Hedlund Law, Monsanto secret documents, www.baumhedlundlaw.com/monsantoroundup-cancer-lawsuits-filed-missouri/, accessed August 5, 2017.

7 Danny Hakim, "Monsanto Weed Killer Faces New Doubts on Safety in Unsealed Documents," New York Times, March 14, 2017.

8 Tiffany Stecker, "Monsanto Pushed EPA to Fast Track Pesticide Report in 2015," Bloomberg News. August 2, 2017, www.bna.com/monsanto-pushed -epa-n73014462610/, accessed August 5, 2017.

9 Gilles-Éric Séralini and Jérôme Douzelet, The Great Health Scam. (New Dehli: Natraj Publishers, 2016).

10 Rudolfo Saracci and Christopher P. Wild, International Agency for Research on Cancer: The first 50 years: 1965–2015, Rudolfo Saracci and Christopher P. Wild. World Health Organization. (Lyon, France: 2015), 32, http://www.iarc .fr/en/about/iarc-history.php.

11 http://publications.iarc.fr/Non-Series-Publications/The-History-Of -Iarc/International-Agency-For-Research-On-Cancer-The-First-50 -Years-1965%E2%80%932015

12 IARC 2017, 140.

13 IARC. Statistical Methods in Cancer Research Vol. 1: The Analysis of Case Control Studies. (Lyons, France, 1980); Statistical Methods in Cancer Research Vol II: The Analysis of Cohort Studies (Lyons, France, 1987).

14 Saracci, 142.

15 IARC, Carcinogens in the Human Environment, 143, www.iarc.fr/en /publications/books/iarc50/IARC_Ch4.2.2_web.pdf, accessed August 4, 2017.

16 András Székács Béla Darvas, "Forty Years with Glyphosate,"

17 Gerald M Dill, R. Douglas Sammons, Paul C.C. Feng et al., "Glyphosate: Discovery, Development, Applications, and Properties," in Glyphosate Resistance in Crops and Weeds: History, Development, and Management, ed. Vijay K. Nandula (New York: John Wiley, 2010).

18 Gerald M Dill, R. Douglas Sammons, Paul C.C. Feng et al., "Glyphosate: Discovery, Development, Applications, and Properties," in Glyphosate Resistance in Crops and Weeds: History, Development, and Management, ed. Vijay K. Nandula (New York: John Wiley, 2010).

19 Environmental Protection Agency (EPA), Office of Prevention, Pesticides, and Toxic Substances, Office of Pesticide Programs, Registration Eligibility Decision (RED): Glyphosate (Washington, DC: EPA, 2017), 738-R-93-014, www3.epa.gov/pesticides/chem_search/reg_actions/reregistration/red_PC -417300_1-Sep-93.pdf.

20 Jerry M. Green and Michael D. K. Owen, "Herbicide-Resistant Crops: Utilities and Limitations for Herbicide-Resistant Week Management," Journal of Agriculture and Food Chemistry 59, no. 11 (2011): 5819–29

21 IARC 2017.

22 United States District Court, Northern District of California, Roundup Products Liability Litigation, MDL No. 2741, Case No. 16-md-02741-VC, videotaped deposition of Aaron Earl Blair (Washington DC, March 20, 2017).

23 IARC, monograph 112, 35

24 Environmental Protection Agency, Office of Pesticide Programs, Revised Glypho-sate Issue Paper: Evaluation of Carcinogenic Potential (December 12, 2017).

25 National Institutes of Health, Agricultural Health Study, http://aghealth.nih.giv /collaboration/questionnaire.html, accessed December 30, 2017.

26 "No association of glyphosate with cancer of the brain in adults was found in the Upper Midwest Health case-control study. No associations in single case-control studies were found for cancers of the oesophagus and stomach, prostate, and soft-tissue sarcoma. For all other cancer sites (lung, oral cavity, colorectal, pancreas, kidney, bladder, breast, prostate, melanoma) investigated in the large AHS, no association with exposure to glyphosate was found," IARC, 396.

27 IARC, monograph 112, 75.

28 Ibid., 78.

29 Ibid.

30 Jose V. Tarazona, Daniele Court-Marques, Manuela Tiramani, Hermine Reich, Rudoff Pfeil, Frederique Istace, Frederica Crivellente, "Glyphosate Toxicity and Carcinogenicity: A Review of the Scientific Basis of the European Union Assessment and Its Differences with IARC, Archives of Toxicology (Published online April 3, 2017). Tarazona, J.V., Court-Marques, D., Tiramani, M. et al. Arch Toxicol (2017) 91: 2723. https://doi.org/10.1007/s00204-017-1962-5.

31 Ibid., 1.

32 Ibid., 16, 18.

33 Kate Kelland, "Cancer Agency Left in the Dark over Glyphosate Evi-dence," Reuters, June 14, 2017, www.reuters.com/investigates/special-report /glyphosate-cancer-data/, accessed July 21, 2017.

34 U.S. District Court, videotaped deposition. 293.

35 International Agency for Research on Cancer, World Health Organization, "Q&A on Glyphosate," March 1, 2016, www.iarc.fr/en/media-centre/iarcnews /pdf/Q&A_Glyphosate.pdf. 34 IARC, 2016 Q&A.

36 Ibid.

37 EPA, 2017, 68.

38 Ibid.

39 Sheldon Krimsky, Science in the Private Interest (Lanham, MD: Rowman & Littlefield Publishers, 2003).

40 Sheldon Krimsky, "The Unsteady State and Inertia of Chemical Regulation under the U.S. Toxic Substances Control Act," PLOS Biology 15, no. 12 (2017): e2002404.

Chapter 14: Reuters vs. U.N. Cancer Agency: Are Corporate Ties Influencing Science Coverage?

1   http://www.iarc.fr/en/media-centre/iarcnews/pdf/MonographVolume112.pdf.

2   http://www.huffingtonpost.com/carey-gillam/iarc-scientists-defend-gl_b _12720306.html.

3   http://www.lemonde.fr/planete/article/2017/06/01/monsanto-operation -intoxication_5136915_3244.html.

4   https://fair.org/home/gmo-crops-are-tools-of-a-chemical-agriculture -system/.

5   http://www.nature.com/news/science-media-centre-of-attention-1.13362.

6   http://www.sciencemediacentre.org/wp-content/uploads/2012/09/Science -Media-Centre-consultation-report.pdf.

7   https://usrtk.org/our-investigations/science-media-centre/.

8   https://ec.europa.eu/energy/sites/ener/files/documents/Risk%20Communication %20with%20the%20media_0.pdf.

9   https://storify.com/Paraphyso/wellcome-trust-science-media-centre.

10 https://blogs.fco.gov.uk/sunilkumar/2013/11/26/engaging-science-and -media/.

11 http://www.reuters.com/investigates/special-report/glyphosate-cancer-data/.

12 https://www.niehs.nih.gov/research/atniehs/labs/epi/studies/ahs/index.cfm.

13 http://www.motherjones.com/environment/2017/06/monsanto-roundup -glyphosate-cancer-who/.

14 https://www.bna.com/monsanto-cancer-study-n73014453449/.

15 https://usrtk.org/pesticides/reuters-kate-kelland-iarc-story-promotes-false -narrative/.

16 http://www.reuters.com/investigates/special-report/glyphosate-cancer-data/.

17 http://www.huffingtonpost.com/entry/monsanto-spin-doctors-target-cancer -scientist-in-flawed_us_594449eae4b0940f84fe2e57.

18 http://governance.iarc.fr/ENG/Docs/IARC_responds_to_Reuters_15 _June_2017.pdf.

19 http://www.stltoday.com/opinion/editorial/editorial-in-a-scientific-dispute -over-roundup-monsanto-gets-a/article_dca54b1d-6025-5c8a-b765 -cf286adac9ec.html.

20 http://www.stltoday.com/opinion/mailbag/editorial-relied-on-inaccurate -reporting-about-glyphosate-study/article_754e8e59-c63a-5925-bdc0 -0006af6cc05c.html.

21 https://twitter.com/USRightToKnow/status/885194196839129088.

22 https://morningconsult.com/opinions/cancer-data-suppressed-international -organization/.

23 https://www.facebook.com/pg/Science-News-Today-1927007787570501 /reviews/.

24 https://twitter.com/USRightToKnow/status/885192740677595137.

25 http://www.reuters.com/investigates/special-report/health-who-iarc/.

26 http://news.cision.com/the-investor-relations-group/r/environmental-health -trust-questions-new-study-claiming-no-cell-phone-brain-cancer-link-among -childr,c9147452.

27 http://science.time.com/2010/09/27/health-a-cancer-muckraker-takes-on -cell-phones/.

28 https://usrtk.org/hall-of-shame/why-you-cant-trust-the-american-council-on -science-and-health/.

29 http://www.acsh.org/news/2016/04/18/whats-ailing-iarc-another-round -of-criticism-ensues.

30 http://www.reuters.com/investigates/special-report/health-who-iarc/.

31 http://jech.bmj.com/content/early/2016/03/03/jech-2015-207005.

32 https://geneticliteracyproject.org/2016/04/18/glyphosate-battles-different -european-agencies-came-different-cancer-conclusions/.

33 https://usrtk.org/hall-of-shame/jon-entine-the-chemical-industrys-master -messenger/.

34 http://www.reuters.com/article/us-health-cancer-iarc-exclusive -idUSKCN12P2FW.

35 https://www.desmogblog.com/energy-environment-legal-institute.

36 http://www.iarc.fr/en/media-centre/iarcnews/pdf/Reuters_questions_and _answers_Oct2016.pdf.

37 http://www.iarc.fr/en/media-centre/iarcnews/pdf/Reuters_Readmore _Oct2016.pdf.

38 http://governance.iarc.fr/ENG/Docs/IARC_responds_to_Reuters_15 _June_2017.pdf.

39 http://archives.cjr.org/the_observatory/science_media_centers_the_pres_1 .php.

40 http://www.sciencemediacentre.org/about-us/funding/.

41 http://www.sciencemediacentre.org/working-with-us/for-journalists /roundups-for-journalists/.

42 http://www.sciencemediacentre.org/working-with-us/for-journalists/briefings -for-journalists/.

43 http://www.sciencemediacentre.org/film/.

44 http://www.emfacts.com/download/Science_media_centre_spin.pdf.

45 http://www.nature.com/news/science-media-centre-of-attention-1.13362.

46 http://www.nature.com/news/science-media-centre-of-attention-1.13362.

47 http://www.scidev.net/global/journalism/feature/uk-s-science-media-centre
-lambasted-for-pushing-corporate-science.html.

48 http://www.sciencemediacentre.org/expert-reaction-to-carcinogenicity
-classification-of-five-pesticides-by-the-international-agency-for-research-on
-cancer-iarc/.

49 https://monsanto.com/company/media/responses-iarc-glyphosate
-classification/.

50 https://www.efsa.europa.eu/en/press/news/151112.

51 https://www.efsa.europa.eu/en/press/news/151112.

52 https://echa.europa.eu/-/glyphosate-not-classified-as-a-carcinogen-by-echa.

53 http://www.sciencemediacentre.org/expert-reaction-to-efsas-conclusions-on
-glyphosate-safety/.

54 http://www.sciencemediacentre.org/expert-reaction-to-the-european-chemicals
-agency-echa-committee-for-risk-assessment-not-classifying-glyphosate-as-a
-carcinogen/.

55 http://www.reuters.com/article/us-health-monsanto-glyphosate
-idUSKCN0T61QL20151117.

56 http://www.sciencemediacentre.org/expert-reaction-to-efsas-conclusions-on
-glyphosate-safety/.

57 https://www.nrdc.org/experts/jennifer-sass/glyphosate-iarc-got-it-right-efsa
-got-it-monsanto.

58 http://www.sciencemediacentre.org/expert-reaction-to-the-european-food
-safety-authority-efsa-draft-scientific-opinion-on-the-safety-of-the-artificial
-sweetener-aspartame/.

59 http://www.sciencemediacentre.org/expert-reaction-to-press-release-from-fote
-and-gm-freeze-about-glyphosate-in-urine/.

60 http://www.sciencemediacentre.org/expert-reaction-to-paper-on-insecticides
-and-birth-defects-to-be-published-in-occupational-and-environmental
-medicine-a-bmj-specialist-journal-2-2/.

61 http://www.sciencemediacentre.org/expert-reaction-to-research-on-alcohol
-and-cancer/.

62 http://www.sciencemediacentre.org/expert-reaction-to-efsa-report-conclusion
-that-seralini-study-conclusions-were-not-supported-by-data/.

63 http://www.sciencemediacentre.org/expert-reaction-to-study-of-trace-metals
-and-lower-or-higher-cancer-risk-as-published-in-gut-2-2/.

64 http://www.sciencemediacentre.org/expert-reaction-to-study-investigating
-content-of-and-contaminants-in-laboratory-rodent-diets/.

65 http://www.sciencemediacentre.org/expert-reaction-to-the-european-chemicals -agency-echa-committee-for-risk-assessment-not-classifying-glyphosate-as-a -carcinogen/.

66 http://www.sciencemediacentre.org/expert-reaction-to-carcinogenicity -classification-of-five-pesticides-by-the-international-agency-for-research-on -cancer-iarc/.

67 https://www.theguardian.com/environment/2016/may/17/unwho-panel-in -conflict-of-interest-row-over-glyphosates-cancer-risk.

68 https://usrtk.org/sweeteners/ilsi-wields-stealthy-influence-for-the-food-and -agrichemical-industries/.

69 http://www.who.int/foodsafety/jmprsummary2016.pdf?ua=1.

70 http://ilsi.eu/about-us/.

71 https://usrtk.org/sweeteners/ilsi-wields-stealthy-influence-for-the-food-and -agrichemical-industries/.

72 https://www.theguardian.com/environment/2016/may/17/unwho-panel-in -conflict-of-interest-row-over-glyphosates-cancer-risk.

73 http://www.who.int/foodsafety/faq/en/.

74 http://www.reuters.com/article/us-health-who-glyphosate-idUSKCN0Y71HR.

75 http://www.sciencemediacentre.org/experts-respond-to-the-news-that-high -levels-of-dioxins-have-been-found-in-slaughtered-pigs-in-the-republic-of -ireland-2/.

76 http://www.reuters.com/article/us-ireland-food-recall-idUSL751490720081207.

77 https://www.theguardian.com/environment/2016/may/17/unwho-panel-in -conflict-of-interest-row-over-glyphosates-cancer-risk.

78 https://usrtk.org/pesticides/mdl-monsanto-glyphosate-cancer-case-key -documents-analysis/.

79 https://www.bloomberg.com/news/features/2017-07-13/does-the-world-s -top-weed-killer-cause-cancer-trump-s-epa-will-decide.

80 https://www.bloomberg.com/news/features/2017-07-13/does-the-world-s -top-weed-killer-cause-cancer-trump-s-epa-will-decide.

81 https://www.reuters.com/investigates/special-report/who-iarc-glyphosate/.

82 https://www.americanchemistry.com/Media/PressReleasesTranscripts/ACC -news-releases/ACC-Calls-Upon-Global-Leaders-to-Take-Action-Against -IARC-Over-Deliberate-Manipulation-of-Data.html.

83 https://usrtk.org/pesticides/carey-gillams-presentation-to-european-parliament -hearing-on-the-monsanto-papers-glyphosate/.

84 http://www.environmentalhealthnews.org/ehs/news/2017/june/of-mice -monsanto-and-a-mysterious-tumor.

85 http://fair.org/home/reuters-vs-un-cancer-agency-are-corporate-ties-influencing
-science-coverage/.

86 http://baumhedlundlaw.com/pdf/monsanto-documents/72-Document-Details
-Monsantos-Goals-After-IARC-Report.pdf.

87 https://usrtk.org/pesticides/how-monsanto-manufactured-outrage-at-iarc-over
-cancer-classification/.

88 https://usrtk.org/our-investigations/science-media-centre/.

89 https://usrtk.org/pesticides/reuters-kate-kelland-iarc-story-promotes-false
-narrative/.

90 https://usrtk.org/pesticides/monsanto-spin-doctors-target-cancer-scientist-in
-flawed-reuters-story/.

Chapter 15: Genetic Engineering, Pesticides, and Resistance to the New Colonialism

1 Javiera Rulli, No Spray Coalition forum, Brooklyn, 2005.

2 Javiera Rulli, "More on the Massacre in Paraguay," *GMWatch*, June 28, 2005,
http://www.gmwatch.org/en/news/archive/2005/2621-more-on-massacre-in
-paraguay-2862005.84

3 Mitchel Cohen, *The Politics of World Hunger;* also, *Somalia, and the Cynical
Manipulation of Hunger;* Silvia Federici, *Africa, the IMF and the New Enclosures,*
Red Balloon Collective; and Midnight Notes, *One No, Many Yeses*, Box 204,
Jamaica Plain, MA 02130, December 1997.

4 Michael Donnelly, "Wall Street's Failed 1934 Coup," *Counterpunch*, December
2, 2011.

5 Diamond v. Chakrabarty, 447 U.S. 303 (1980), in which Chakrabarty, a micro-
biologist, sought to patent a human-made, genetically engineered bacterium
capable of breaking down crude oil, a property that is possessed by no naturally
occurring bacteria. In a 5–4 decision, the US Supreme Court ruled that a live,
human-made microorganism is patentable under US statutes. Justices in favor:
Chief Justice Burger, Stewart, Blackmun, Rehnquist, and Stevens. Opposed:
Brennan, White, Marshall, and Powell.

6 Jonah Raskin, *A Terrible Beauty: The Wilderness of American Literature*, Berkeley:
Regent Press, 2014.

7 Fiji Water is an interesting example, as the people of Fiji don't have clean
drinking water for themselves and yet water from Fiji is purified and sold all
over the rest of the world. (See http://www.eastbaypesticidealert.org/lbam
.html#THEPUSHERS). In addition, Aquafina (owned by Pepsi-Cola) and
Dasani (owned by Coca-Cola) have finally admitted that they do not draw

water from natural springs despite the pictures on the labels and have been drawing water to sell from tap water—or, as they call it, a "Public Water Source" (PWS). Nestle owns Poland Springs bottled water, but the water "is drawn from underground streams and not from the mountain stream depicted on the label."

8   "The use of Agent Orange was an experimental form of chemical and biological warfare, designed to strip foliage and deny the enemy jungle cover—and to deprive enemy forces of their food supply (directly spraying rice-fields, for instance). Experimental in this instance meaning no idea of the long-term effects of this deadly herbicide, which can release dioxin—one of the most potent toxins known to mankind." Mick Grant, *The Ecologist*, "First Agent Orange, Now Roundup: What's Monsanto up to in Vietnam? Ecologist Special Investigation," October 10, 2016.

9   Jeff Nesmith, "Monsanto Altered Dioxin Study, EPA Memo Says," *Indianapolis Star*, March 23, 1990, A3.

10 Admiral Elmo Zumwalt, Affidavit in Agent Orange Case, Ivy v. Diamond Shamrock Chemicals Co. et al., CV-89-03361 (E.D.N.Y.) 8.

11 Sustainable Pulse, "Vietnam Bans Paraquat and 2,4-D over Human and Environmental Damage," February 27, 2017.

12 Michael Hansen, "Possible Human Health Hazards of Genetically Engineered Bt Crops," Consumer Policy Institute/Consumers Union, reported in Organic Consumers Association, October 2000, https://www.organicconsumers.org/old_articles/ge/btcomments.php.

13 http://articles.mercola.com/sites/articles/archive/2011/11/06/aspartame-most-dangerous-substance-added-to-food.aspx. See also A.V. Krebs, "Monsanto's Misdeeds," *The Multinational Monitor*, July/August 1990: "Dr. Louis Elias, director of medical genetics at Emory University Medical School, argues that NutraSweet was not properly tested. 'They never asked the right questions about what it does to brain function in humans. They decided without data that you had to have enormous amounts of phenylalanine in your blood before it becomes a problem. We don't know that's the case.' A 1987 study by a University of Illinois scientist indicated that using NutraSweet appeared to heighten chances of behavioral disturbances and birth defects. Dr. Reuben Matalona, a pediatrician and geneticist, argues that his test results suggested that large amounts of NutraSweet could affect small children and millions of people unaware of their body's inability to fully process phenylalanine. High concentrations of the chemical, he suggests, can cause reduced attention span and concentration and memory loss."

14 https://www.epa.gov/pcbs/learn-about-polychlorinated-biphenyls-pcbs.

15 Greenpeace, "Monsanto: Greenpeace Corporate Criminal Report."

16 A.V. Krebs, "Monsanto's Misdeeds," *The Multinational Monitor,* July/August 1990. "EPA officer Robert Taylor has told Greenpeace that the EPA did not approve Monsanto's application for butachlor registration due to 'environmental, residue, fish and wildlife and toxicological concerns.' Butachlor can cause skin and eye irritation, as well as decreased body weights, organ weight changes, reduced brain size and lesions, according to Monsanto's safety data sheet for the chemical. Nevertheless, butachlor appears in the U.S. food supply; it is used in Argentina, Brazil, China, India, the Philippines, Taiwan, Thailand and Venezuela, which together produce 97.5 percent of U.S. rice imports, reports the U.S. Department of Agriculture Economic Research Service."

17 Ibid.

18 See Mitchel Cohen, *Somalia, and the New World Order: You Provide the Collateral, We'll Provide the Damage* (New York: Red Balloon Publications, 1994). See also GM Watch, "GM Cassava 'Our Only Hope,'" www.gmwatch.org/en/gm-cassava-our-only-hope.

19 See "Why Iraqi Farmers Might Prefer Death to Paul Bremer's Order 81," GM Watch, September 19, 2008.

20 Monsanto was forced to pull its genetically engineered canola off the market in June 1997. But US patent laws allowed the company to conceal the reasons why, despite the threat to public health and safety.

21 S'ra DeSantis, "Using Free Trade Agreements to Contaminate Indigenous Corn," *Synthesis/Regeneration: A Magazine of Green Social Thought,*" Winter 2004. S'ra DeSantis was writing on behalf of the Institute for Social Ecology's Biotechnology Project.

22 Ibid.

23 Ibid.

24 To keep track of these sorts of movements, contact OrganicWatch, (202) 547–9359, owatch@icta.org; or Pure Food Campaign, 860 Hwy 61, Little Marais, MN 55614, (800)-253-0681, purefood@aol.com.

25 Janet N. Abramovitz, "Imperiled Waters, Impoverished Future: The Decline of Freshwater Ecosystems" (Worldwatch Paper 128, Washington, DC, Worldwarch Institute, 1996).

26 In 2013, the European Union General Court overruled an earlier decision in Germany to grow genetically engineered potatoes, called *Amflora,* developed by German chemical corporation BASF. "EU Court bans BASF's 'Amflora' GM Potato, Annuls Commission Approval," *Deutsche Welle,* Dec. 13, 2013. Three years earlier, the European Union had approved its first genetically modified plant, the Amflora potato, since 1998. "The World from Berlin GM Potato Approval 'A Big Step for Germany,'" *Der Spiegel,* March 3, 2010. Mark Lynas, political director of the Cornell Alliance for Science, wrote: "[H]ypocrisy rules: Europe imports over

30 million tons per year of corn and soy-based animal feeds, the vast majority of which are genetically modified, for its livestock industry."

27   Paul Farmer, *AIDS and Accusation: Haiti and the Geography of Blame.*1992 University of California Press.

28   Mary Shelley's nineteenth century novel was one of many warnings against the dangers of runaway technology. Her *Frankenstein* is nowhere more apropos than it is in pointing to genetic engineering.

29   "Genetic Seed Sterilization is 'Holy Grail' for Ag Biotechnology Firms: New Patents for 'Suicide Seeds' Threaten Farmers and Food Security Warns RAFI." Berkeley College of Natural Resources, https://nature.berkeley.edu/srr/Alliance /novartis/sterile.htm. "The latest version of Monsanto's suicide seeds won't even germinate unless exposed to a special chemical, while AstraZeneca's technologies outline how to engineer crops to become stunted or otherwise impaired if not regularly exposed to the company's chemicals."

30   Although I occasionally use the terms "pests" and "weeds" as commonly understood in this book, those are characterizations advanced as part of a context. What might be a "pest"—a disruptive annoyance—in one context is another organism's food, or symbiotic partner in another. The same is true for what some call "weeds"—a negative term if you're growing an immaculate grassy lawn, for example, but if you're an artist appreciating vibrant colors or a practitioner of holistic medicines, so-called weeds would be far more important than suburban lawns—a different context. *Anything* can be considered a "pest," "weed," "foreigner," or a "disruptor of the natural ecosystem"—it depends on the lens through which one is looking.

31   http://www.gmo-free-regions.org/gmo-free-regions.html.

32   Layla Katiraee and Kavin Senapathy, "Gerber Formula Goes Non-GMO, But Not Really," *Forbes,* February 22, 2016. The authors, who are strongly pro-GMO, raise questions about Gerber's claims as to being non-GMO.

33   See Jimmy Carter's letter to the *New York Times,* August 26, 1998, in favor of genetic engineering as a solution for feeding people in Africa.

34   "As Crisis Hits, Seed Giant Monsanto Sees Business in Russia and Ukraine," *The Moscow Times,* January 23, 2015.

35   US Department of Agriculture GAIN report, https://gain.fas.usda.gov/Recent %20GAIN%20Publications/Russia%20Bans%20Cultivation%20and% 20Breeding%20of%20GE%20Crops%20and%20Animal_Moscow_Russian% 20Federation_7-12-2016.pdf. For more information see FAS/Moscow GAIN report GMO Registration for Cultivation Postponed 6/27/2014.pdf; Russian "Producers Consider It Reasonable to Ban GMO products," 05/07/2016, http:// ria.ru/economy/20160705/1459098131.html.

36 "Russia to Ban Genetically Modified Organisms in Food Production," *Moscow Times,* Sept. 20, 2015.

37 *Moscow Times,* January 23, 2015, *op cit.* "Ukraine is the world's sixth largest grain grower this season, and Goncalves said the region remained a priority for Monsanto."

38 Eduard Korniyenko, "Putin Wants Russia to Become World's Biggest Exporter of Non-GMO Food," *Reuters,* December 3, 2015.

39 Seminar on Socialist Renewal and the Capitalist Crisis—A Cuban-North American Exchange," Havana, Cuba, June 16–30, 2013, in association with the Center for Global Justice, San Miguel de Allende, Mexico, and the Radical Philosophy Association.

40 "An Open Letter to the Cuban People against Genetic Engineering of Agriculture," Mitchel Cohen, in *A Talk in Havana,* Red Balloon Collective, 2013.

41 Organic Consumers Association, https://www.organicconsumers.org/campaigns/millions-against-monsanto.

42 See Mitchel Cohen, *Got Pus? Bovine Growth Hormone & the New World Order,* (New York: Red Balloon Collective, 1999). Updated and reissued, 2013.

43 Dr. Samuel Epstein, *Cancer Research,* June 1995.

44 *New York Times,* "Synthetic Hormone in Milk Raises New Concerns," Jan. 19, 1999.

45 Ibid.

46 "Milk, rBGH, and Cancer," *Rachel's Environment and Health Weekly,* no. 593, April 9, 1998.

47 See especially Brian Tokar, ed., *Redesigning Life: The Worldwide Challenge to Genetic Engineering,* Zed Books, 2001, distributed in the U.S. by St. Martin's Press; Sheldon Krimsky & Jeremy Gruber, eds., *The GMO Deception: What you need to know about the food, corporations, and government agencies putting our families and our environment at risk,* Skyhorse publishing, 2014; and Jeffrey Smith, *Seeds of Deception: Exposing Industry and Government Lies About the Safety of the Genetically Engineered Foods You're Eating,* 2003.

48 TruthWiki, "Glyphosate," www.truthwiki.org/glyphosate/. See also Elena Keates, "EPA Sued for Approving Herbicide to Treat Agent Orange Crops," *Greenpeace,* October 2014.

49 Isabella Kenfield, "Monsanto's Man in the Obama Administration," Counterpunch.org, August 14, 2009.

50 Mitchel Cohen, "Biotechnology and the New World Order," in Brian Tokar, ed., *Redesigning Life? The Worldwide Challenge to Genetic Engineering* (London: Zed Books, 2001). I cited Dan Fagin and Marianna Lavelle, *Toxic Deception: How the Chemical Industry Manipulates Science, Bends the Law and Endangers*

Your Health, Monroe, ME: Common Courage Press, 1999), reviewed by Russell Mokhiber and Robert Wiessman: "Soon after this visit, and after a lot of lobbying pressure from the industry, Gore directly ordered the EPA to slow down its implementation of these tougher pesticide standards that were required by the FQPA," Fagin told us. "He also told the EPA to make a special effort to consider the needs of agribusiness and the views of the U.S. Department of Agriculture. A new advisory committee, which included many of the key chemical manufacturers and their consultants, was set up to review implementation of the new law. That committee is still meeting, and the EPA still hasn't moved against organophosphates."

51 Full disclosure: The pro-GMO BIO corporation was founded by Lisa Raines, a former colleague and friend of mine at SUNY Stony Brook who participated in her younger years in many anti-war protests with the Red Balloon Collective and wrote a front-page article for the Collective's newspaper featuring her observations on her trip through Europe and the kidnapping and murder of politician Aldo Moro. A decade later, she had moved dramatically to the right and when we got back in touch with each other after many years of silence, I was already deeply involved in organizing protests against genetic engineering and pesticides. Lisa had taken the opposite path. Lisa Raines was killed on September 11, 2001—a passenger on the plane that was shot down by the U.S. Air Force over Pennsylvania. Her husband, Steve Push, went on to be a spokesperson for a time for some of the families whose relatives had been killed on 9/11.

52 Dave Murphy, Food Democracy Now, http://www.fooddemocracynow.org /campaign/hillary-s-monsanto-how-clinton-state-department-became-global -marketing-arm-monsanto.

53 "U.S. Taxpayer Money Used to Help Promote Monsanto GMO Products Overseas, Documents Reveal," *New York Daily News*, May 14, 2013.

54 Tom Philpott, "Taxpayer Dollars Are Helping Monsanto Sell Seeds Abroad," *Mother Jones*, May 18, 2013.

55 According to Snopes.com, Hillary Clinton was interviewed by Jim Greenwood, president and CEO of the Biotechnology Industry Organization. Here is an excerpt from the Snopes report:

Clinton endorsed the use of genetically modified organisms such as by engineering them for drought resistance. She suggested the biotech industry stress these characteristics instead of focusing on the term GMOs.

Clinton said the biotech industry "should continue to try to make the case to those who are skeptical that they may not know what they are eating already, because the question of genetically modified foods or hybrids has gone on for many many years, and there is a big gap between what the facts are and what the perceptions are."

"If you talk about drought-resistant seeds, and I have promoted those all over Africa, by definition they have been engineered to be drought-resistant," Clinton said. "That's the beauty of them. Maybe somebody can get their harvest done and not starve, and maybe have something left over to sell."

The Monsanto Company is also listed among the entities who have donated between $1 million and $5 million to the Clinton Foundation, a nonprofit corporation established by former President Bill Clinton to "strengthen the capacity of people throughout the world to meet the challenges of global interdependence."

Additionally, some sources have posited that the rumor arose from a connection between the Clinton campaign and Jerry Crawford, an Iowa lawyer and Democratic party leader. Crawford worked on Hillary Clinton's 2016 campaign in Iowa and is often described as a "Monsanto lobbyist."

Crawford was brought on to help her win Iowa. According to Opensecrets. org, his lobbying firm has represented Monsanto, as well as the Humane Society. This shouldn't be a surprise, considering the fact that Iowa is a major state for agriculture, and a number of seed companies do business with farmers there.

Chapter 16: Big Science and the Curious Notion of "Progress"

1 The Institute for Social Ecology, whose leading luminary was Murray Bookchin, existed for many years at Goddard College in Plainfield, Vermont. It continues to be headed by Brian Tokar, a contributor to this book.

2 Werner Heisenberg, *Physics and Philosophy* (New York: Harper & Row), 1962.

3 Stephen Jay Gould, *Ever Since Darwin*, (New York, NY: W.W. Norton & Co., 1977).

4 Dana Bramel and Ron Friend, "The Theory and Practice of Psychology," in Bertell Ollman and Edward Vernoff, *The Left Academy: Marxist Scholarship on American Campuses*, (New York, NY: McGraw-Hill, 1982).

5 Richard Levins and Richard Lewontin, *The Dialectical Biologist*, (Cambridge, MA: Harvard University Press, 1985).

6 Ibid.

7 Harry Braverman, *Labor and Monopoly Capital: The Degradation of Work in the Twentieth Century* (New York: Monthly Review Press, 1974), 73.

8 The dehumanization of the labor process into fragmentary pieces was intensified under the "tyranny of the clock" and rigid work schedules that are enforced eight to twelve hours per day fifty weeks a year and do not leave time for people to have a life, as Ellen Buff points out in private correspondence, in 2004.

Atomization of work fostered reductionist ways of seeing, which echoed throughout the culture. Radical artists such as Duchamps attempted to reflect and critique on the canvas the rise of reductionist ways of seeing in society at large. (See "Nude Descending a Staircase," for example.

9 See, among others who challenge reductionist constructs, Stuart Newman, "Idealist Biology," *Perspectives in Biology and Medicine* 31, no. 3, (Spring 1988), 353–368; Paul Weiss, "The Living System: Determinism Stratified," in *Beyond Reductionism: New Perspectives in the Life Sciences,* ed. Arthur Koestler and J.R. Smythies (Boston: Beacon Press, 1971); Martha Herbert, *Incomplete Science, the Body and Indwelling Spirit* (Online journal, 2000) Metanexus Institute http://www.metanexus.net/incomplete-science-body-and-indwelling-spirit/; Joel Kovel, *The Enemy of Nature* (London: Zed Books, 2002); Richard Levins and Richard Lewontin, *The Dialectical Biologist* (Cambridge, MA: Harvard University Press, 1985); Brian Tokar, ed., *Redesigning Life? The Worldwide Challenge to Genetic Engineering* (London: Zed Books, 2001); Mae Wan-Ho, *Living with the Fluid Genome* (London: Institute of Science and Society, 2003); and Stuart Kauffman, *At Home in the Universe: The Search for the Laws of Self-Organization and Complexity* (New York: Oxford University Press, 1995).

10 This essential relationship between parts and wholes, individual and environment, is generally given short shrift by many scientists, even ignored. Instead, they pursue a reductionist unidirectional causality—the parts, pieced together (they say) determine the whole in cause and effect sequence. Their linear framework provides the basis for the mechanistic formulations (such as reductionism, positivism, empiricism, and behaviorism) that, I would argue, are incorrect when applied to genes or nanolevel interactions in a determinist manner. These meditations on wholes and parts, holism and reductionism, freedom and determinism, grew out of discussions of a paper I presented at the Radical Philosophy Association conference in Havana in 1992 titled "A Call for a Revolutionary Science."

11 Evelyn Fox Keller, *A Feeling for the Organism: The Life and Work of Barbara McClintock,* (New York and San Francisco: W.H. Freeman & Co., 1983).

12 Ibid.

13 Ibid., 199.

14 Science Daily, July 14, 2014.

15 Stephen Jay Gould, "Humbled by the Genome's Mysteries," *The New York Times,* February 19, 2001.

16 Keller, *A Feeling for the Organism,* 199–200.

17 For one of many research papers on this theme, see "Phase Transitions in a Gene Network Model of Morphogenesis," *Journal of Biosciences* 17, no. 3, 193–215.

18 Chambon, 1981.Organization and expression of eurcaryotic split genes coding for proteins Ann.Rev.Bioche. 1981 50–3 49–83

19 Holderedge, Craig (2005) The gene a needed revolution; The Nature Institute In Context #14 (Fall, 2005, pp. 14–17); http://natureinstitute.org/pub/ic/ic14 /gene.htm.

20 Stuart Newman, "Generic Physical Mechanisms of Morphogenesis and Pattern Formation as Determinants in the Evolution of Multicellular Organization," *Journal of Biosciences*, 17, no. 3 (September 1992), 193–215.

21 Stuart Newman, "Dynamic Balance in Living Systems," *GeneWATCH*, November-

22 See, for instance, Morris Berman, *The Reenchantment of the World* (Ithaca, NY) Cornell University Press 1981). The book is a truly great historical *tour de force* through the history of dualistic thinking on which modern science is based.

23 Ibid.

24 Karl Marx and Fredrick Engels, *The German Ideology*. (original date of writing, unpublished in German, 1845–46) International Publishers, Chicago 1947.

Chapter 17: When Rights Collide: Genetic Engineering & Preserving Biocultural Integrity

1 *Rights and Liberties in the Biotech Age: Why We Need a Genetic Bill of Rights.* Krimsky, Sheldon and Peter Shorrett, eds., Rowman and Littlefield, 2005.

2 The present chapter is an update and revision of Chapter 6 of Krimsky and Shorrett, ibid.

3 Ivan Illich, *Tools for Conviviality* (New York: Harper, 1980).

4 Margaret Mellon, "The Wages of Hype: Agricultural Biotechnology After 25 Years," Arthur Miller Lecture presented at MIT (October 3, 2003); Marc Lappé and Britt Bailey, *Against the Grain* (Monroe, ME:Common Courage Press, 1998); Mae-Won Ho, *Genetic Engineering: Dream or Nightmare* (Bath, UK: Gateway Books, 1998).

5 Andrew Kimbrell (ed.), *Fatal Harvest:The Tragedy of Industrial Agriculture* (Covelo, CA: Island Press, 2002).

6 M. S. Prakash and Gustavo Esteva, *Grassroots Post-Modernism: Remaking the Soil of Culture* (London: Zed Books, 1998); Gustavo Esteva, "Re-Embedding Food in Agriculture," *Culture & Agriculture* (Winter 1994): 2–12.

7 Frances Moore Lappe, Joseph Collins, and Peter Rossett, with Luis Esparza, *World Hunger: 12 Myths* (London: Earthscan, 1998). Miguel A. Altieri and Peter Rosset, "Ten Reasons Why Biotechnology Will Not Ensure Food Security, Protect the Environment and Reduce Poverty in the Developing World," *AgBioForum* 2 (1999): 155–62, www.agroeco.org/doc/10reasonsbiotech1.pdf.

8 Richard Levins, "When Science Fails Us," www-trees.slu.se/newsl/32/32levin .htm (1996).

9 Barry Commoner, "Unraveling the DNA Myth: The Spurious Foundation of Genetic Engineering," *Harper's* (February 2002): 39–47.

10 Les Levidow and Susan Carr, "Unsound Science? Trans-Atlantic Regulatory Disputes over GM Crops," *International Journal of Biotechnology* 2(2000): 257–73; B.Vogel and B. Tappeser, Der Einfluss der Sicherheitsforschung und Risikoabschätzung bei der Genehmigung von Inverkehrbringung und Sortenzulassung transgener Pflanzen, Öko-Institut e.V., study commissioned by the German Technology Assessment Bureau Auftrag, Berlin, 2000, available as pdf-file under www.oeko.de (only german). Also, see Jane Anne Morris, "Sheep in Wolf's Clothing," *By What Authority* (Fall 1998), www.poclad.org/bwa/fall98.htm.

11 Miguel Altieri, "Agroecology: The Science of Natural Resource Management for Poor Farmers in Marginal Environments," *Agriculture, Ecosystems and Environment* 93 (December 2002): 1–24, www.agroeco.org/doc/NRMfinal.pdf.

12 See USAID bilateral assistance programs, for example. Alan P. Larson, "The Future of Agricultural Biotechnology in World Trade," remarks at the Agricultural Outlook Forum 2002, www.state.gov/e/rls/rm/2002/8447.htm.

13 See Ian Sample, "Naïve, Narrow, and Biased," *The Guardian*, Op-Ed, July 24, 2003; Sujatha Byravan, "Genetically Engineered Plants: Worth the Risk?" plenary lecture at Viterbo University, February 3, 2004; Sheldon Krimsky and Tim Schwab, Conflicts of interest among committee members in the National Academies' genetically engineered crop study. *PLoS One*, 2017 Feb 28; e0172317.

14 Lily Kay, *The Molecular Vision of Life: Caltech, the Rockefeller Foundation, and the Rise of the New Biology.* Oxford University Press, 1996.

15 See, for example, the "oncomouse" decision of the U.S. Patent and Trademark Office, U.S. patent No. 4,736,866 (1988).

16 L. LaReesa Wolfenbarger and Paul R. Phifer, "The Ecological Risks and Benefits of Genetically Engineered Plants," *Science* 290 (2000): 2088–93.

17 Don Westfall, food industry marketing strategies consultant formerly with Promar International, quoted in Stuart Laidlaw, "Starlink Fallout Could Cost Billions," *Toronto Star* (January 9, 2001).

18 Emmy Simmons, assistant administrator, USAID, quoted in Philip Bereano, "Engineered Food Claims Are Hard to Swallow," *Seattle Times* (November 19, 2001).

19 Evelyn Fox Keller, *The Century of the Gene* (Cambridge, MA: Harvard University Press, 2000).

20 Richard Lewontin, *Triple Helix: Gene, Organism, and Environment* (Cambridge, MA: Harvard University Press, 2000); Richard Lewontin, *It Ain't Necessarily So* (New York: New York Review Books, 2000).

21 Commoner,"Unraveling the DNA Myth"; Ruth Hubbard and Elijah Wald, *Exploding the Gene Myth* (Boston: Beacon Press, 1993).

22 See Michael Hansen,"Genetic Engineering Is not an Extension of Conventional Breeding," Consumers Union Discussion Paper (2000), www.consumersunion .org/food/widecpi200.htm. Also see David Schubert, "A Different Perspective on GM Food," *Nature Biotechnology* 20 (October 2002): 969.

23 Richard C. Strohman, "Organization Becomes Cause in the Matter," *Nature Biotechnology* 18 (June 2000): 575–6; Richard C. Strohman, "Five Stages of the Human Genome Project," *Nature Biotechnology* 17 (February 1999): 112.

24 Sui Huang, "The Practical Problems of Post-Genomic Biology," *Nature Biotechnology* 18 (May 2000): 471–2.

25 Federation of American Scientists, Agricultural Biotechnology: Safety, Security and Ethical dimensions: Methods of Gene Transfer in Plants, https://fas.org /biosecurity/education/dualuse-agriculture/2.-agricultural-biotechnology /methods-of-gene-transfer-in-plants.html, 2011.

26 See G. Riddihough and E. Pennisi, "The Evolution of Epigenetics," *Science* 293 (2001): 1063.

27 See, for example, P. Meyer, F. Linn, I. Heidmann, H. Meyer, I. Niedenhof, and H. Saedler, "Endogenous and Environmental Factors Influence 35S Promoter Methylation of a Maize A1 Gene Construct in Transgenic Petunia and Its Colour Phenotype," *Molecular Genes and Genetics* 231 (1992): 345–52.

28 Sheldon Krimsky, "Biotechnology at the Dinner Table: FDA Oversight of Transgenic Food," *Annals of the American Academy of Political and Social Science* 584 (November 2002): 80–96.

29 Erik Millsone, Eric Brunner, and Sue Mayer, "Beyond Substantial Equivalence," *Nature* 401 (October 7, 1999): 525–6.

30 Hansen, "Genetic Engineering Is not an Extension of Conventional Breeding." Also see Meyer et al., "Endogenous and Environmental Factors Influence 35S Promoter Methylation"; and A N. E. Birch, I. E. Geoghegan, D. W. Griffiths, and J.W. McNichol, "The Effect of Genetic Transformations for Pest Resistance on Foliar Solandine-based Glycoalkaloids of Potato (*Solanum tuberosum*)," *Annals of Applied Biology* 140 (2002): 134–49.

31 Jimmy Carter, "Who's afraid of Genetic Engineering?" New York Times, Opinion, Aug. 26, 1998.

32 Sheldon Krimsky, *Science in the Private Interest* (New York: Rowman & Littlefield, 2003).

33 Kimbrell, *Fatal Harvest*; Wes Jackson and Wendell Berry, *New Roots for Agriculture* (Lincoln: University of Nebraska Press, 1985).

34 Miguel A. Altieri, *Agroecology: The Science of Sustainable Agriculture* (Boulder, CO: Westview Press, 1995).

35 See Pesticide Action Network North America, www.panna.org.

36 Altieri, *Agroecology.*

37 Jane Rissler and Margaret Mellon, *The Ecological Risks of Genetically Engineered Crops* (Cambridge, MA: MIT Press, 1996).

38 Wendell Berry, *The Unsettling of America: Culture and Agriculture* (San Francisco: Sierra Club Books 1977).

39 Margaret Mellon, "The Wages of Hype."

40 Stephen B. Brush and Doreen Stabinsky (eds.), *Valuing Local Knowledge* (Covelo, CA: Island Press, 1996); Vandana Shiva, *Biopiracy* (Boston: South End Press, 1996).

41 Levins, "When Science Fails Us"; Steve Lerner, *Eco-Pioneers: Practional Visionaries Solving Today's Environmental Problems* (Cambridge, MA: MIT Press, 1997); Kenny Ausubel, *The Bioneers: Declarations of Interdependence* (South Burlington, VT: Chelsea Green, 2001); Alan Weisman, *Gaviotas: A Village to Reinvent the World* (South Burlington, VT: Chelsea Green, 1995).

42 Perry ED, Ciliberto F, Hennessy DA, Moschini G.,"Genetically engineered crops and pesticide use in U.S. maize and soybeans. *Science Advances,* 2016 August 31; 2(8):e1600850.

43 Doug Gurian-Sherman, *Failure to Yield: Evaluating the Performance of Genetically Engineered Crops.* Union of Concerned Scientists, April, 2009.

Chapter 18: Glyphosate Acting as a Glycine Analogue: Slow Insidious Toxicity

1 N. L. Swanson, A. Leu, J. Abrahamson, and B. Wallet, "Genetically Engineered Crops, Glyphosate and the Deterioration of Health in the United States of America," *Journal of Organic Systems* 9 (2014): 6–37.

2 G. E. Seralini, E. Clair, R. Mesnage, S. Gress, N. Defarge, M. Malatesta, D. Hennequin, and J. S. de Vendomois, "Retracted Long Term Toxicity of a Roundup Herbicide and a Roundup-Tolerant Genetically Modified Maize," *Food and Chemical Toxicology* 50 (2012): 4221–31, *retracted.*

3 G. E. Seralini E. Clair, R. Mesnage, N. Defarge, M. Malatesta, D. Hennequin, J. Spiroux de Vendômois, "Long-Term Toxicity of a Roundup Herbicide and a Roundup-Tolerant Genetically Modified Maize. Environmental Sciences Europe 26 (2014):14, *republished.*

4 D. Cattani, V. L. de Liz Oliveira Cavalli, C. E. Heinz Rieg, J. T. Domingues, T. Dal-Cim, C. I. Tasca, F. R. Mena Barreto Silva, and A. Zamoner, "Mechanisms Underlying the Neurotoxicity Induced by Glyphosate-Based Herbicide in Immature Rat Hippocampus: Involvement of Glutamate Excitotoxicity," *Toxicology* 320 (2014) : 34–45.

5 Q. Li, M. J. Lambrechts, Q. Zhang, S. Liu, D. Ge, R. Yin, M. Xi, and Z. You. "Glyphosate and AMPA Inhibit Cancer Cell Growth through Inhibiting Intracellular Glycine Synthesis," *Drug Design Development and Therapy* 7 (2013): 635–43. L. M. Kitchen, W. W. Witt, and C. E. Rieck, "Inhibition of δ-Aminolevulinic Acid Synthesis by Glyphosate," *Weed Science* 29 (1981): 571–577.

6 L. M. Kitchen, W. W. Witt, and C. E. Rieck, "Inhibition of δ-Aminolevulinic Acid Synthesis by Glyphosate," *Weed Science* 29 (1981): 571–577.

7 I. Astner, J. O. Schulze, J. van den Heuvel, D. Jahn, W. D. Schubert, and D. W. Heinz, "Crystal Structure of 5-Aminolevulinate Synthase, the First Enzyme of Heme Biosynthesis, and Its Link to XLSA in Humans," *EMBO Journal* 24 (2005): 3166–77.

8 A. Samsel and S. Seneff, "Glyphosate, Pathways to Modern Diseases IV: Cancer and Related Pathologies," *Journal of Biological Physics and Chemistry* 15 (2015):121–59.

9 C. Lowrie, *Metabolism of [14C]-N-Acetyl-Glyphosate (IN-MCX20) in the Lactating Goat* (Charles River Laboratories Project no. 210583, submitted by E. I. du Pont de Nemours and Company), DuPont Report No. DuPont-19796 (2007).

10 M. Bajaj, M. D. Waterfield, J. Schlessinger, W. R. Taylor, and T. Blundell. "On the Tertiary Structure of the Extracellular Domains of the Epidermal Growth Factor and Insulin Receptors," *Biochimica Biophysica Acta* 916 (1987): 220–26.

11 M. Topf, P. Varnai, and W. G. Richards, "Ab Initio QM/MM Dynamics Simulation of the Tetrahedral Intermediate of Serine Proteases: Insights into the Active Site Hydrogenbonding Network," *Journal of the American Chemical Society* 124 (2002): 14780–88.

12 J. Walter, W. Steigemann, T. Singh, H. Bartunik, W. Bode, and R. Huber, "On the Disordered Activation Domain in Trypsinogen: Chemical Labeling and Low Temperature Crystallography," *Acta Crystallographica* 38 (1982): 1462–72.

13 C. Chen, J. Ke, X. E. Zhou, W. Yi, J. S. Brunzelle, J. Li, E. L. Yong, H. E. Xu, and K. Melcher, "Structural Basis for Molecular Recognition of Folic Acid by Folate Receptors," *Nature* 500 (2013): 486–89.

14 U. M. Koivisto, J. S. Viikari, and K. Kontula, "Molecular Characterization of Minor Gene Rearrangements in Finnish Patients with Heterozygous Familial Hypercholesterolemia: Identification of Two Common Missense Mutations (Gly823Asp and Leu380His) and Eight Rare Mutations of the LDL Receptor Gene," *The American Journal of Human Genetics* 57 (1995): 789–97.

15 F. Kinose, S. X. Wang, U. S. Kidambi, C. L. Moncman, and D. A. Winkelmann. "Glycine 699 Is Pivotal for the Motor Activity of Skeletal Muscle Myosin," *Journal of Cell Biology* 134 (1996): 895–909.

16 M. Bucciantini, E. Giannoni, F. Chiti, F. Baroni, L. Formigli, J. Zurdo, N. Taddei, G. Ramponi, C. M. Dabsan, and M. Stefani, "Inherent Toxicity of Aggregates Implies a Common Mechanism for Protein Misfolding Diseases," *Nature* 416 (2002): 507–511.

17 H. N. Du, L. Tang, X. Y. Luo, H. T. Li, J. Hu, J. W. Zhou, and H. Y. Hu, "A Peptide Motif Consisting of Glycine, Alanine, and Valine Is Required for the Fibrillization and Cytotoxicity of Human α-Synuclein." Biochemistry 42, no. 29 (July 2003): 8870–78.

18 S. Pesiridis, V. M. Y. Lee, and J. Q. Trojanowski, "Mutations in TDP-43 Link Glycine-Rich Domain Functions to Amyotrophic Lateral Sclerosis," *Human Molecular Genetics* 18 (2009): R156–62.

19 C. F. Harrison, V. A. Lawson, B. M. Coleman, Y. S. Kim, C. L. Masters, R. Cappai, K. J. Barnham, and A. F. Hill, "Conservation of a Glycine-Rich Region in the Prion Protein is Required for Uptake of Prion Infectivity," *Journal of Biological Chemistry* 285 (2010): 20213–23.

20 K. Liu, D. Kozono, Y. Kato, P. Agre, A. Hazama, and M. Yasui, "Conversion of Aquaporin 6 from an Anion Channel to a Water-Selective Channel by a Single Amino Acid Substitution," *Proceedings of the National Academy of Sciences usA* 102, no. 6 (2005): 2192–97.

21 A. Tanuma, H. Sato, T. Takeda, M. Hosojima, H. Obayashi, H. Hama, N. Iino, K. Hosaka, R. Kaseda, N. Imai, M. Ueno, M. Yamazaki, K. Sakimura, F. Gejyo, and A. Saito, "Functional Characterization of a Novel Missense CLCN5 Mutation Causing Alterations in Proximal Tubular Endocytic Machinery in Dent's Disease," *Nephron Physiology* 107, no. 4 (2007): 87–97.

22 M. J. E. Sternberg and W. R. Taylor, "Modelling the ATP-Binding Site of Oncogene Products, the Epidermal Growth Factor Receptor and Related Proteins," *FEBS Letters* 175 (1984): 387–92.

23 A. Zuin, M. Isasa, and B. Crosas, "Ubiquitin Signaling: Extreme Conservation as a Source of Diversity," *Cells* 3, no. 3 (2014): 690–701.

24 M. Bajaj, *op cit.*

25 M. Topf, *op cit.*

26 A. Samsel and S. Seneff, "Glyphosate, Pathways to Modern Diseases IV: Cancer and Related Pathologies," *Journal of Biological Physics and Chemistry* 15 (2015):121–59.

27 S. Seneff, "Cholesterol Sulfate Deficiency and Coronary Heart Disease," Weston A. Price Foundation, August 2, 2017, www.westonaprice.org/health -topics/cholesterol-sulfate-deficiency- coronary-heart-disease/.

28 T. H. Kim, J. Yang J, P. B. Darling, and D. L. O'Connor, "A Large Pool of Available Folate Exists in the Large Intestine of Human Infants and Piglets," *Journal of Nutrition* 134, no. 6 (2004):1389–94.

29 F. M. Asrar and D. L. O'Connor, "Bacterially Synthesized Folate and Supplemental Folic Acid are Absorbed across the Large Intestine of Piglets," *Journal of Nutritional Biochemistry* 16, no. 10 (2005): 587–93.

30 F. Kinose, *op cit.*

31 S. Seneff and G. L. Nigh, "Glyphosate and Anencephaly: Death by A Thousand Cuts," *Journal of Neurology and Neurobiology* 3, no. 2 (2017): 1–15.

32 Washington State Department of Agriculture (2013) "Integrated Pest Management Plan for Freshwater Emergent Noxious and Quarantine Listed Weeds," revised January 2013, http://agr.wa.gov/PlantsInsects/Weeds/NPDESPermits/docs/IPMFreshwaterEmergentNoxiousQuarantine.

33 Samsel and Seneff, "Glyphosate, Pathways to Modern Diseases IV," *op cit.*

34 R. E. Day, P. Kitchen, D. S. Owen, C. Bland, L. Marshall, A. C. Conner, R. M. Bill, and M. T. Conner, "Human Aquaporins: Regulators of Transcellular Water Flow," *Biochimica et Biophysica Acta* 1840, no. 5 (2014): 1492–506.

35 E. Baggaley, S. Nielsen, and D. Marples, "Dehydration-Induced Increase in Aquaporin-2 Protein Abundance Is Blocked by Nonsteroidal Anti-Inflammatory Drugs. *American Journal of Physiology-Renal Physiology* 298, no. 4 (2010): F1051–58.

36 U. Paula Santos, D. M. Zanetta, M. Terra-Filho, and E. A. Burdmann, "Burnt Sugarcane Harvesting Is Associated with Acute Renal Dysfunction," *Kidney International* 87, no. 4 (2015): 792–99.

37 R. Garca-Trabanino, E. Jarqun, C. Wesseling, R. J. Johnson, M. Gonz´alez-Quiroz, I. Weiss, J. Glaser, J, Jos´e Vindell, L. Stockfelt, C. Roncal, T. Harra, and L. Barregard, "Heat Stress, Dehydration, and Kidney Function in Sugarcane Cutters in El Salvador—A Cross-Shift Study of Workers at Risk of Mesoamerican Nephropathy," *Environmental Research* 142 (2015):746–55.

38 O. Devuyst and W. B. Guggino, "Chloride Channels in the Kidney: Lessons Learned from Knock-Out Animals," *American Journal of Physiology-Renal Physiology* 283, no. 6 (2002): F1176–91.

39 S. H. Fatemi, T. J. Reutiman, T. D. Folsom, and P. D. Thuras, "GABA(A) Receptor Downregulation in Brains of Subjects with Autism," *Journal of Autism and Developmental Disorders* 39, no. 2 (2009): 223–30.

40 Sternberg, *op cit.*

41 Samsel and Seneff, "Glyphosate, Pathways to Modern Diseases IV," *op cit.*

42 G. Ruiz-Irastorza, M. Crowther, W. Branch, and M. A. Khamashta, "Antiphospholipid Syndrome," *Lancet* 376, no. 9751 (2010): 1498–1509.

43 K. J. Rodgers and N. Shiozawa, "Misincorporation of Amino Acid Analogues into Proteins by Biosynthesis," *International Journal of Biochemistry and Cell Biology* 40 (2008): 1452–1466.

44 Zuin, *op cit.*

45 S. Seneff S, W. Morley, M. J. Hadden, and M. C. Michener, "Does Glyphosate Acting as a Glycine Analogue Contribute to ALS?" *Journal of Proteomics and Bioinformatics* 2, no. 3 (2016): 1–21.

46 M. M. Newman, N. Lorenz, N. Hoilett, N. R. Lee, R. P. Dick, M. R. Liles, C. Ramsier, and J. W. Kloepper, "Changes in Rhizosphere Bacterial Gene Expression Following Glyphosate Treatment," *Science of the Total Environment* 533 (2016): 32–41.

47 M. Krüger, W. Schrödl, J. Neuhaus, and A.A. Shehata, "Field investigations of glyphosate in urine of Danish dairy cows," *J Environ Anal Toxicol* 2013; 3: 186.

48 E. Rubenstein, "Misincorporation of the Proline Analog Azetidine-2-Carboxylic Acida in the Pathogenesis of Multiple Sclerosis: A Hypothesis. *Journal of Neuropathology & Experimental Neurology* 67 (2008): 1035–40.

49 R. A. Dunlop, P. A. Cox, S. A. Banack, and K. J. Rodgers, "The Non-protein Amino Acid BMAA Is Misincorporated into Human Proteins in Place of L-Serine Causing Protein Misfolding and Aggregation," *PLoS ONE* 8 (2013): e75376.

50 S. Eschenburg, M. L. Healy, M. A. Priestman, G. H. Lushington, and E. Schonbrunn, "How the Mutation Glycine 96 to Alanine Confers Glyphosate Insensitivity to 5-Enolpyruvyl shikimate-3-phosphate Synthase from Escherichia coli. *Planta* 216 (2002): 129–135.

51 J. Krakauer, A. Long Kolbert, S. Thanedar, and J. Southard, "Presence of L-Canavanine in *Hedysarum alpinum* Seeds and Its Potential Role in the Death of Chris McCandless," *Wilderness & Environmental Medicine* (2015): 36–42.

52 Samsel and Seneff, "Glyphosate Pathways to Modern Diseases VI," *op cit.*

53 S. Eschenburg, M. L. Healy, M. A. Priestman, G. H. Lushington, and E. Schonbrunn, "How the Mutation Glycine 96 to Alanine Confers Glyphosate Insensitivity to 5-Enolpyruvyl shikimate-3-phosphate Synthase from *Escherichia coli*. *Planta* 216 (2002): 129–135.

54 T. Funke, H. Han, M. L. Healy-Fried, M. Fischer M, and E. Schönbrunn, "Molecular Basis for the Herbicide Resistance of Roundup Ready Crops," *Proceedings of the National Academy of Sciences usA* 103, no. 35 (2006): 13010–15.

55 D. Zukowski, "Health Advisory: Venison, Elk May No Longer Be Safe to Eat Study: Deadly Chronic Wasting Disease Could be Moving to Humans," *Enviro News*, August 15, 2017, www.environews.tv/081517-venison-elk-may-no-longer-safe-eat-study-deadly-chronic-wasting-disease-moving-humans/.

56 E. M. Norstrom and J. A. Mastrianni, "The AGAAAAGA Palindrome in PrP Is Required to Generate a Productive PrPSc-PrPC Complex That Leads to Prion Propagation," *Journal of Biological Chemistry* 280, no. 29 (2005): 27236–43.

57 "Massive Fish Kill in Nova Scotia is 'Perplexing' but No 'Great Cause of Concern': Federal Scientist," *National Post*, December 30, 2016, http://nationalpost .com/news/canada/no-great-cause-for-concern-in-nova-scotia-fish-kill -federal-scientist-says/wcm/2fb66bd9-ec3f-4885-8565-5a860d10a295.

58 Nova Scotia gives OK to spray hundreds of hectares of woodland with glyphosate. https://www.localxpress.ca/local-news/nova-scotia-gives-ok-to-spray -hundreds-of-hectares-of-woodland-with-glyphosate-378654.

59 Ellis JD, Evans JD, Pettis J. Colony losses, managed colony population decline, and Colony Collapse Disorder in the United States. Journal of Apicultural Research 2010; 49(1): 134–136.

60 D. Huber, "Glyphosate Could Cause Bee Colony Collapse Disorder (CCD)," http://www.gmoevidence.com/dr-huber-glyphosate-and-bee-colony-collapse -disorder-ccd/.

61 L. T. Herbert, D. E. Vazquez, A. Arenas, and W. M. Farina, "Effects of Field-Realistic Doses of Glyphosate on Honeybee Appetitive Behaviour," *Journal of Experimental Biology* 217, pt. 19 (2014): 3457–64.

62 C. M. Handel, L. M. Pajot, S. M. Matsuoka, C. van Hemert, J. Terenzi, S. L. Talbot, D. M. Mulcahy, C. Ul Meteyer, and K. A. Trust, "Epizootic of Beak Deformities Among Wild Birds in Alaska: An Emerging Disease in North America?" *Auk: Ornithological Advances* 127, no. 4 (2010): 882–98.

63 C. M. Handel and C. van Hemert, "Environmental Contaminants and Chromosomal Damage Associated with Beak Deformities in a Resident North American Passerine," *Environmental Toxicology and Chemistry* 34, no. 2 (2015): 314–27.

64 Ibid.

65 M. I. Kang, A. Kobayashi, N. Wakabayashi, S. G. Kim, and M. Yamamoto, "Scaffolding of Keap1 to the Actin Cytoskeleton Controls the Function of Nrf2 as Key Regulator of Cytoprotective Phase 2 Genes," *Proceedings of the National Academy of Sciences usA* 101, no. 7 (2004): 2046–51.

66 K. H. Treiber, D. S. Kronfeld, T. M. Hess, B. M. Byrd, R. K. Splan, and W. B. Staniar, "Evaluation of Genetic and Metabolic Predispositions and Nutritional Risk Factors for Pasture-Associated Laminitis in Ponies," *Journal of the American Veterinary Medical Association* 228, no. 10 (2006): 1538–45.

67 J. Hoy, N. Swanson, and S. Seneff, "The High cost of Pesticides: Human and Animal Diseases," *Poultry, Fisheries and Wildlife Sciences* 3 (2015): 1.

68 N. Martin, and R. Montagne, "U.S. Has the Worst Rate of Maternal Deaths in the Developed World," *National Public Radio*, May 12, 2017, www.npr.org/2017/05/12/528098789/u-s-has-the-worst-rate-of-maternal-deaths-in-the-developed-world.

69 "League of Denial: The NFL's Concussion Crisis," *Frontline*, www.pbs.org/wgbh/pages/frontline/oral-history/league-of-denial/.

70 W. A. Morley and S. Seneff, "Diminished Brain Resilience Syndrome: A Modern Day Neurological Pathology of Increased Susceptibility to Mild Brain Trauma, Concussion, and Downstream Neurodegeneration," *Surgical Neurology International* 5 (2014): 97.

71 K. Hassan, V. Bhalla, M. E. El Regal, and H. H. A-Kader, "Nonalcoholic Fatty Liver disease: A Comprehensive Review of a Growing Epidemic," *World Journal of Gastroenterology* 20, no. 34 (2014): 12082–101.

72 Seralini, Clair, et al. "Long-Term Toxicity . . ." *op cit.*

73 R. Jasper, G. O. Locatelli, C. Pilati, and C. Locatelli, "Evaluation of Biochemical, Hematological and Oxidative Parameters in Mice Exposed to the Herbicide Glyphosate-Roundup," *Interdisciplinary Toxicology* 5, no. 3 (2012): 133–40.

74 V. Kauba, M. Mili, R. Rozgaj, N. Kopjar, M. Mladini, S. Unec, A. L. Vrdoljak, I. Pavii, A. M. M. Ermak, A. Pizent, B. T. Lovakovi, and D. Eljei, "Effects of Low Doses of Glyphosate on DNA Damage, Cell Proliferation and Oxidative Stress in the HepG2 Cell Line," *Environmental Science and Pollution Research International* 24, no. 23 (2017): 19267–81.

75 R. Mesnage, G. Renney, G. E. S'eralini, M. Ward, and M. N. Antoniou, "Multiomics Reveal Non-alcoholic Fatty Liver Disease in Rats Following Chronic Exposure to an Ultra-low Dose of Roundup Herbicide," *Scientific Reports* 7 (2017): 39328.

76 Seneff, Morley, et al. "Does Glyphosate Acting as a Glycine Analogue Contribute to ALS?" *op cit.*

77 M. Basaranoglu, G. Basaranoglu, and E. Bugianesi, "Carbohydrate Intake and Nonalcoholic Fatty Liver Disease: Fructose as a Weapon of Mass Destruction," *Hepatobiliary Surgery and Nutrition* 4, no. 2 (2015): 109–16.

78 M. B. Vos and J. E. Lavine, "Dietary Fructose in Nonalcoholic Fatty Liver Disease," *Hepatology* 57, no. 6 (2013): 2525–31.

79 S. Thongprakaisang A. Thiantanawat, N. Rangkadilok, T. Suriyo, and J. Satayavivad, "Glyphosate Induces Human Breast Cancer Cells Growth via Estrogen Receptors," *Food and Chemical Toxicology* 59 (2013): 129–36.

80 Swanson, et al., *op cit.*

81 M. Hollstein, D. Sidransky, B. Vogelstein, and C. C. Harris, "p53 Mutations in Human Cancers," *Science* 253 (1991): 49–53.

82 Y. Higashimoto, Y. Asanomi, S. Takakusagi, M. S. Lewis, K. Uosaki, S. R. Durell, C. W. Anderson, E. Appella, and K. Sakaguchi, "Unfolding, Aggregation, and Amyloid Formation by the Tetramerization Domain from Mutant p53 Associated with Lung Cancer," *Biochemistry* 45, no. 6 (2006): 1608–19.

83 R. S. Legro, A. R. Kunselman, W. C. Dodson, and A. Dunaif, "Prevalence and Predictors of Risk for Type 2 Diabetes Mellitus and Impaired Glucose Tolerance in Polycystic Ovary Syndrome: A Prospective, Controlled Study in 254 Affected Women," *Journal of Clinical Endocrinology & Metabolism* 84, no. 1 (1999): 165–69.

84 K. Graham, "Argentina Study: 85% of Tampons Contaminated with Glyphosate," *Digital Journal*, Oct 27, 2015, www.digitaljournal.com/life/health/85-percent-of-tampons-found-to-contain-glyphosate/article/447733.

85 F. Qu, F. F. Wang, X. E. Lu, M. Y. Dong, J. Z. Sheng, P. P. Lv, G. L. Ding, B. W. Shi, D. Zhang, and H. F. Huang, "Altered Aquaporin Expression in Women with Polycystic Ovary Syndrome: Hyperandrogenism in Follicular Fluid Inhibits Aquaporin-9 in Granulosa Cells through the Phosphatidylinositol 3-Kinase Pathway," *Humam Reproduction* 25, no. 6 (2010): 1441–50.

86 C. Barratt, "Most Men in the U.S. and Europe Could Be Infertile by 2060, According to a New Study," *Quartz*, July 28, 2017, https://qz.com/1040302/most-men-in-the-us-and-europe-could-be-infertile-by-2060-according-to-a-new-study/.

87 Y. Yang, and H. Lockett, "Sperm Crisis in China as Fertility Slides," *Financial Times*, November 28, 2016, www.ft.com/content/d4b5325c-abad-11e6-ba7d-76378e4fef24.

88 J. Nardi, P. B. Moras, C. Koeppe, E. Dallegrave, M. B. Leal, and L. G. Rossato-Grando, "Prepubertal Subchronic Exposure to Soy Milk and Glyphosate Leads to Endocrine Disruption," *Food and Chemical Toxicology* 100 (2017): 247–52.

89 V. L. de Liz Oliveira Cavalli, D. Cattani, C. E. Heinz Rieg, P. Pierozan, L. Zanatta, E. Benedetti Parisotto, D. Wilhelm Filho, F. R. Mena Barreto Silva, R. Pessoa-Pureur, and A. Zamoner, "Roundup Disrupts Male Reproductive Functions by Triggering Calcium-Mediated Cell Death in Rat Testis and Sertoli Cells," *Free Radical Biology and Medicine* 65 (December 2013): 335–46.

90 S. Reddy, "A Striking Rise in Serious Allergy Cases," *Wall Street Journal*, August 21, 2017, www.wsj.com/articles/a-striking-rise-in-serious-allergy-cases-1503327581.

91 B. Han, W. M. Compton C. Blanco, E. Crane, J. Lee, and C. M. Jones, "Prescription Opioid use, Misuse, and use Disorders in U.S. Adults: 2015 National Survey on Drug use and Health," *Annals of Internal Medicine* 167, no. 5 (2017).

92 I. J. Wallace, S. Worthington, D. T. Felson, R. D. Jurmain, K. T. Wren, H. Maijanen, R. J. Woods, and D. E. Lieberman, "Knee Osteoarthritis Has Doubled in Prevalence Since the Mid-20th Century," *Proceedings of the National Academy of Sciences usA* 114, no. 35 (2017), e-pub ahead of print.

93 Samsel and Seneff, "Glyphosate Pathways to Modern Diseases VI," *op cit.*

94 A. Vasin, S. Klotchenko, and L. Puchkova, "Phylogenetic Analysis of Six-Domain Multi-copper Blue Proteins. *PLOS Currents* (March 2013) 5, doi: 10.1371/cur- rents.tol.574bcb0f133fe52835911abc4e296141.

95 V. Dubljevic, A. Sali, and J. W. Goding, "A Conserved RGD (Arg-Gly-Asp) Motif in the Transferrin Receptor Is Required for Binding to Transferrin," *Biochemical Journal* 341, pt. 1 (1999): 11–14.

96 A. Chauhan, V. Chauhan, W. T. Brown, and I. Cohen, "Oxidative Stress in Autism: Increased Lipid Peroxidation and Reduced Serum Levels of Ceruloplasmin and Transferrin—the Antioxidant Proteins," *Life Sciences* 75, no. 21 (2004): 2539–49.

97 A. Samsel and S. Seneff, "Glyphosate, Pathways to Modern Diseases III: Manganese Neurological Diseases and Associated Pathologies," *Surgical Neurology International* 6 (2015): 45.

98 G. Wang , X.N. Fan, Y.Y. Tan, Q. Cheng, S.D. Chen, "Parkinsonism after chronic occupa- tional exposure to glyphosate," *Parkinsonism Relat Disord* 2011; 17: 486–7.

99 E.R. Barbosa, Leiros da Costa MD, L.A. Bacheschi, M. Scaff, C.C. Leite, "Parkinsonism after glycine-derivate exposure," *Mov Disord* 2001; 16: 565–8.

100 R. Negga, J.A. Stuart, M.L. Machen, J. Salva, A.J. Lizek, S.J. Richardson, et al, "Exposure to glyphosate-and/or Mn/Zn-ethylene-bisdithiocarbamate-containing pesticides leads to degeneration of γ-aminobutyric acid and dopamine neurons in Caenorhabditis elegans," *Neurotox Res* 2012; 21: 281–90.

101 Hernández-Plata I, Giordano M, Daz-Muñoz M, Rodrguez VM. The herbicide glyphosate causes behavioral changes and alterations in dopaminergic markers in male Sprague-Dawley rat. Neurotoxicology 2015; 46: 79–91.

102 M. Barbariga, F. Curnis, A. Andolfo, A. Zanardi, M. Lazzaro, A. Conti, G. Magnani, M.A. Volontè, L. Ferrari, G. Comi, A. Corti, M. Alessio, "Ceruloplasmin functional changes in Parkinsons disease-cerebrospinal fluid," *Molecular Neurodegeneration* 2015; 10: 59.

103 C.W. Olanow, P. Brundin, "Parkinson's disease and alpha synuclein: is Parkinson's disease a prion-like disorder?" *Mov Disord* 2013; 28(1): 31–40.

104 R. Chandra, A. Hiniker, Y.M. Kuo, R.L. Nussbaum, R.A. Liddle, "α-Synuclein in gut endocrine cells and its implications for Parkinson's disease," *JCI Insight* 2017; 2(12):e92295.

105 H.N. Du, L. Tang, XY Luo, H.T. Li, J. Hu, J.W. Zhou, H.Y. Hu, "A peptide motif consisting of glycine, alanine, and valine is required for the fibrillization and cytotoxicity of human alpha-synuclein." *Biochemistry* 2003; 42(29): 8870–8.

106 D. I. Finkelstein, J.L. Billings, P.A. Adlard, S. Ayton, A. Sedjahtera, C.L. Masters, S. Wilkins, D.M. Shackleford, S.A. Charman, W. Bal, I.A. Zawisza, E. Kurowska, A.L. Gundlach, S. Ma, A.I. Bush, D.J. Hare, P.A. Doble, S. Crawford, E.C. Gautier, J. Parsons, P. Huggins, K.J. Barnham, R.A. Cherny, "The novel compound PBT434 prevents iron mediated neurodegeneration and alpha-synuclein toxicity in multiple models of Parkinson's disease," *Acta Neuropathol Commun* 2017; 5(1): 53.

107 C. Jayasumana, S. Gunatilake, P. Senanayake, "Glyphosate, Hard Water and Nephro-toxic Metals: Are They the Culprits Behind the Epidemic of Chronic Kidney Disease of Unknown Etiology in Sri Lanka?" *Int J Environ Res Public Health* 2014; 11(2): 2125–2147.

108 A. Mirza, A. King, C. Troakes, C. Exley. "Aluminium in brain tissue in familial Alzheimer's disease," *Journal of Trace Elements in Medicine and Biology* 2017; 40: 30–36.

109 C.A. Shaw, S.D. Kette, R.M. Davidson, S. Seneff. "Aluminums Role in CNS-immune System Interactions leading to Neurological Disorders." *Immunome Res* 2013; 9:1000069.

110 M. Purgel, Z. Takács, C.M. Jonsson, L. Nagy, I. Andersson, I. Bányai, I. Pápai, P. Persson, S. Sjöberg, I.Tóth, "Glyphosate complexation to aluminium(III). An equilibrium and structural study in solution using potentiometry, multinuclear NMR, ATR-FTIR, ESI-MS and DFT calculations," *J Inorg Biochem* 2009; 103(11): 1426–38.

111 S. Seneff, N. Swanson, C. Li, "Aluminum and Glyphosate Can Synergistically Induce Pineal Gland Pathology: Connection to Gut Dysbiosis and Neurological Disease," *Agricultural Sciences* 2015; 6: 42–70.

112 J.M. Purdey, "Ecosystems supporting clusters of sporadic TSEs demonstrate excesses of the radical-generating divalent cation manganese and deficiencies of antioxidant co factors Cu, Se, Fe, Zn," *Med Hypotheses* 2000; 54: 278–306.

113 Samsel and Seneff, "Glyphosate, Pathways to Modern Diseases III: Manganese Neurological Diseases and Associated Pathologies," *op cit.*

114 Seneff, Morley, et al. *op cit.*

115 Samsel and Seneff, "Glyphosate, Pathways to Modern Diseases IV," *op cit.*

116 Ibid.

117 P.G. Ince, G.A. Codd, "Return of the cycad hypothesis: Does the amyotrophic lateral sclerosis/parkinsonism dementia complex (ALS/PDC) of Guam have new implications for global health?" *Neuropathol Appl Neurobiol* 2005; 31: 348, 345–353.

INDEX